Books are to be returned on or before
the last date below.

British civil engineering 1640–1840

British civil engineering 1640–1840:

A Bibliography of Contemporary Printed Reports, Plans and Books

A. W. Skempton

Mansell Publishing Limited
London and New York

First published 1987 by Mansell Publishing Limited
(A subsidiary of The H. W. Wilson Company)
6 All Saints Street, London N1 9RL, England
950 University Avenue, Bronx, New York 10452, U.S.A.

British Library Cataloguing in Publication Data

Skempton, A. W.
  British civil engineering 1640–1840:
  a bibliography of contemporary printed reports,
  plans and books.
  1. Civil engineering—Great Britain—History
  —Bibliography
  I. Title
  016.624′0941   Z5855.G7

  ISBN 0-7201-1746-1

Library of Congress Cataloging-in-Publication Data

Skempton, A. W.
  British civil engineering, 1640–1840.

  Includes index.
  1. Civil engineering—Great Britain—Bibliography.
2. Civil engineering—Great Britain—History
—Bibliography.   I. Title.
Z5851.S54 1986 [TA15]   016.624′0941   85-29819
ISBN 0-7201-1746-1

Printed in Great Britain by
Whitstable Litho Ltd., Whitstable, Kent

Contents

Illustrations

All illustrations are title-pages, unless otherwise indicated.

1. Report by Vermuyden on draining the Fens, 1642.
2. Perry's Account of Stopping Dagenham Breach, 1721.
3. First page, with caption title, of Armstrong's report on King's Lynn harbour, 1724.
4. Report by John Grundy senior and junior on the River Witham, 1744.
5. Labelye's Description of Westminster Bridge, 1751.
6. Plate from Smeaton's book on the Eddystone Lighthouse; engraved 1763, published 1791.
7. Plan of the Birmingham Canal, 1767. James Brindley engineer, drawn by Robert Whitworth.
8. Smeaton's Review of Several Matters relative to the Forth and Clyde Canal, 1768.
9. Reports by Jessop and Mylne on the Thames Navigation, 1791.
10. Part of a plan of the Thames from Lechlade to Abingdon, annexed to the 1791 report.
11. Docket title of Hudson's report on the Eau Brink Cut, 1792.
12. Report by Chapman on draining the low grounds adjacent to the River Derwent in Yorkshire, 1800.
13. Part of a plan of the Caledonian Canal by Telford and Downie, 1805.
14. Report by Rennie on the South and Middle Levels, 1810.
15. Elevation and details of Bonar Bridge by Telford, 1813.
16. Report by Stevenson on Dundee Harbour, 1814.
17. Reports by Telford on the Holyhead Road, 1820.
18. First page of report by Grainger and Miller on a railway from Edinburgh to Glasgow, 1831.
19. Printed wrapper of Walker's report on Aberdeen Harbour, 1838.
20. Warrington Viaduct: engraved plate from Roscoe *Grand Junction Railway*, 1839.
21. New Cross Cutting: lithograph plate from Trotter *Croydon Railway*, 1839.
22. Offprint of Eaton Hodgkinson's paper on the strength of iron columns, 1840.

Preface

This bibliography gives details of books, printed reports and plans published in Great Britain and Ireland, before 1840, on civil engineering matters: fen drainage, river navigations, canals, docks and harbours, bridges, roads, railways, and water supply; with a selection of related pamphlets, and works on mechanics, strength of materials and hydraulics. It begins with the first notable engineering publication in England, Vermuyden's *Discourse* on draining the Fens (1642), and continues through the period represented by such masters as Smeaton, Rennie and Telford: a period which saw the canal system virtually completed, the creation of many harbours, major improvements in the trunk roads, and the opening of the first thousand miles of our railways.

References to parts of this literature exist (*see* p. xx) but few of them are satisfactory from a bibliographic point of view, or give locations for what are often quite rare items, and in any case they leave wide areas of the field uncovered: no catalogue of English canal plans, for example, almost nothing on harbours, and above all there are scarcely any lists of publications under authors' names.

In connection with my biographical studies of early civil engineers the lack of such information became very apparent by 1971. I then began filling in the gaps and soon found an unexpected richness of material; not so much in the books, which were relatively well known, but in a wealth of printed reports.

It emerged that these reports, often finely printed and accompanied by engraved or lithographed plans, were far more numerous than could have been supposed and constituted a principal form of publication before the 1840s; before, that is to say, the time when they were to an increasing extent replaced in importance by papers published in engineering journals such as the *Transactions* of the Institution of Civil Engineers, first issued in 1836.

Reports were of course by no means unknown. But the discovery, if so it can be called, of their dominant role brings the subject of civil engineering bibliography into proper perspective. To take an example: Robert Mylne, John Grundy, Robert Whitworth and Thomas Yeoman, distinguished men of their day and among the founder members in 1771 of the Society of Civil Engineers, can now be associated with an appropriate number of publications. But if reports (and plans) were to be omitted from consideration their names, and many others, would not appear in any list of contemporary printed material.

Encouraged by the progress being made in this

research, I compiled in 1977 a list of pre-1850 reports in the library of the Institution and in 1979 decided to embark on more systematic work leading towards a substantially complete bibliography. The main concentration of effort has been in Edinburgh at the National Library, in Oxford at the Bodleian, and in London at the British Library (British Museum), the Institution of Civil Engineers and the University of London Goldsmiths' Library. Other sources used in London include the Royal Geographical Society, the Science Museum Library and the House of Lords Record Office. Visits have been made, usually on more than one occasion, to several county record offices (notably at Beverley, Cambridge and Lincoln), to university libraries at Cambridge and Hull, and to various city libraries.

With few exceptions, which are duly noted, at least one copy of each item in the bibliography has been examined. When more than one location is given, the other copies have been seen or unambiguously identified by correspondence or entries in a library catalogue.

The ideal of completeness obviously cannot be achieved. The collections used provide a representative but not complete coverage. For this reason alone some items must have been missed, and no doubt there are others which I did not succeed in tracking down. Conversely there were items which on examination proved too trivial for inclusion. Also there is the problem of deciding what is relevant. Here, on the boundary between engineering and commerce, in arguments on the relative merits of railways and canals, in the host of pamphlets attacking or supporting fen drainage schemes, and so on, the principle adopted has been to include a selection of the better-known and worthwhile contributions.

Finally, the bibliography includes only 'separate' publications; not articles in journals or encyclopaedias, unless in the form of offprints.

Acknowledgements

A work of this nature could not be undertaken without positive encouragement. I am therefore grateful to Mansell Publishing Ltd. for agreeing at an early stage to handle the publication, and to the Royal Society for a generous grant towards travel expenses.

It is a pleasure to record my indebtedness to Mr. Roland Paxton of Edinburgh, Miss Julia Elton of B. Weinreb Ltd. and Mr. Michael Chrimes of the Institution of Civil Engineers for responding to numerous enquiries and finding items on my behalf. The unfailing help and courtesy of librarians and archivists has been greatly appreciated, and particular thanks are due to Miss Anna Hikel for typing the entire set of entries with impeccable accuracy.

Permission to reproduce photographs has kindly been given by Manchester Central Library (pl. 7), the Institution of Civil Engineers (pls. 3, 5, 9, 10, 21), B. Weinreb Ltd. (pl. 15) and the Royal Society (pl. 6). The other plates are from items in the author's collection.

Guide to the Literature

It may be helpful to give a broad classification of the material and to mention some examples in each category.

Books

Perhaps the books of greatest value are those giving a detailed account of actual jobs. Early examples are Perry's description of his work at Dagenham Breach (1721) and Labelye on Westminster Bridge (1751). Three classics of engineering literature in this category, fully illustrated by engraved plates, are Smeaton's Eddystone Lighthouse (1791 and two later editions), Stevenson's Bell Rock Lighthouse (1824) and the account of the design and construction of Telford's Menai suspension bridge (1828) by the resident engineer W. A. Provis.

Telford's autobiography, edited and enlarged by his friend John Rickman (1838), with a splendid atlas of plates, is essentially a description of many of his works. In a rather similar style, giving illustrated details of a wide range of jobs, are Brees on Railway Practice (1837) and Simms on the Public Works of Great Britain (1838).

Widely read textbooks, all passing into several editions, include McAdam (1816) and Parnell (1833) on roads, Wood (1825) on railways, and the series of monographs on strength of materials, etc. by Barlow (1817) and Tredgold (1820).

Histories, of a less technical nature, begin with books on fen drainage by Dugdale (1662, second edition 1772) and Badeslade (1725, reprinted 1766), the latter with beautifully engraved maps and plans. An anonymous work traditionally attributed to Brindley (1766) gives the first account of English canals; it was followed by enlarged editions in 1769 and 1779, and the popular book by Phillips (1792 with five subsequent editions). Later histories include outstanding contributions by Wells on Bedford Level (1828–30), Priestley on inland waterways (1831) and Whishaw on railways (1840).

Books illustrated with views of engineering works form a separate, valuable category. They include Bury's Liverpool & Manchester Railway (1831, the plates finely executed by Ackermann's aquatint engravers), George Rennie on London Bridge with etchings by E. W. Cooke (1833), drawings by J. W. Carmichael and descriptions by Blackmore of the Newcastle & Carlisle Railway (1836), and Roscoe's Grand Junction Railway (1839) with steel engravings by Radclyffe after David Cox, George Dodgson and others.

In a class of its own is the book of lithographs by John C. Bourne (1839, with text by John Britton) faithfully depicting work in progress on the London & Birmingham Railway.

Also in an unique category are the three large quarto volumes of Smeaton's reports, published in 1812 by the Society of Civil Engineers from manuscript copies in his report books. These were highly regarded by the profession; a second printing appeared in 1837.

Pamphlets and Tracts

Pamphlets mostly relate to a particular scheme and tend to be biased or contentious. Beginning with Burrell's strictures on Vermuyden's work in the Fens (1642), pamphlets of this type continue throughout the period under consideration and, as mentioned before, only a selection is recorded here. At their best they can be thoughtful and productive. Prominent are the pamphlets by Vaughan (1793–96) on the case for providing enclosed docks in the Port of London; Bentley's *View of the Advantages* of a canal linking the Trent and Mersey (1765) is an important document in canal history; an expert criticism of the designs for Blackfriars Bridge is given by the anonymous author of *Observations on Bridge Building* (1760); a forceful but objective account of the state of Holderness Drainage (1786) by 'A Friend of the Undertaking' had the beneficial effect of leading to improvements made under Jessop's direction, and Chapman's *Observations* (1824) formed a starting point for planning the Newcastle & Carlisle Railway.

Somewhat different in character are the replies by engineers to criticism of their works or plans, a famous example being Smeaton's *Review of Several Matters* relative to the Forth & Clyde Canal (1768).

Tracts come under the same broad definition as pamphlets but nevertheless can be distinguished as serving a more academic or general purpose. Indeed, tracts differ from books chiefly in being shorter (and unbound), and there are instances in which a tract by accretion of material in later editions becomes of sufficient size (say eighty pages or more) to rank as a book, Gray's essay on railways (1820–25) being an example. In its purest form, then, a tract is a short treatise such as Atwood's *Dissertation* on arches (1801), Macneill's account of experiments on the resistance of water to the passage of boats on canals (1833) and Buck's *Essay* on oblique arches (1839). In the same category can be placed the rare but highly important offprints of major scientific papers, ranging in date from Smeaton's pioneer contribution on experimental fluid mechanics (1759) to Eaton Hodgkinson's research on the strength of cast iron columns (1840).

More mundane works include tracts on road construction, for example by 'A.E.' (1774), and comparisons between canal, road and rail transport: Maclaren (1825), Page (1832) and Grahame (1835).

Printed Reports

In format these range from a short report printed on a folded folio sheet, with a docket title on the back page, to quite substantial documents having a title page and often with engraved or lithographed folding plans. In most cases they were stitched, not bound, with or without paper covers; but reports of exceptional length or containing several large plans might be issued, like a book, in 'boards'.

Reports were commissioned (as they still are today) by the client: a canal or railway company, a 'committee of survey' for some proposed scheme, or a parliamentary commission, etc. Sometimes the reports would be published on sale, but usually they were distributed by the client free of charge (postage as always before 1840 being paid by the recipient) to landowners, local authorities, members of Parliament and the committee or commissioners themselves; also to

Examples of numbers printed

1729	Badeslade	*Scheme* for draining the Fens	150
1748	Labelye	Report, Sunderland harbour	100
1768	Smeaton	*Review* of Forth & Clyde Canal	500
1771	Brindley	Report + plan, River Thames	1000
1772	Grundy	Report + plan, Walling Fen	300
1773	Whitworth	Report + plan, Moorfields Canal	1000
1791	Mylne	Report + plan, Thames Navigation	500
1803	Jessop	Plan, Bristol Docks. 1st & 2nd state	800
1807	Rennie	Report, Holderness Drainage	500
1811	Chapman	Report + plan, Scarborough Harbour	250
1812	Smeaton	*Reports of the late John Smeaton*	500
1813	Smeaton	*Eddystone Lighthouse*, 3rd edn.	300
1824	Stevenson	*Bell Rock Lighthouse*	300
1829	Walker	Report, Liverpool & Manchester Railway	500

the engineer for his own records and for his friends and colleagues, much as offprints are circulated.

Reports were always submitted by the engineer in manuscript, a copy having been kept in his report book, and not infrequently a few further copies would suffice for the client's needs. However, when a decision was made to have the report printed (and the plan or plans engraved), an order would be placed typically for several hundred copies—numbers wholly comparable to those printed in an edition of a book: *see* the above Table. In some cases it seems that the order was excessively large, as a batch of twenty-five or fifty reports may on occasion turn up unexpectedly in some archive. But there are many instances in which the first printing proved insufficient; for example, both of Smeaton's reports on the Forth & Clyde Canal went into second editions within weeks of first publication. Mylne's report on the Eau Brink Cut was printed twice, in London and King's Lynn, while Chapman's reports on the proposed canal from Newcastle to Maryport attracted so much interest that they were reprinted a few months later in a collected second edition by J. Whitfield and 'sold by him and the other booksellers in Newcastle, Hexham, Carlisle, etc'.

Most engineer's reports appear as separate documents but sometimes were printed along with a statement by the commissioning authority, especially in the case of reports submitted to Parliament.

Printed reports first appear with regularity in the second quarter of the eighteenth century. Examples include Armstrong on King's Lynn harbour (1724), Perry on the South Holland fens (1727), Labelye on the foundations of Westminster Bridge (1739) and on the Fens (1745), John Grundy senior and junior on the River Witham (1744) and Brett and Desmaretz on Ramsgate harbour (1755), the latter being one of the earliest reports to Parliament on an engineering scheme.

After 1760 what had been a small stream of publications rapidly turns into a flood. Here it is possible to mention only a small sample from the leading exponents. Smeaton, renowned for his logical and lucid style, could be represented by reports on the River Lea (1766), the Forth & Clyde Canal (1767), Hatfield Chase (1776) and Ramsgate Harbour (1791). Of the fifteen printed reports by his friend John Grundy, the one on Walling Fen (1772) might be chosen. Jessop wrote dozens of reports, usually short and direct and very practical; many were printed, those on the upper Thames navigation (1791) and on Sunderland harbour (1808) being good examples.

The Eau Brink Cut, a scheme for improving the Great Ouse near King's Lynn, gave rise in 1791 and 1792 to several reports, notably by Mylne.

James Golborne and John Hudson were among the others who reported and, in these notes, they have to represent the many engineers with a smaller or more local practice whose names are associated with perhaps no more than about four or five printed reports. When eventually the Eau Brink job was carried out, Rennie took charge. He is widely recognized for his bridges and canals, but as it happens two of his most memorable reports are on fen drainage: the East, West and Wildmore Fens in Lincolnshire (1800) and the South Level (1810), the latter with nine engraved plans and sections from surveys by his assistant, Anthony Bower. Another prolific writer was William Chapman. His reports, especially those on Scarborough harbour (for which he acted as consulting engineer for many years), show how ideas have to be modified in searching for an optimum solution to design problems. But some of his best reports, like those of Rennie, are on the drainage of low grounds, an excellent example being on the Vale of Pickering (1800).

Some of R. L. Stevenson's literary talent may have been inherited from his grandfather Robert, the Edinburgh engineer, whose masterpiece on the Bell Rock Lighthouse has been mentioned above. The thirty printed reports by him (some written jointly with his son Alan) are models of clarity and presentation. Two harbour reports, on Dundee (1814) and Granton (1834), might be singled out together with the first of a series of railway reports written in 1818 and published early in 1819.

In Thomas Telford one encounters an engineer whose talents were employed during a long career of the utmost distinction. Those talents are reflected in his output of over one hundred printed reports, many being to Parliament in connection with his great public works: the Caledonian Canal, roads and bridges in the Highlands, and the road from London to Holyhead which includes the Menai suspension bridge. Nowhere is his mastery of a grand project more clearly displayed than in the *Survey and Report of the Coasts and Central Highlands of Scotland* (1803), and if a single report had to be selected as characteristic of his attention to detail in the context of a vast

undertaking it might well be one written in 1820 on the 'English Part' of the Holyhead Road (i.e. from London to Shrewsbury) accompanied by nine plans and sections and a short appendix on the rules for repairing roads, later to be issued as a small separate book.

Of reports by Telford's near contemporaries, apart from those by Rennie, Chapman and Stevenson, there are, to mention a very few from this most prolific period, Sir Marc Brunel's first report on the Thames Tunnel (1823), describing his brilliant concept of the tunnelling shield, and his *Exposition* to the King on the progress of this work (1833); the reports by George Stephenson and by Grainger and Miller on the Edinburgh & Glasgow Railway (1831); reports by Sir John Rennie on the River Nene (1836), and those by James Walker on Aberdeen Harbour (1838).

Plans

Engineering plans (and more rarely maps) were used to illustrate a report, either being sewn in with the printed text or issued separately, or they could be published in their own right with or without short explanatory notes. The engravers and lithographers employed in producing the plans and maps include eminent craftsmen: W. H. Toms, Thomas Jefferys, William Faden, John Cary, Charles Hullmandel and others.

Sir Jonas Moore's accurate map of the Fens (1684, reprinted 1705) was ahead of its time. River navigation plans of the Mersey & Irwell by Steers (1712) and of the Don (1722) and Yorkshire Ouse (1725) by Palmer are interesting but, from a cartographic point of view, rather primitive. A distinct advance in clarity is displayed in Whittenbury's plan of the Lea (1740), Smeaton's more elaborate plan of the Calder (1757), showing the navigation locks, and Grundy's plan of the Witham (1762).

The eight canal plans by Brindley and his associates, all dating from the 1760s, are admirably clear, while retaining a little of the nature of a sketch, and delineate property boundaries adjacent to the canal. The Birmingham Canal plan (1767) is an example and, like most of those in the mid eighteenth century, is decorated with a cartouche title.

A new standard of accuracy was achieved in 1770 with the large plan and profile of the Thames, from Boulter's Lock to Mortlake, surveyed and drawn by Brindley's assistant Robert Whitworth and engraved by Jefferys with hand colouring. Many of the canal plans of the later eighteenth century are splendid examples of engineering and cartography: Whitworth's Thames & Severn (1783), Jessop's Grantham Canal (1791), the plan and section by James Barnes of the Grand Junction (1792) and Rennie's Kennet & Avon (1794) must suffice to illustrate the point.

In the early nineteenth century engraved plans are so numerous that a choice of examples becomes arbitrary; they can be represented by Jessop's Bristol Floating Harbour (1803), Bower's large plan and section of the Witham (1804), the East India Docks by Rennie and Ralph Walker (1804) and the Caledonian Canal by Telford and Downie (1805). The first accurate map of Scotland is by Arrowsmith (1807), showing Telford's survey of the Highland roads and bridges.

Engraving of plans, and especially of maps with their close detail, continued at least to the mid century, but from about 1820 lithography was used to an increasing extent. This technique proved particularly applicable where plans are large in size and numbers, as for deposited plans (required by Parliament) which, from the 1830s, were usually on a scale of 4 inches to a mile (see for example the plans and sections of I. K. Brunel's Great Western Railway, 1835), though the laborious procedure of copying by hand continued to a surprising extent.

Characteristic of this period, also, are folding and sectioned maps issued in a slip case. Examples include Bradshaw's set of canal maps (1828–32 and later editions) with levels all related to a single datum at Liverpool, Wells's

map of the Fens (1829) and several important railway maps such as Robert Stephenson's London & Birmingham (1835) and Locke's Grand Junction (1836).

In a special category are engraved plans and elevations of bridges. These may be straightforward, if elegant, engineering or architectural drawings (for example the engraving of Labelye's Westminster Bridge (1751) issued to accompany his book of the same date), or they may be embellished with landscape as in Mylne's design for Blackfriars Bridge (1760), James Walker's Vauxhall Bridge (1816) and the Union suspension bridge at Berwick by Sir Samuel Brown (1823). Artistic views of engineering works, other than those in illustrated books, are not listed in the bibliography.

Specifications

Not more than about thirty printed specifications (with or without forms of tender) have survived, and in all cases only one or two copies have been found. They have considerable interest from an historical engineering point of view. Examples are the specifications by Rennie for a river wall at Deptford dockyard (1815) and by Francis Giles for two large masonry bridges or viaducts on the Newcastle & Carlisle Railway (1830).

Minutes of Evidence

Verbatim transcripts of evidence presented to parliamentary committees often provide biographical details, as the engineers respond to questions on their previous experience, and give an opportunity to hear (as it were) them speaking on their own subject. Two volumes of printed evidence on the Eau Brink Bill (1794, 1795) may specially be mentioned, as also the evidence on the Port of London (1796) and the state of London Bridge (1821). Following the famous minutes of evidence on the original and revised lines for the Liverpool & Manchester Railway (1825, 1826) come a whole series of enquiries on railway bills including the London & Birmingham (1832), London & Southampton (1834), Great Western (1835), South Eastern (1836) and the London & Brighton (1836, 1837). Scarcely any engineer of note failed to give evidence on one or more of these schemes.

Explanatory Notes

The majority of entries are listed under authors' names, in chronological order, and indexed under subject headings. Second editions or reprints, however, follow immediately after the first issue. In cases of joint authorship cross-references are given under each name. Anonymous or collective items are listed and indexed under subject.

The author's signature and the place and date of composition are always recorded, as are the names of engravers or lithographers of plans.

Dimensions refer to the page height or height × width for plans, measured to the border or plate-mark if there is no border.

Pagination applies to printed pages only, following the printer's numbering; printed pages without a number are given in brackets.

Notes have been added concerning the original covers (e.g. printed wrappers or boards), the published price and numbers printed, when known.

Locations of copies are shown by symbol or name below each item. For the sake of completeness items seen only in private collections or antiquarian booksellers have been recorded. Locations are not given for parliamentary papers issued from the Commons or Lords; these can be found in the British Library, the House of Lords Record Office and, with few exceptions, in the Institution of Civil Engineers.

List of Symbols

B	Antiquarian bookseller
BDL	Bodleian Library, Oxford
BM	British Library, British Museum
CE	Institution of Civil Engineers
CLRO	Corporation of London Record Office
CRO	Cambridgeshire Record Office
CUL	Cambridge University Library
HLRO	House of Lords Record Office
HRO	Humberside Record Office, Beverley
IC	Imperial College, London
ISE	Institution of Structural Engineers
LCL	Lincoln City Library
LAO	Lincolnshire Archives Office, Lincoln
MCL	Manchester Central Library
NLS	National Library of Scotland, Edinburgh
P	Private collection
PLA	Port of London Authority
RGS	Royal Geographical Society
RIBA	Royal Institute of British Architects
RS	Royal Society
SML	Science Museum Library
Sutro	Sutro Library, San Francisco
ULG	Goldsmiths' Library, University of London

Other, less frequently used, sources are named in full. And it must be emphasized that locations recorded under any given item do not constitute a complete census; copies may well exist elsewhere.

Collections

Particular copies of items in the bibliography are known to have originated in personal collections. The more interesting of these are recorded, under the following abbreviations.

MCL A contemporary collection of more than forty canal plans, mostly dating from 1758 to 1774, now in Manchester Central Library. The provenance has not been established.

RS (Smeaton) The folios of drawings by John Smeaton (1724–1792) in the Royal Society, include about a hundred engraved plans of bridges, river navigations, harbours, etc. They were part of his working collection.

BDL (Gough) Among a massive collection formed by the eminent topographer Richard Gough (1735–1809), and bequeathed by him to the Bodleian Library, are numerous reports, plans, etc. on civil engineering matters.

Sutro A collection made by Sir Joseph Banks (1743–1820) and partly inherited from his father, including reports and plans on fen drainage, harbours and river navigations chiefly in Lincolnshire and neighbouring counties. In the Sutro Library, San Francisco; but reduced photocopies are in the Natural History Museum, London.

CE (Vaughan) Five bound volumes in the Institution of Civil Engineers of reports on canals, docks and harbours collected by William Vaughan (1752–1850). His two large folios of London dock plans have been transferred to the Port of London Authority, the items being noted as PLA (Vaughan).

CE (Telford) Thomas Telford (1757–1834) donated his library to the Institution. Items recorded as above bear his signature or are endorsed 'Telford', but probably most of his own reports in CE come from this source.

CE (Page) Numerous pamphlets and printed reports on inland navigation, and a folio of thirty canal plans, collected by Frederick Page (1769–1834) and given to the Institution by his widow.

ULG (Rastrick) The working library of J. U. Rastrick (1780–1856), purchased by the University of London Goldsmiths' Library after the death of his son.

CE (Rennie) Two large portfolios of prints and drawings collected by John Rennie (1761–1821) and George Rennie (1791–1866). Presented to the Institution by Mrs. J. A. Rennie in 1949.

Elton The collection formed by Sir Arthur Elton (1906–1973) of books, drawings and prints illustrating the Industrial Revolution; now in the Ironbridge Gorge Museum.

Though not specifically referenced, all CE items by William Chapman (1749–1832) are from an almost complete collection of his printed reports given to the Institution by his brother in 1835.

Selected References

1. Richard Gough. *British Topography*. 2 vols. London, 1780.
    Indispensable for early plans; partially superseded by Hanson (item 11 in this list) and Higgs (item 7) for reports.
2. Hansard's *Catalogue of Parliamentary Reports 1696–1834*. House of Commons, 1834. With *A Continuation from 1835 to 1837*.
    Valuable but not complete catalogue; includes brief analysis of contents.
3. W. H. Wheeler. *A History of the Fens of South Lincolnshire*. 2nd edn. Boston, 1896.
    Gives titles of all relevant reports and plans.
4. *List of Manuscripts, Maps and Plans, and Printed Books and Pamphlets . . . from the Collections of John Urpeth Rastrick and his son Henry Rastrick.*

University of London Goldsmiths' Library, 1908.

5. W. T. Jackman. *The Development of Transportation in Modern England.* 2 vols. Cambridge, 1916.
   Copious references in the text to contemporary literature including parliamentary reports, with a good list of books and pamphlets.

6. R. A. Peddie. *Railway Literature 1556–1830.* London, 1931.
   Hand-list, virtually complete, of books, reports, pamphlets and plans. *See also* Haskell, 1955 (item 10).

7. Henry Higgs. *Bibliography of Economics 1751–1775.* Cambridge, 1935.
   Includes books, pamphlets and reports on fen drainage and transport. *See also* Hanson, 1963 (item 11).

8. Edward Lynam. Maps of the Fenland. (In) *Victoria County History of Huntingdonshire.* Vol. 3. London, 1936.
   Full bibliography of fenland maps and plans with very few omissions.

9. H. C. Darby. *The Draining of the Fens.* Cambridge, 1940.
   Includes a useful list of books, reports and pamphlets.

10. D. C. Haskell. *A Tentative Check-List of Early European (and British) Railway Literature 1831–1848.* Boston, Mass., 1955.
    A continuation of Peddie, 1931 (item 6).

11. L. W. Hanson. *Contemporary Printed Sources for British and Irish Economic History 1701–1750.* Cambridge, 1963.
    Companion volume to Higgs, 1935 (item 7).

12. George Ottley. *A Bibliography of British Railway History.* London, 1965.

Books, pamphlets and (non-parliamentary) reports.

13. Blackwell's Catalogue 865. *Rare and Interesting Books on Science Mainly Related to Transport and Technology.* Oxford, 1969.
    Bibliographic details of many books and reports, chiefly on railways and canals.

14. A. W. Skempton. *Early Printed Reports and Maps (1665–1850) in the Library of the Institution of Civil Engineers.* London, 1977.
    A finding-list arranged under author with a subject index and notes on production, distribution, etc. of printed reports; illustrated.

15. L. G. Booth. Tredgold's Publications. *Trans. Newcomen Soc.* **51** (1981 for 1979) 86–92.
    A complete bibliography of Thomas Tredgold.

16. A. W. Skempton (ed.) *John Smeaton, F.R.S.* London, 1981.
    With an account of Smeaton's working library, manuscript report books, plans, etc. and a list of his printed reports.

17. Julia Elton. *Bridges, Docks and Harbours with related works.* B. Weinreb Ltd. Catalogue 45. London, 1982.
    Annotated bibliographic descriptions of more than 500 items, many pre-1850.

18. Royal Scottish Geographical Society. *The Early Maps of Scotland to 1850.* Vol. 2. Edinburgh, 1983.
    Full bibliography including all Scottish road, canal and railway plans and maps.

19. Mark Baldwin. A Bibliography of British Canals, 1623–1950. (In) *Canals: a New Look* (ed. Mark Baldwin and Anthony Burton). Chichester, 1984.
    A useful list of books and pamphlets.

The Bibliography

ABERCROMBIE, Charles [d. 1817]

1 *Report of Mr Charles Abercrombie.*
Signed: Charles Abercrombie. Doonmouth, near
Ayr, July 1813.
pp. (3)–17.
[Printed with]
Report by Bryce MACQUISTON, 20 Aug 1813
[q.v.].
[In]
*Reports respecting a New Line of Road, from Hamilton
to Elvanfoot.*
Glasgow: Printed by J. Hedderwick and Co . . .
1813.
4to   28 cm   26 pp.   Folding engraved plan.
*Plan of different Lines of Roads, between Hamilton and
Elvan-Foot . . . from the Respective Surveys of Messrs
Abercrombie and Macquiston.*
Drawn by Peter Fleming, engraved by J. Hal-
dane.
NLS, CE

ABERDEEN HARBOUR

2 *The Reports by Smeaton, Rennie, and Telford, upon
the Harbour of Aberdeen.*

Aberdeen: Printed by G. Cornwall . . . 1834.
8vo   22 cm   56 pp.   6 folding hand-coloured
litho plans.
CE

AINSLIE, John [1745–1828]

See WHITWORTH, Robert. Plan of Forth &
Clyde Canal 1785, and Plan of proposed Leith
Docks 1786, drawn and engraved by Ainslie.

See AINSLIE and WHITWORTH. Report on
proposed canal betwixt Edinburgh and Glasgow [1794].

See RENNIE. Plan of different lines proposed
for canal between Edinburgh and Glasgow
1798, Plan of proposed canal between Glasgow
and Saltcoats 1803, and Plan of harbour at
Burntisland 1809, surveyed by Ainslie.

3 *Comprehensive Treatise on Land Surveying, comprising the Theory and Practice in all its branches; in which
the use of the various instruments employed in surveying,
levelling, &c, is clearly elucidated by Practical Examples.*
By John Ainslie, Land Surveyor.
Edinburgh: Printed for Silvester Doig & Andrew
Stirling . . . 1812.

4to 26 cm xvi + 248 + (1) pp. 40 engraved
  plates, some hand-coloured.
BM, NLS, IC

AINSLIE, John and WHITWORTH, Robert *junior*

4 *Report of John Ainslie and Robert Whitworth jun.
concerning the Practicability and Expence of Making the
Different Tracks proposed for a Canal betwixt the Cities
of Edinburgh and Glasgow.*
[?Edinburgh, 1794].
Signed: John Ainslie, Rob. Whitworth junior.
4to 26 cm (i) + 39 [i.e. 43] pp. Folding
  engraved plan.
*Plan of all the different Proposed Tracts for a Canal
between Leith and Glasgow.*
  Reduced to a small scale from the Large Plans
  Survey'd and Drawn by John Ainslie.
ULG (Rastrick)

AIRD, John

5 *Copies of Letters from Mr John Aird, to John
Rennie, Esquire, Engineer* [on Howth Harbour].
Signed: John Aird. Howth, 14 and 19 Dec 1816.
pp. 52–55.
[In]
*Fourth Report from the Select Committee on the Roads
from Holyhead to London.*
  House of Commons, 1817.

See RENNIE and AIRD. Plan of proposed harbour at Dunleary, 1817.

ALEXANDER, Daniel Asher
[1768–1846]

6 *The London Docks* [inset title: *Plan of the River
Thames with the Proposed Docks and Cut*].
  Dan.¹ Alexander, Surveyor. J. Cary, Engraver.
  1794.
Engraved plan 29 × 51 cm (scale 1 inch to
  1,200 feet).
PLA (Vaughan)
  The second scheme, including a canal from
  Wapping to Blackwall. For an earlier plan by
  John Powsey *see* VAUGHAN, 1794.

7 Second issue
[In] *Report from the Committee . . . on the Port of
London.*
  House of Commons, 1796.
  With a valuation of the lands by Daniel Alex-
  ander, 22 April 1796 (pp. 227–230) and
  engineering estimates by John RENNIE [q.v.].
  The plan also in *A Collection of Tracts on Wet
  Docks*, 1797. *See* VAUGHAN.

8 Another issue
The plan (on a smaller scale), valuation and
  estimates reprinted in *Reports from Committees
  of the House of Commons* Vol. 14 (1803)
  pp. 437–438 and pl. 12(b).

9 Revised scheme
The London Docks. D. Alexander, Surveyor. Nov.ʳ
  1796. J. Cary, Engraver.
Engraved plan, hand coloured 29 × 51 cm
  (scale 1 inch to 1,200 feet).
PLA (Vaughan)
  Also in *A Collection of Tracts on Wet Docks*, 1797.
  See VAUGHAN.

10 Second issue
[In] *Minutes of the Evidence taken at the Committee of
the Bill for making Wet Docks . . . within the Port of
London. Merchants Plan.*
  House of Commons, 1799.
  With evidence given by Daniel Alexander and
  John Rennie, 16–18 April 1799.

11 *Plan of the proposed London Docks*.
Daniel Alexander, Surveyor. Engraved by W. Faden . . . 1796.
Engraved plan, hand coloured 37 × 47 cm (scale 1 inch to 220 feet).
PLA
The revised scheme drawn to a larger scale with detailed local topography.

12 Another edition
Similar to No. 11 but 'Engraved by W. Faden 1797'.
BM, PLA (Vaughan)

13 Another edition
Similar to No. 11 but 'Engraved by W. Faden 1799'.
[In] *Second Report of the Select Committee on the Port of London*.
House of Commons, 1799. Pl. 13.

14 Another edition
[In] *Reports from Committees of the House of Commons* Vol. 14 (1803) pl. 27.

15 *A Correct Plan of the London Docks, shewing the Houses & other Premises, within the Limits granted by Act of Parliament, 1800*.
Published Apr.¹ 26th 1802 by John Fairburn. Engraved by Robert Rowe.
Engraved plan, hand coloured 42 × 58 cm (scale 1 inch to 220 feet).
PLA
From a drawing by Daniel Alexander, May 1801, following revisions proposed by Rennie, Mylne and Huddart. The plan as finally adopted.

16 *Report relative to Rye Harbour*.
By Daniel Alexander, Esq.
London: G. Cooke, Printer.
Signed: Daniel Alexander. London, 18 Nov 1813.
fol 28 cm 4 pp.
East Sussex R.O.

See DANCE, CHAPMAN *et al*. Report on London Bridge, 1814.

17 *Rye Harbour. Plan of Works proposed therein. Sep ¹ 1816*.
Daniel A. Alexander, Engineer.
Engraved plan 36 × 66 cm (scale 5 inches to 1 mile).
BM

ALLNUTT, Zachary

18 *Considerations on the Best Mode of improving the present imperfect state of the Navigation of the River Thames from Richmond to Staines. Showing the advantages to the Public . . . by improving the Navigation of the River in preference to the making any Canal*.
By Zach. Allnutt, Superintendent of the 2nd and 3rd Districts.
Henley: Printed for the Author . . . 1805.
8vo 19 cm 46 pp. 3 hand-coloured printed plans, 2 of which are folding.
[1] *A map of the Rivers Thames and Isis*.
[2] *A Plan of an Opening Weir across the Thames*.
[3] *A Profile of the Thames from Staines to Richmond*.
BM, ULG, BDL

19 *Useful and Correct Accounts of the Navigation, of Rivers and Canals West of London . . . with Tables of Distances; Time of Navigating; and Prices of Carriage on each River and Canal*.
By Mr Z. Allnutt. Superintendant and Receiver &c. on the Thames Navigation. The Second Edition, much improved.
Henley: Printed for the Author . . . [c. 1810].
8vo 22 cm 20 pp. Folding hand-coloured map.
BM, CE (lacking map)
The first edition not found.

ANDERSON, James [d. 1856] F.R.S.E.

See JARDINE. Plan of Callendar Park, 1818.

20 *Report relative to a Design for a Chain Bridge, proposed to be thrown over the Frith of Forth at Queensferry* [etc.].

By James Anderson, Civil Engineer and Land Surveyor.
Edinburgh: Printed by Balfour and Clarke, 1818.
Signed: James Anderson. Edinburgh, Jan 1818.
4to   27 cm   (i) + 31 pp.
CE (Telford)

21 *Report on the Present State of Leith Harbour, and the Practicability of rendering available its Wet Dock, by means of a Deep-Water Entrance at Newhaven, and a communicating Dock or Ship-Canal.*

By James Anderson, Civil Engineer.
Edinburgh: Adam and Charles Black . . . 1834.
Signed: James Anderson. Edinburgh, 6 Nov 1834.
4to   28 cm   (i) + 28 pp.   Folding engraved plan.
Reduced Plan of Leith Harbour & Docks exhibiting a Deep Water Entrance at Newhaven.
By James Anderson . . . April 1834.
BM, NLS

22   Second, revised edition
Title and publisher as No. 21 but *1836*.
Signed: James Anderson. Edinburgh, 29 Nov 1835.
4to   28 cm   (i) + 28 pp.   Folding engraved plan (title as in No. 21 but 1836).
CE

ARCHER, Thomas

23 *A Letter to the Right Honourable the Earl of Hardwicke, K.G. on the Use of the . . . Steam Dredging Engine in Deepening the Rivers in the Bedford Level; with Observations on the Effects of the Eau-Brink Cut.*

By Thomas Archer, Clerk to the South Level Commissioners.
Ely: Printed and Sold by J. Clements . . . 1829. Price One Shilling.
Signed: Thomas Archer. Ely, 7 Aug 1829.
8vo   22 cm   (i) + 21 pp.
CE, CRO

ARMSTRONG, *Colonel* John [1674–1743] F.R.S.

24 *Colonel Armstrong's Report, with Proposals for Draining the Fens and Amending the Harbour of Lynn, 1724.*
Signed: John Armstrong, 12 May 1724.
pp. 1–4.
[Printed with]
FULLER *et al.* A Report touching Lynn Navigation, 22 Jan 1723/4 [q.v.].
[and]
The Complaints of the Merchants, Mariners and Watermen of Lynn about Navigation, [August] 1724.
n.p.
fol   33 cm   8 pp.
CE, BDL (Gough)

25   Reprint
[In] An Answer to Mr Bridgman [December] 1724. pp. 15–17. *See* KING'S LYNN.

26   Reprint
[In] BADESLADE. Ancient and Present State of the Navigation of King's Lynn, 1725 [q.v.]. pp. 109–111.

ARROWSMITH, Aaron [1750–1823] Assoc.M.Inst.C.E.

See TELFORD. Plans of the Caledonian Canal, drawn by A. Arrowsmith, 1806–1820.

See TELFORD. Map of Scotland exhibiting the Highland Roads and Bridges, drawn by A. Arrowsmith 1807–1821.

ATWOOD, George [1745–1807] F.R.S.

27 *Answers [to the Questions respecting the Construction of a Cast Iron Bridge, of a Single Arch, 600 Feet in the Span, and 65 Feet Rise].*
By Mr Atwood.
Signed: G. Atwood. Sloane-street, Knightsbridge, [April 1801].
[In]
Report from the Select Committee upon Improvement of the Port of London.
1801. pp. 48–52.

28 Second printing
In *Reports from Committees of the House of Commons*
Vol. 14 (1803) pp. 621–623.

29 *A Dissertation on the Construction and Properties of Arches.*
By G. Atwood, Esq. F.R.S.
London: Printed by W. Bulmer and Co . . . 1801.
4to 28 cm viii + 51 pp. 7 engraved plates, 2 double-page and 1 folding.
BM, RS, CE, IC, BDL

30 *A Supplement to a Tract, entitled a Treatise on the Construction and Properties of Arches . . . to which is added, a Description of Original Experiments to verify and illustrate the Principles in this Treatise [etc].*
Part II. By the author of the first part.
London: Printed by W. Bulmer and Co . . . 1804.
Signed: G.A. London, 29 Nov 1803.
4to 28 cm xii + 47 + (11) pp. 5 engraved plates, 2 double-page and 1 folding.
BM, CE, IC

BADESLADE, Thomas

See FULLER *et al.* A Report touching Lynn Navigation, 1724.

See KING'S LYNN. An Answer to Mr Bridgman, 1724.

31 *The History of the Ancient and Present State of the Navigation of the Port of King's Lyn, and of Cambridge . . . and of the Navigable Rivers that have their Course through the Great Level of the Fens, called Bedford Level. Also the History of the Ancient and Present State of Draining in that Level . . . From Authentick Records . . . and from Observations and Surveys carefully made upon the Spot these Three Years last past.*
London: Printed by J. Roberts for the Author . . . 1725.
Signed: Tho. Badeslade.
fol 40 cm (xii) + 148 pp. 7 hand-coloured engraved plans (T. Badeslade del. S. Parker sc.).
These include:
[6] *The Upright & Plan of Denver Sluices.*
and four folding maps:
[2] *A Mapp of the River of Great Ouse.* Surveyed by Tho. Badeslade 1723.
[4] *A Map of Lynn-Haven & of the River Ouse to Germans.*
[5] *A Mapp of the great Level of ye Fens.* By Tho. Badeslade 1723.
[7] *A Plan and Description of the Fens . . . Survey'd* by Wm. Hayward A.D. 1604. Copied by T. Badeslade 1724.
BM, LCL, BDL, CUL
The Appendix (pp. 109–141) includes reports by Sir Clement Edmonds 1618 and Lord (Richard) Gorges 1682 [?printed here for the first time] and a reprint of Colonel ARMSTRONG's report of 1724 [q.v.].

32 Second edition
Title as No. 31.
London: Printed for L. Davis and C. Reymers . . . 1766.
fol 40 cm (vi) + 148 pp. 7 engraved plans (as in first edition but mostly uncoloured).
BM, CE, ULG, CUL
Using original copper plates, and apparently with identical typesetting of the main text.

33 *A Map of the North Level &c and of the Marshes a Drain is propos'd to be made through to convey the Waters yᵗ now Overflow that Level to Sea.*
T. Badeslade del. S. Parker sc. [1727].
Engraved plan, hand coloured 16 × 30 cm (scale 1 inch to 2½ miles).

RGS, BDL (Gough)

Issued at about the same time as PERRY's report of February 1727 on South Holland [q.v.].

34 *A Scheme for Draining the Great Level of the Fens, called Bedford-Level; and for Improving the Navigation of Lyn-Regis. Founded upon self-evident Principles in experimental Philosophy and practical Mathematicks, and upon Historical Facts* [etc].

By Tho. Badeslade.

London: Printed and sold by J. Roberts . . . 1729. No more than 150 copies printed. Price Two Shillings.

fol   40 cm   10 + (1) pp.   Folding hand-coloured engraved plan (No. [5] from The Ancient and Present State . . . 1725).

CE, ULG, BDL (Gough), CUL, Sutro

Some copies include an additional leaf in different type with the caption title *An Abstract of the Ancient and Present State of the Navigation of Lyn . . . and of Draining in the Fens.*

35 *Reasons humbly offer'd to the Consideration of the Publick; shewing how the Works now executing . . . to recover and preserve the Navigation of the River Dee, will destroy the Navigation; and occasion the Drowning of all the Low Lands adjacent to the said River. From Observations made on the Spot; and from Instances of the ruinous Effects like Works have had at the Ports of Lyn, Rye, Wisbech and Spalding. Illustrated with a Map of each of those Rivers, to compare with a Map of the River Dee, all drawn by Hand.*

Chester: Printed by Roger Adams. No more than 100 copies printed. Price Two Shillings and Sixpence with the Maps.

Signed: Thomas Badeslade, 15 Nov 1735.

fol   36 cm   14 pp.   4 plans in brown ink.

[1] A Map of the River Dee.
[2] Map of Lyn River & Harbour.
[3] Map of Rye Harbour & River.
[4] Map of Wisbech & Spalding Rivers.

BM, CUL

36 Second edition

Reasons humbly offer'd to the Consideration of the Publick [etc]. *Illustrated with a map of the River Dee. The Second Edition.*

Chester: Printed by Roger Adams. Price 2s. 6d.

fol   36 cm   14 pp.   4 plans, copied by hand from those in the first edition.

ULG, BDL (Gough)

37 *The New Cut Canal, intended for Improving the Navigation of the City of Chester, with the Low Lands adjacent to the River Dee, compared with the Welland, alias Spalding River, now Silted up, and Deeping-Fens adjacent, now Drowned.* [etc]

Chester: Printed by Roger Adams. Price Two Shillings.

Signed: Thomas Badeslade, 25 March 1736.

fol   36 cm   22 pp.

BM, ULG, BDL (Gough), CUL

Includes (pp. 17–22) a reprint, without Preface, of GRUNDY's *Philosophical and Mathematical Reasons*, 1736 [q.v.] and (on p. 22) *Mr John Reynolds, his Remarks on the Dam and Sluice making on the River Dee, to turn the Waters into the New Cut.*

BAGE, Charles [c. 1752–1822]

38 *Answers* [*to the Questions respecting the Construction of a Cast Iron Bridge, of a Single Arch, 600 Feet in the Span, and 65 Feet Rise*].

By Mr Charles Bage.

Signed: Chas Bage. Shrewsbury, 23 April 1801.
[In]
Report from the Select Committee upon Improvement of the Port of London.

1801. pp. 73–75.

39 Second printing

In *Reports from Committees of the House of Commons* Vol. 14 (1803) pp. 631.

BAIRD, Hugh [1770–1827]

40 *Report on the proposed Edinburgh and Glasgow Union Canal.*

By Hugh Baird, Civil Engineer.

Glasgow: Printed by D. Prentice & Co.

Signed: H. Baird. Kelvinhead, near Kilsyth, 20 Sept 1813.

4to   27 cm   24 + (2) pp.   Folding engraved plan, the lines entered in colours.

General Plan of the Communications that will be Formed by the Proposed Edinburgh & Glasgow Union Canal with other Navigations.

W. & D. Lizars sc.

NLS

General Meeting of Subscribers, 8 Oct 1813: ordered that the plan, report, and estimates, made by Mr Baird, be immediately printed. Also issued from the Caledonian Mercury Office, 1813.

41   Another edition

Title as No. 40.

Edinburgh: Printed by Oliver & Boyd . . . 1814.

4to   27 cm   24 pp.   with *General Plan of the Communications* [etc].

CE

42   *Subsidiary Report to the General Committee of Subscribers to the intended Edinburgh and Glasgow Union Canal.*

By Hugh Baird, Engineer.

Signed: H. Baird. Edinburgh, 24 Jan 1814.

pp. 1–11   with folding engraved plan.

Reduced Plan of the Proposed Edinburgh & Glasgow Union Canal from Lock Nº 16 on the Forth & Clyde Canal Navigation to the City of Edinburgh.

Surveyed by Hugh Baird Civil Engineer and Francis Hall Surveyor 1813. W. & D. Lizars sc.

[Printed with]

Report by Mr John RENNIE, 15 Jan 1814 [q.v.].

Edinburgh: Printed by Oliver & Boyd . . . 1814.

4to   27 cm   15 pp.   Folding engraved plan (as above).

NLS, CE

43   *Additional Report. To the Committee of Subscribers to the Edinburgh and Glasgow Union Canal.*

By Hugh Baird, Engineer.

n.p.

Signed: H. Baird. Kelvinhead, 17 Oct 1814.

4to   27 cm   12 pp.   Folding engraved plan.

Reduced Plan of the Proposed Edinburgh & Glasgow Union Canal from Lock Nº 16 . . . to the City of Edinburgh.

Surveyed by Hugh Baird Civil Engineer and Francis Hall Surveyor 1813 & 14. W. & D. Lizars sc.

CE

Second state of the plate with some revisions in the line of canal.

44   *Remarks, &c. To the Subscribers to the Intended Edinburgh and Glasgow Union Canal.*

n.p.

Signed: H. Baird. Edinburgh, 16 Nov 1814.

4to   27 cm   6 pp.

NLS, CE

45   *Report of Mr Hugh Baird on the proposed Plan for making the Dike or Track-Path to Carron Water Mouth.*

Signed: H. Baird, 8 Sept 1814.

Appendix pp. 11–12.

[In]

State of Facts and Observations relative to the Affairs of the Forth & Clyde Navigation.

1816. *See* FORTH & CLYDE CANAL.

46   *Reduced Plan of the Proposed Edinburgh & Glasgow Union Canal from Lock Nº 16 on the Forth & Clyde Canal Navigation to the City of Edinburgh.*

Surveyed by Hugh Baird Civil Engineer and Francis Hall Surveyor. 1813 & 16. Engraved by W. & D. Lizars.

Engraved plan   29 × 88 cm   (scale 1 inch to 1 mile).

NLS

Third state of the plate with further revisions in the line of canal.

See SPECIFICATIONS. Edinburgh & Glasgow Union Canal, 1818.

47   *Reports on the Improvements of the River Leven and Loch Lomond.*

By H. Baird, Civil Engineer.

n.p.

Signed: H. Baird. Kelvinhead, by Kilsyth, 25 Aug 1824 [and] Edinburgh, 20 Oct 1824.

4to   26 cm   18 pp.   Folding engraved plan.

Plan (and Section) of the River Leven shewing the Improvements Proposed by H. Baird.
W. H. Lizars sc.
ULG (Rastrick)

48 *Report on the proposed Railway, from the Union Canal at Ryal, to Whitburn, Polkemmet, and Benhar: or, the West Lothian Railway.*
By Hugh Baird, Civil Engineer.
Edinburgh: Printed by James Auchie.
Signed: H. Baird, 18 Dec 1824.
4to   27 cm   14 + (4) pp.   Folding hand-coloured engraved plan.
Plan (and Section) of the Proposed West Lothian Railway.
By H. Baird, Civil Engineer. Surveyed 1824.
W. H. Lizars sc.
BM, NLS

BALD, Robert [c. 1778–1861]
F.R.S.E., M.Inst.C.E.

49 *Report of a Mineral Survey of the Country through which the North or Level Line of Canal passes, as projected by John Rennie, Esq. Anno 1798.*
Signed: Robert Bald. Edinburgh, 28 May 1814.
pp. (1)–25.
[Printed with]
Observations occasioned by the Report of Mr Robert Bald . . . by John PATERSON, 12 Sept 1814 [q.v.].
[In]
Report of a Mineral Survey along the Track of the Proposed North or Level Line of Canal betwixt Edinburgh and Glasgow.
Leith: W. Reid & Co. Printers. [1814]
4to   25 cm   (iii) + 25 + 9 + 2 pp.
NLS, CE, ULG (Rastrick)

BALD, William [c. 1789–1857]
F.R.S.E., M.R.I.A.

50 *The Report (and Second and Third Reports) of Mr William Bald on Three large Districts in the County of Mayo.*
Signed: William Bald. Castlebar, 9 Dec 1811, n.d., and 28 Jan 1813.
With 4 folding hand-coloured engraved plans. (James Basire sc.).
[In]
Third Report of the Commissioners on the practicability of draining and cultivating the Bogs in Ireland.
House of Commons, 1814. pp. 131–152 and pls. 14–17.

51 *A Map of Lakes Corrib, Mask, Carra, Castlebar or Rahins, Cullin & Conn, showing the proposed navigable communications between them and the sea at Galway, Westport and Ballina.*
By William Bald, Civil Engineer, F.R.S.E., M.R.I.A. 1835. Day & Haghe lith.
Litho map, hand coloured   57 × 51 cm   (scale 1 inch to $4\frac{1}{4}$ miles).
[In]
Reports from the Select Committee on Connaught Lakes.
House of Commons, 1835. pl. 1.

See WALKER and BURGES. Belfast harbour. Sections of channel and quays, by William Bald, 1836.

52 *Report on the Harbour of Drogheda.*
By William Bald, Civil Engineer.
Printed at the Drogheda Journal Office, 1837.
Signed: William Bald. Dublin, February 1837.
12mo   12 pp.
CE

53 *General Report on a Part of the River Clyde, between Jamaica-Street Bridge, and the Glasgow Water-Works.*
By William Bald, F.R.S.E., M.R.I.A., Civil Engineer.
Glasgow: Printed by James Hedderwick & Son. 1839.
Signed: William Bald. Glasgow, 31 July 1839.
8vo   21 cm   12 pp.
CE

BALDWIN, Robert

54 *The New Bridge Building at Black Friars London.*
R.ᵗ Baldwin del. et sculp. Published . . . 15 Dec 1766.
Engraved elevation, including foundations, and plan   31 × 120 cm   (scale 1 inch to 25 feet).
BM, RIBA
  Blackfriars Bridge, designed by and built under the direction of Robert Mylne. This engraving is by one of his assistants.

55 *Plans, Elevations, and Sections of the Machines and Centering used in erecting Black-Friars Bridge.*
Drawn and engraved by R. Baldwin, Clerk of the Work.
London: Printed for I. and J. Taylor . . . [1787].
Oblong folio   41 × 54 cm   Title-page + 7 engraved plates. The plates dated 1 March 1787.
RIBA
  Pile engine, underwater pile-saw, caissons, and timber centering for the arches.

BANKS, *Sir* Edward [1770–1835]

56 *The Report of Edward Banks, on the Practicability and Expence of a Navigable Canal, proposed to be made between the Grand Southern Canal, near Copthorne Common and Merstham, to communicate with the River Thames at Wandsworth, by means of the Surry Iron Railways.*
London: Printed by E. Blackader . . . 1810.
Signed: Edward Banks. London, 3 July 1810.
4to   28 cm   8 pp.   Folding engraved plan.
Plan of a Canal proposed to be made . . . Surveyed by Edward Banks 1810.
BM (plan only), CE, ULG, BDL

BANKS, John

57 *A Treatise on Mills, in four parts.*
By John Banks, Lecturer in Experimental Philosophy.
London: Printed for W. Richardson . . . and W. Pennington, Kendal. 1795.
Dated: 29 Jan 1795.
8vo   22 cm   xxiv + 172 + (4) pp.   3 folding engraved plates.
BM, IC, BDL

58 Second edition
As No. 57 but *Second Edition*.
London: Printed for Longman, Hurst . . . and W. Grapel, Liverpool. 1815.
8vo   22 cm   vii + 172 + (3) pp.   3 folding engraved plates.
BM, SML

59 *On the Power of Machines, including . . . horizontal water wheel, centrifugal pump, common pump, etc., with the method of computing their force. Description of a simple instrument for measuring the velocity of air . . . Observations on wheel carriages . . . Experiments on the strength of Oak, Fir, and Cast Iron* [etc].
By John Banks, Lecturer on Philosophy.
Kendal: Printed by W. Pennington . . . 1803.
8vo   22 cm   viii + 127 pp.   3 folding engraved plates.
BM, ULG, IC, BDL

BARLOW, *Professor* Peter [1776–1826] F.R.S., Hon.M.Inst. C.E.

60 *An Essay on the Strength and Stress of Timber, founded on experiments performed at the Royal Military Academy . . . preceded by an Historical Review of former Theories and Experiments . . . also an Appendix on the Strength of Iron, and other materials.*
By Peter Barlow.
London: Printed for J. Taylor . . . 1817.
Dated: Royal Military Academy, Oct 1817.

8vo 22 cm xvi + 258 pp. 6 folding engraved plates.
BM, IC

61 Second edition
Title as No. 60 but *Second Edition, Corrected.*
London: Printed for J. Taylor . . . 1824.
Dated: Royal Military Academy, 16 Jan 1824.
8vo 22 cm xviii + 294 pp. 6 folding engraved plates (as in first edition).
RS, IC

62 Third edition
Title as No. 60 but *Third Edition, Corrected.*
London: Printed for J. Taylor . . . 1826.
Dated: Royal Military Academy, 16 Jan 1826.
8vo 22 cm xviii + 306 pp. 6 folding engraved plates (as in first edition).
BM
Price 16*s.* in boards.

63 *Calculation of the Stress and Strength of the pro-jected Iron Hanging Bridge over the Menai Strait.*
Signed: Peter Barlow. [May 1818].
[In]
Papers Relating to the building a Bridge over the Menai Strait, near Bangor Ferry.
House of Commons, 1819. pp. 13–14.

64 *Experiments on the Transverse Strength and other properties of Malleable Iron, with reference to its uses for railway bars; and a Report founded on the same, addressed to the Directors of the London and Birmingham Railway Company.*
By Peter Barlow, F.R.S. Cor. Mem. Inst. of France . . . [etc].
London: B. Fellowes . . . 1835.
Dated: Woolwich, 25 March 1835.
8vo 22 cm 97 pp. Woodcuts in text.
BM, NLS, SML, RS, ULG

65 *Second Report addressed to the Directors and Pro-prietors of the London and Birmingham Railway, foun-ded on an inspection of, and experiments made on the Liverpool and Manchester Railway.*
By Peter Barlow, F.R.S. Cor. Mem. Inst. France . . . [etc].
London: B. Fellowes . . . 1835.
Dated: Woolwich, 29 Oct 1835.

8vo 22 cm (iv) + 67 + 82–116 pp. Wood-cuts in text.
BM, NLS, RS, ULG, Elton

66 Second edition
Text exactly as No. 65 but new title page.
Report on the Weight of Rails . . . the Distance of Sup-ports . . . of the Liverpool and Manchester Railway.
By Peter Barlow . . . Second Edition.
London: John Weale . . . 1837.
ULG (Rastrick)

67 *A Treatise on the Strength of Timber, Cast Iron, Malleable Iron, and other materials; with Rules for Application in Architecture, Construction of Suspension Bridges, Railways, etc. with an Appendix on the Power of Locomotive Engines and the effect of inclined planes and gradients.*
By Peter Barlow, F.R.S. Mem. Inst. of France . . . and Hon. Mem. Inst. Civil Engineers.
London: John Weale . . . 1837.
Dated: 10 May 1837.
8vo 23 cm xii + 492 pp. 5 folding engraved plates.
BM, ISE, ULG, IC
A new edition was published in 1845.

See DRUMMOND *et al.* Two reports of the Commissioners on Railways in Ireland, 1837 and 1838.

See MAHAN. Course of Civil Engineering, edited by Barlow, 1838.

See TREDGOLD. Principles of Carpentry, edited by Barlow, 1840.

BARLOW, William Henry
[1812–1902] F.R.S.L. & E., M.Soc. C.E., M.Inst.C.E.

68 *On the Adaptation of different Modes of Illuminat-ing Light-Houses.*
By William Henry Barlow, Esq.
London: Printed by R. and J. E. Taylor . . . 1837.
4to 28 cm (i) + [221–235] pp.
CE (from the author)

Offprint, with new title-page, from *Phil. Trans.*, Vol. 127 (1837).

BARNARD, John

69 *To [the] Prince . . . and Princess . . . of Wales. This Design of a Bridge over the River Thames, from Kew, in the County of Surrey, to the opposite Shore, in the County of Middlesex . . .*
> is most Humbly Dedicated . . . by Their Royal Highnesses most Dutiful and Obedient Servant John Barnard Arch*!*

Printed for Rob*!* Sayer . . . Price 2*s*. 6*d*.
Engraved plan and elevation   29 × 72 cm (scale 1 inch to 20 feet).
RS (Smeaton), BDL (Gough)
> This timber-arch bridge was built 1758–59.

BARNES, James [*c*. 1740–1819]

70 *A Plan of the Proposed Canal from the Oxford Canal at Braunston, in the County of Northampton, to join the River Thames at New Brentford, in the County of Middlesex, to be called the Grand Junction Canal; with the collateral Cuts or Branches, from the said Canal to Daventry, Northampton, and Old Stratford, in the County of Northampton, and to Watford in the County of Hertford. [1792].*
Engraved plan   34 × 136 cm   (scale 1 inch to 1⅓ miles).
BM
> Survey by Barnes, approved with minor amendments (incorporated in this plan) by Jessop, October 1792.

> [Issued with]:

71 *A Section of the Proposed Canal . . . to be called the Grand Junction Canal* [etc, as above].
Engraved section   36 × 342 cm   (scale 1½ inches to 1 mile and 1 inch to 80 feet)
BM

The plan and accompanying section constitute perhaps the finest of all canal surveys.

72 *A Report and Estimate, made by Mr James Barnes, Engineer: under the Directors of the Company of Proprietors of the Leicestershire and Northamptonshire Union Canal.*
Banbury: Printed by J. Cheney . . . 1802.
Signed: James Barnes. Market Harborough, 26 July 1802.
4to   24 cm   20 pp.
CE

BATESON, *Reverend* Peter

73 *Some Papers relating to the General Draining of Marsh-Land in the County of Norfolk. With Mr Berner's Objections and Proposals. As also, an Answer to those Objections and Proposals.*
> By Peter Bateson.

n.p.   Printed in the Year 1710.
4to   23 cm   24 pp.   Folding engraved plan.
*A Map of Marsh-Land in Norfolk by S*ʳ *W*ᵐ *Dugdale with Additions and Amendments.*
CE, CRO, BDL (Gough), Sutro

See FULLER *et al.* Report touching Lynn Navigation, 22 Jan 1724.

BAYLIS, Thomas

74 *Plan (and Section) of the intended Rail or Tram Road from Stratford upon Avon in the County of Warwick to Moreton in the Marsh in the County of Gloucester with a Branch to Shipston upon Stower in the County of Worcester.*
> By Thomas Baylis, 1820.   T. Radclyffe sc.

Engraved map   22 × 43 cm   (scale 1 inch to 1 mile).
BM

BEDFORD LEVEL

75 *A Discourse concerning the Great Benefit of Drayning and imbanking, and of transportation by water within the Country.*
Presented to the High Court of Parliament by I.L.
n.p.   Printed by G.M., 1641.
Signed: J.L.
4to   20 cm   (iv) + 16 pp.   Folding engraved map of Cambridgeshire.
BM, BDL (Gough)

76 *The Drayner Confirmed, and the Obstinate Fen-Man Confuted. In a Discourse concerning the Drayning of Fennes, and surrounded Grounds in the Six Counties of Norfolk, Suffolk, Cambridge, Huntington, Northampton and Lincolne.*
London: Printed Anno Dom, 1647.
4to   19 cm   23 pp.
BM, BDL (Gough)
Second edition of *A Discourse concerning the drayning of the Fennes* by H.C., 1629.

77 *The State and Condition of the River Ouse, before and after the Erecting of Denver Sluces: and since their Fall.*
n.p.
Single sheet   40 × 25 cm   Docket title on verso.
CE
Endorsed in contemporary hand 'printed 1721'.

78 *Remarks upon that Part of the Great Bedford Level, called the North Level: in which the Causes of its present ruinous Condition are considered; with some Proposals for the better Draining the said Level. First published in 1748; and now Corrected, and Published, for the Perusal of the Publick, as there is a Bill depending in Parliament, for the better Draining of the North Level, with Porsand.*
n.p.   [1753]
fol   38 cm   3 pp.   Docket title on p. (4).
BM, CRO
Refers to the Bill for the first North Level Act of 1753.

79 *An Inquiry into Facts, and Observations thereon. Humbly submitted to the Candid Examiner into the Principles of a Bill intended to be offered to Parliament, for the Preservation of the Great Level of the Fens* [etc].
London: Printed for W. Owen . . . 1777.
8vo   19 cm   (iii) + 115 pp.
CE, ULG
Relates to the South and Middle Levels.

See EVIDENCE. Bedford Level Petition, etc. 1777.

See EAU BRINK CUT for pamphlets published 1793–94 on improvement of the South and Middle Levels.

80 *An Historical Account of the North Level of the Fens* . . . *1809.*
London: Strahan & Preston, Printers.
4to   24 cm   23 pp.
CE

BENTHAM, *Sir* Samuel [1757–1831] Hon.M.Soc.C.E.

81 *Answers* [*to the Questions respecting the Construction of a Cast Iron Bridge, of a Single Arch, 600 Feet in the Span, and 65 Feet Rise*].
By General Bentham. Inspector General of the Naval Works of the Admiralty.
Signed: Sam.ᴶ Bentham, 25 April 1801.
pp. 76–83   with folding engraved plate.
[In]
Report from the Select Committee upon Improvement of the Port of London.
House of Commons, 1801.

82 Second printing
[In] *Reports from Committees of the House of Commons*
Vol. 14 (1803) pp. 631–634 and pl. 56.

83 *Minute or Paper, from Mr Bentham, to the Navy Board, on the Subject of the Breakwater in Plymouth Sound.*

House of Commons, 1812.
Signed: S. Bentham, 4 Oct 1811.
fol   33 cm   5 pp.   Docket title on p. (6).

BENTINCK, *Lord* William Cavendish [1774–1839]

See RENNIE. Reports on the proposed bridge over the River Nene and a letter thereon from Lord William Cavendish Bentinck, 1819.

84 *A Letter from Lord William Bentinck to the Eau Brink Commissioners, on the Expediency of immediately widening the Cut, as recommended by the Engineers for Drainage and Navigation, and of Enlarging Denver Sluice.*

London: Printed for J. Booth . . . 1822. One Shilling.
Signed: William Cavendish Bentinck, 1822.
8vo   21 cm   29 + (2) pp.
CE

Includes (pp. 25–29) the report by Thomas TELFORD and John RENNIE, 15 April 1822 [q.v.].

BENTLEY, Thomas [1731–1780]

85 *A View of the Advantages of Inland Navigations: with a Plan of a Navigable Canal, intended for a Communication between the Ports of Liverpool and Hull.*

London: Printed for Becket and De Hondt . . . Parsons and Smith, in Newcastle under Lyne; and J. Gore, in Liverpool. 1765.
8vo   22 cm   (iii) + 40 pp.   Folding engraved plan (untitled) showing proposed canal from Wilden to Harecastle and thence to Northwich or the Bridgewater Canal.   Hugh Henshall del. Tho. Kitchin sc.
BM, CE, ULG, BDL (Gough)
Known to be written by Bentley.

86 Second edition
Title, etc as in No. 85 but *The Second Edition . . .* 1766.
8vo   22 cm   43 pp.   Folding engraved plan (as in first edition).
SML, CE, ULG
New setting of type.

BEVAN, Benjamin [*c.* 1772–1833]

87 *Plan of the Proposed Grand Union Canal. Survey'd 1808–1809.*

By B. Bevan.
Engraved plan, the line entered in colour
  17 × 49 cm   (scale 1 inch to 1 mile).
                    [Printed with]
Prospectus of the Intended Grand Union Canal.
[London]: S. Gosnell, Printer . . . 1809.
BM

88 *A Map of the intended Line of Canal, from Market Harborough to Stamford.*

Survey'd by B. Bevan 1810. J. Cary sc.
Engraved plan, the line entered in colour
  23 × 74 cm   (scale 1 inch to 1 mile).
BM

89 *A Practical Treatise on the Sliding Rule: in Two Parts. Part the First being an Introduction to the Use of the Rule . . . Part the Second containing Formulae for the Use of Surveyors, Architects, Civil Engineers, and Scientific Gentlemen.*

By B. Bevan, Civil Engineer and Architect.
London: Printed for the Author; and Sold by Longman, Hurst . . . also by W. Cary . . . 1822.
Signed: B. Bevan, 13 Nov 1821.
8vo   23 cm   101 + (6) pp.
BM

90 Second issue

Title as No. 89.

London: Printed for the Author; and sold by W. Cary . . . 1838.

BM

A reissue with new title page.

91 *Report of Benjamin Bevan, Esq. Civil Engineer, upon the Stour Navigation and Sandwich Harbour.*

Canterbury: Printed by Cramp and Co . . . 1824.

Signed: B. Bevan. Leighton Buzzard, 9 Dec 1824.

8vo   21 cm   8 pp.

Kent A. O.

BIDDER, George Parker
[1806–1878] M.Inst.C.E.

See STEPHENSON, George and Robert. Map of proposed railway between London and Brighton, 1836. G. P. Bidder, acting engineer.

See STEPHENSON and BIDDER. London & Blackwall Railway report, 1838.

BIEDERMANN, Henry Augustus
[*c.* 1744–1816]

92 *A Plan of Wells Harbour with the adjoining Salt Marshes and the several Imbankments of Wells & Warham &c. 1780.*

Engraved plan, hand-coloured   34 × 43 cm (scale 5 inches to 1 mile).

RS (Smeaton), ULG

This plan should have been annexed to MYLNE's printed report of 28 April 1781. The original drawing, signed by Biedermann, is with Mylne's papers on Wells Harbour in the University of London Library.

BLACKADDER, William
[1789–1860] M.Inst.C.E.

See STEVENSON, Robert. Plan of proposed railway from Montrose into Strathmore, 1826.

93 *Report relative to the Strathmore Railway, being the extension of the Dundee and Newtyle Railway from Newtyle along Strathmore between Coupar and Forfar.*

By William Blackadder.

Dundee: Printed . . . by D. Hill, 1833.

Signed: Wm. Blackadder. Glamis, Dec 1832.

4to   29 cm   26 pp.   Folding engraved plate.

Map of the Strathmore Railway from near Nethermill to Coupar.

1834. Designed & Surveyed by W. Blackadder. Vallentine & Son sc.

CE

BLACKFRIARS BRIDGE

94 *The Expedience, Utility, and Necessity of a New Bridge at or near Blackfryars.*

London: Printed for M. Cooper . . . 1756. Price Sixpence.

8vo   20 cm   (i) + 22 pp.

BM, CE

95 *Observations on Bridge Building, and on the several Plans offered for a New Bridge. In a Letter addressed to the Gentlemen of the Committee, appointed by the Common-Council of the City of London, for putting in Execution, a Scheme for building a new Bridge across the Thames, at or near Black Friars.*

London: Printed and Sold by J. Townsend . . . 1760.

Signed: Publicus.

8vo   21 cm   (i) + 50 pp.

BM, IC, BDL

This has been attributed to Robert Mylne.

96 *Postscript* [to Observations on Bridge Building . . .]
[London, 1760].
8vo   21 cm   pp. (51)–58.
BM, IC
   Published as an addendum to *Observations*, in response to SMEATON's *Answer* 9 Feb 1760 [q.v.].

97 [*Report from*] *The Committee appointed to carry into Execution the Act of Parliament, for building the Bridge at Blackfriars.*
   18 Jan 1771.
   pp. 5–19.
                  [Printed with]
Various papers including Petitions by Robert Mylne dated 13 Dec 1770, 21 Jan 1771 and 15 April 1774.
                  [In]
Bull, Mayor. A Common-Council holden . . . on Wednesday the 27th Day of April 1774.
   Corporation of London, 1774.
fol   30 cm   23 pp.
CLRO, CE

BLACKMORE, John [*c.* 1801–1844] M.Inst.C.E.

98 *Views on the Newcastle and Carlisle Railway, from original drawings by J. W. Carmichael, with details by John Blackmore, Engineer to the Company.*
Newcastle: Currie and Bowman . . . 1836 [–1838].
4to   28 cm   Engraved vignette title page + (25)pp.   23 steel-engraved plates (J. W. Archer, T. E. Nicholson, J. T. Willmore and others after John Wilson Carmichael).
BM, NLS, SML, ULG, Elton
   Issued in parts 1836, 1837 and 1838. Also published 1839 in blind and gilt-stamped cloth: price 21*s*.

99 *Great Inland Junction Railway, or Inland Line of Railway from Newcastle-upon-Tyne to Edinburgh.*

[*Report*] *To the Directors of the Newcastle-upon-Tyne and Carlisle Railway.*
Newcastle: Printed by John Hernaman.
Signed: John Blackmore. Newcastle, Dec 1838.
fol   33 cm   4pp.
BM

BOOTH, Henry [1788–1869]

100 *An Account of the Liverpool and Manchester Railway, comprising a History of the Parliamentary Proceedings, preparatory to the passing of the Act, a Description of the Railway . . . and a Popular Illustration of the Mechanical Principles applicable to Railways. Also, an Abstract of the Expenditure . . . with Observations on the same.*
By Henry Booth, Treasurer to the Company.
Liverpool: Printed by Wales and Baines . . . [1830].
Dated: Liverpool, June 1830.
8vo   24 cm   (i) + 104 pp.   Litho frontispiece (Hullmandel) + engraved plate + 2 folding litho plans: a map and section of the line (Miller).
BM, NLS, SML, ULG

101 Second edition
As No. 100 but *Second Edition . . .* 1831.
Dated: 20 Jan 1831.
8vo   23 cm   104 pp.   Frontispiece, plate and plans as in first edition.
BM, SML, CE, ULG
   New setting of type throughout, with some additional material.

BOURNE, John Cooke [1814–1896]

102 *Drawings of the London and Birmingham Railway.*

By John C. Bourne. With an Historical and Descriptive Account by John Britton F.S.A.

London: Published by J. C. Bourne . . . Ackermann and Co . . . and C. Tilt . . . 1839

fol 54 cm 26 pp. Tinted litho title page + map + 29 litho plates (J. C. Bourne del et lith).

BM, SML, CE, ULG, Elton

Issued in four parts at £1 1s. each, between September 1838 and July 1839, then published with map and Account by John Britton (1771–1857) as above; price £4 14s. 6d. in half morocco. Seventeen of the plates are from drawings made in 1836–37 during construction of the railway.

BOWDLER, Astley

103 Deposited Plan
Plan of the Severn & Wye Railway and Canal, from Bishop's wood and Lidbrook on the River Wye, to Nass point on the Severn, in the County of Gloucester.

Astley Bowdler, Surveyor and Engineer, 1810. Matlow sc.

Engraved plan, lines entered in colour 73 × 42 cm (scale 2 inches to 1 mile).

HLRO

BOWER, Anthony

104 *A Plan of the Holderness Drainage in the County of York.*

By A. Bower, Surveyor 1781.

Engraved plan 74 × 58 cm (scale 1 inch to 2,000 feet).

HRO, Hull City Archives

105 *A Plan of the Town & Harbour of Kingston upon Hull.*

By A. Bower, 1786. Published 1 March 1787 by Rob.ᵗ Wilkinson.

Engraved plan 23 × 36 cm (scale 1 inch to 360 feet).

Hull City Archives, RS (Smeaton), RGS, Sutro

See RENNIE. Plan and profile of the River Witham between Boston and Lincoln, surveyed by Bower 1803 and 1804.

106 *A Plan of the River Foss Navigation with the low lands adjoining, shewing the Intended Improvement of the Drainage.*

Anthony Bower, 1804. W. Faden sc.

Engraved plan 48 × 118 cm (scale 4 inches to 1 mile).

BM

107 *Mr Bower's Report respecting the Improvement of the River Witham, from the Grand Sluice at Boston, to Stamp End Lock, at Lincoln.*

Sleaford: Thornhill, Printer.

Signed: Anthony Bower. Lincoln, 11 Nov 1806.

fol 32 cm 2 pp.

Sutro

See RENNIE. Longitudinal and transverse sections of the River Ouse, the Hundred Foot and Old Bedford and the River Cam from Kings Lynn to Earith and Clayhithe, surveyed by Bower, 1810.

An outstanding survey of these fenland rivers.

See RENNIE. Plan of the River Witham between Lincoln and Boston, with intended improvements of the navigation, surveyed by Bower, 1812.

108 *Mr Bower's Statement as to the Drainage and Levels of the Fens North of Boston, and Comparison with the Levels and Drainage of the Low Lands of South Holland and the Bedford Level.*

Signed: Anthony Bower. Lincoln, 23 Feb 1814.

pp. 23–32 with folding hand-coloured engraved plan.

Plan of the East, West and Wildmore Fens, near Boston. With the respective Drains which have been executed for the Drainage thereof.

Anth⸢ʳ⸣ Bower 24 Sept 1811.
[Printed with]
RENNIE's reports on drainage of South Holland
&c. 28 Jan 1813 and 26 Jan 1814 [q.v.].
[In]
Reports as to Wisbech Outfall, and the Drainage of the
North Level and South Holland.
By John Rennie, Civil Engineer.
With a Statement &c. as to the Drainage of the Fens
North of Boston.
By Anthony Bower.
London: Printed by Richard and Arthur Taylor
. . . 1814.
4to   28 cm   (i) + 33 pp.   2 folding hand-
coloured engraved plans.
LCL, CRO, CE, ULG, CUL
Issued in blue wrappers.

BRADFORD, Henry

109 *A Plan of the Rivers Trent and Tame from Burton*
to Tamworth in the County of Stafford.
Surveyed by Henry Bradford. 1758. Engraved
by Tho⸢ˢ⸣ Jeffreys 1759.
Engraved plan   30 × 26 cm   (scale 2 inches to
1 mile).
RS (Smeaton), RGS, MCL, BDL (Gough)

BRADSHAW, George [1801–1853]
Assoc.Inst.C.E.

Note that Bradshaw published three large-scale
maps of the canals, navigable rivers
and railways in the Northern, Midland and
Southern parts of England. Their publication history, covering a period of about
seven years before *c.* 1835, can be clarified by
recognizing two distinct 'editions' of both the
Northern and Midland maps and a single edi-

tion of the Southern map, and then distinguish-
ing several 'states' within each edition dependent upon relatively minor variations. It is
remarkable that levels throughout all three
maps are related to a single fixed datum (at
Liverpool) before the Ordnance Survey
adopted a similar practice.

110   Northern map. First edition, first state
[*c.* 1828]
G. Bradshaw's Map of Several Canals, situated in the
Counties of Lancaster, York, Derby & Chester; shewing
the Heights of their Pools above the Level of the Sea at
Low Water, from Levels taken by William Johnson and
Son, Manchester.
Manchester Published by G. Bradshaw & Sold by
Mr J. Gardner . . . London.
Engraved map, hand-coloured   86 × 124 cm
(scale 1 inch to 2 miles).
CE
With 'Parliamentary Line of the Manchester &
Liverpool Railway'. According to a note in an
early MS. catalogue of maps and plans in the
Institution of Civil Engineers the levels were
taken in 1825.

111   First edition, second state [*c.* 1829]
G. Bradshaw's Map of Canals, situated in the Counties of
Lancaster, York, Derby & Chester; shewing the Heights
of their Pools from a Level of 6ft. 10in. under the Old
Dock Sill at Liverpool. From Levels taken by William
Johnson and Son, Manchester.
And Dedicated by Permission to Thomas Tel-
ford, F.R.S.L. & E. President of the Institution
of Civil Engineers by his Obliged G. Bradshaw.
SML
The map and levels appear to be identical with
the first state. This is traditionally the 'First
Edition'.

112   Second edition, first state [1833]
Title as No. 111 but map has a wider border
(88 × 126 cm), much additional topography in
Yorkshire, and the Liverpool & Manchester
and other railways as detailed in *Lengths and*
Levels 1833 (No. 123).
BM, SML

113 Second edition, second state [*c.* 1834]
As No. 112 but with line of the Leeds & Selby
  Railway.
RGS

114 Midland map. First edition, first state.
1829
*G. Bradshaw's Map of Canals, Navigable Rivers, Rail
Roads &c in the Midland Counties of England. From
Actual Survey shewing the Heights of the Pools on the
Lines of Navigation above the Low Water at Liverpool.
From Levels taken by Twyford & Wilson Surveyors &
Engineers, Manchester.*
Published by G. Bradshaw Manchester, Feb-
  ruary 12th 1829, and Sold by Mr Jas. Gardner
  . . . London. Price mounted on rollers
  £2 2s.—in sheet £1 11s. 6d. Engraved &
  Printed by W. R. Gardner, 13 Harper Street,
  London.
Engraved map, hand-coloured   131 × 121 cm
  (scale 1 inch to 2 miles).
ULG

115 First edition, second state [*c.* 1829]
*G. Bradshaw's Map of Canals, Navigable Rivers, Rail
Roads &c in the Midland Counties of England. From
Actual Survey shewing the Heights of the Ponds on the
Lines of Navigation from a level of 6 feet 10 inches under
the Old Dock Sill at Liverpool. From Levels taken by
Twyford & Wilson Surveyors & Engineers, Manchester.*
Published by G. Bradshaw Manchester, Feb-
  ruary 12th 1829, and Sold by Mr Jas. Gardner
  . . . London. Price mounted on rollers
  £2 2s.—in sheets £1 11s. 6d. Engraved &
  Printed by W. R. Gardner, Harper Street, Lon-
  don. To Thomas Telford . . . This Map as a
  Tribute of Respect is by Permission dedicated
  by his obliged G. Bradshaw.
SML, CE
  The map and levels appear to be identical with
  the first state. This is traditionally the 'First
  Edition'. The CE copy is a slightly earlier version with the (printed) dedication pasted on.

116 First edition, third state. 1830
As No. 115 but 'Published . . . February 12th
  1830', and price not stated.
P, B

117 Second edition, first state [*c.* 1832]
Title and date as No. 116 but the Grand Junction
  Canal is extended from Stoke Hammond to
  Tring with branches to Aylesbury and Wen-
  dover, and the tabulated lengths and widths of
  locks are re-engraved further to the left.
BM, SML, ULG
  The extension, cut into part of the border, fills
  what would otherwise be a gap in the Grand
  Junction Canal between the Midland and
  Southern maps.

118 Second edition, second state [1835]
As No. 117 but with lines of the London & Birmingham and Leicester & Swannington Railways.
RGS

119 Southern map, first state [1832]
*G. Bradshaw's Map of Canals, Navigable Rivers Rail-
ways &c in the Southern Counties of England. From
Actual Survey shewing the Heights of the Pools on the
Lines of Navigation also the Planes on the Railways from
a Level of 6ft. 10in. under the Old Dock Sill at Liver-
pool.*
  Dedicated by Permission to Thomas Telford,
  F.R.S.L. & E. President of the Institution of
  Civil Engineers.
Engraved map, hand-coloured   95 × 185 cm
  (scale 1 inch to 2 miles)   Issued in three
  sheets.
P
  The map was presumably published at the
  same time as *Lengths and Levels* for the Southern
  Counties (No. 122) in 1832. Trinity High
  Water Mark is recorded as 21 ft above datum
  at Liverpool.

120 Second state [1833]
Title and map as in No. 119 but with line of the
  'Proposed London & Birmingham Rail Way'.
BM, SML, CE (from the author March 1833),
  RGS, ULG

121 *Appendix to G. Bradshaw's Map of the Canals
and Navigable Rivers of the Midland Counties of England.*

Manchester: Printed by A. Prentice . . . 1829.
Dated: Manchester, 29 Feb 1829.
8vo   21 cm   20 pp.
BM, CE (Page), ULG

Lengths and falls of the canals and rivers in the first edition of the Midland map.

122 *Lengths and Levels to Bradshaw's Maps of Canals, Navigable Rivers, and Railways.*

From Actual Survey. Taken from a Datum of Six Feet Ten Inches under the Sill of the Old Dock Gates at Liverpool. Dedicated to Thomas Telford, F.R.S.L. & E. President of the Institution of Civil Engineers.

London: T. G. White and Co. Printers . . . 1832.
8vo   21 cm   15 pp.
BM, CE (Page), RGS, ULG

For the Southern Counties only, but includes that part of the Grand Junction Canal and its branches added in the second edition of the Midland map. *See also* No. 124.

123 *Lengths and Levels to Bradshaw's Maps of the Canals, Navigable Rivers, and Railways, in the Principal Part of England.*

Dedicated to Thomas Telford, F.R.S.L. & E. President of the Institution of Civil Engineers. Datum, Six Feet Ten inches under the Sill of the Old Dock Gates at Liverpool.

London: Printed by E. Ruff . . . 1833.
8vo   22 cm   15 + [5–20]pp.
BM, RGS, ULG

First part: Canals, rivers and railways in the Northern map. Second part: a reissue (without prelims) of the 1829 *Appendix*.

124 Collected edition
Nos. 122 and 123 bound together, with folding engraved plate.
Continuation of the Lea Navigation . . . [and] of the Grand Junction Canal.
SML, CE (Telford), ULG (Rastrick)

As issued in publisher's cloth.

125 *Map & Sections of the Railways of Great Britain.*

Dedicated by Permission to James Walker,

F.R.S.L. & E. President of the Institution of Civil Engineers by George Bradshaw. Engraved by I. Dower . . . London. Published 14 Jan 1839 by G. Bradshaw, Manchester, and Sold by James Gardner . . . London.

Engraved map, hand coloured, with gradient profiles   162 × 100 cm   (scale 1 inch to 10 miles).
BM, SML

Also issued mounted and bound integrally with *Table of the Gradients*; price £3 3*s*. There are several later editions.

126 *Table of the Gradients to Bradshaw's Map of the Railways, of Great Britain, containing particulars of the Lengths, Levels, and Gradients, of all the Principal Railways in the Kingdom.*

Dedicated to James Walker, F.R.S.L. & E. President of the Institution of Civil Engineers.
Manchester: George Bradshaw . . . 1839, and sold by James Gardner . . . London.
Dated 14 Jan 1839.
8vo   24 cm   26 + (2) pp.
CE

127 Second edition
Title, date, etc as No. 126 but with 3 extra pages of text.
[Manchester ?1840].
8vo   24 cm   29 + (1 blank) + (2) pp.
SML

BRAITHWAITE, John [1797–1870] M.Inst.C.E.

128 *Eastern Counties Railway.*

John Braithwaite, Engineer. 1 November 1834. Waterlow & Moreland, lithog.
Litho map   48 × 32 cm   (scale 1 inch to 6 miles).

[In]

Prospectus of the Eastern Counties Railway. From Lon-

don to Ipswich, Norwich and Yarmouth, by Romford, Chelmsford, Colchester, &c.
1835.
[London]: Cunningham and Solnar, Printers.
fol   50 cm   (4) pp.
BM

Another Prospectus (1834) with a similar map in BM and CE.

129   Deposited plan
Map of the Intended Eastern Counties Railway from London to Norwich and Yarmouth.
1835. On Stone by C. F. Cheffins.
Litho plans on 21 sheets, each 41 × 70 cm (scale 4 inches to 1 mile).
[and]
Enlarged Map of that portion of the intended Eastern Counties Railway extending through the County of Middlesex.
John Braithwaite, Esq. Engineer. John Doyley, Surveyor.
Litho plans on 2 sheets, each 41 × 70 cm   (scale 16 inches to 1 mile)   Oblong folio   Title page + 23 sheets of plans.
BM

130   *Central Kent Railway (Direct Line from London to Dover).*
Consulting   Engineer,   John   Braithwaite. Engineer, J. W. Bazalgette. Allens, Lith. [?1837].
Litho map   32 × 38 cm   (scale 1 inch to 4½ miles).
BM

BREES, Samuel Charles [*c.* 1800–1865]

131   *Railway Practice. A Collection of Working Plans and Practical Details of Construction in the Public Works of the most Celebrated Engineers,* [etc].
By S. C. Brees, C.E.
London: John Williams . . . 1837.

Signed: S. C. Brees, 40 Ely Place [London].
4to   28 cm   xxxii + 108 pp.   Tinted litho frontispiece, 65 + 12 double-page and folding litho plates (W. Clerk and Day & Haghe); the plates issued plain or hand coloured.
BM, SML, CE, ULG.

Frontispiece (from a drawing by Brees) depicts work in progress at North Church Tunnel. There are 65 plates of working drawings. These include 40 showing works on the London & Birmingham Railway, and 7 on the Leeds & Selby Railway, and two drawings of the Lancaster Canal aqueduct over the R. Lune.

132   Second edition
Title as No. 131 but *Second Edition, corrected and improved.*
London: John Williams . . . 1838.
4to   28 cm   xiv + [xix]–xxxvi + 106 pp. Tinted litho frontispiece + 80 double-page and folding litho plates (W. Clerk and Day & Haghe); the plates issued plain or hand coloured.
CE, ULG (lacks plates)

With revisions and three extra plates of working drawings. As in the first edition the last 12 plates are of 'designs' by Brees. French translations were published in 1841 (Paris and Brussels) and a third English edition in 1847.

133   *Appendix to Railway Practice, containing a copious abstract of the whole of the Evidence given upon the London and Birmingham, and Great Western Railway Bills . . . to which is added A Glossary of Technical Terms, used in Civil Engineering . . . and Details of Hawthorne's Celebrated Locomotive Engine, for the Paris and Versailles Railway.*
By S. C. Brees, C.E.
London: John Williams . . . 1839.
4to   28 cm   vii + errata   leaf + 373 pp.   6 engraved plates of which 3 are folding (W. Kelsall sc.).
BM, CE, ULG

All the plates (drawn by John Hodgson junior) illustrate the Hawthorne locomotive.

134 *Second Series of Railway Practice: a Collection of Working Plans and Practical Details of Construction in the Public Works of the most Celebrated Engineers: [etc].*

By S. C. Brees. C.E.

London: John Williams . . . 1840.

Dated: 12 South Square, Gray's Inn, 1 May 1840.

4to   28 cm   viii + [9]–124 pp.   Engraved frontispiece (pl. 1) + 2–61 double-page and folding engraved plates (B. R. Davies and B. Winkles).

BM, CE, ULG

Frontispiece (from a drawing by Brees) shows the Great Western Railway bridge over the Thames at Maidenhead. The plates illustrate many railway works and swing bridges at the London and St. Katharine Docks, works at Grangemouth harbour, etc. A second edition was published in 1847, as also the *Third Series* and *Fourth Series*.

135 *A Glossary of Civil Engineering, comprising Theory and Modern Practice; and the Subjects of Field and Office Work, and Mechanical Engineering connected with that science.*

By S. C. Brees, C.E., Author of 'Railway Practice'.

London: John Weale . . . 1840.

Dated: 12 South Square, Gray's Inn, 20 May 1840.

8vo   22 cm   vi + errata leaf + [7]–310 pp. Wood-engraved frontispiece [after J. C. Bourne] and numerous woodcut illustrations.

P

Issued in publisher's blind-stamped cloth. Second and third editions appeared in 1841 (BM, CE) and 1844.

BRETT, *Sir* Peircy [1709–1781]

See BRETT and DESMARETZ. Plan and report on Ramsgate Harbour, 1755.

BRETT, *Sir* Peircy and DESMARETZ, *Captain* John

136 *A Plan for making a Harbour at Ramsgate. Survey'd Sep.ʳ 1755.*

By S.ʳ Peircy Brett and Captain Desmaretz.

Engraved plan   35 × 33 cm   (scale 1 inch to 200 feet).

BM, RS (Smeaton), Sutro

137 *Report and Estimate subjoined, relating to the Harbour of Ramsgate.*

[House of Commons]: Printed 1756.

Signed: Peircy Brett, J. P. Desmaretz. 24 Dec 1755.

fol   35 cm   9 pp.

BM

BRIDGEMAN, Charles [?1685–1738]

138 *A Report of the Present State of the Great Level of the Fenns called Bedford-Level. And of the Port of Lynn; and of the Rivers Ouse and Nean . . . with Considerations on the Scheme propos'd by the Corporation of Lynn . . . And also, a Scheme humbly propos'd for the Effectual Draining those Fenns, and Reinstating that Harbour or Port. From a Survey thereof made in August, 1724.*

n.p.   [Printed November 1724].

Signed: Charles Bridgeman.

fol   33 cm   15 pp.

CE, BDL (Gough)

139 Reprint

[In] An Answer to Mr Bridgman, [December] 1724. pp. 1–14 recto. *See* KING'S LYNN.

Includes an engraved plan illustrating Bridgeman's scheme.

BRINDLEY, James [1716–1772]

140 *A Plan for a Navigation chiefly by a Canal from Longbridge near Burslem in the County of Stafford to Newcastle, Lichfield and Tamworth, and to Wilden in the County of Derby.*

By James Brindley. Revis'd and approv'd by John Smeaton F.R.S. 1760. R. W. Seale sc.
Engraved plan   53 × 74 cm   (scale 1 inch to 2 miles)
RS (Smeaton)

141 *A Plan of the Rivers Irwell and Mersey from Manchester to Runcorn Gap and from Longford Bridge to the Hempstones in the County of Chester.*

Taken in·Novemb.ʳ 1761 by Hugh Oldham.
Engraved plan   29 × 55 cm   (scale 1 inch to 1 mile)
MCL, BDL (Gough)
    Land-survey for the proposed extension of the Bridgewater Canal. Brindley took the levels along this line in Sept–Nov 1761.

142 *A Plan of the Duke of Bridgewater's Navigable Canal already made, with the Extention proposed from Longford Bridge to Liverpool.*

Engraved plan   18 × 33 cm   (scale 1 inch to 2 miles).
MCL, BDL (Gough)
    A 'popular' edition of Oldham's plan. A later version is annexed to [BRINDLEY] *History of Inland Navigations*, Part I, 1766 [q.v.].

143 *A Plan of the Great Navigable Canal, intended to be made from the Trent to the Mersey; and also of the Duke of Bridgewaters Navigations, proposed to Communicate therewith.*

Hugh Henshall del.   Tho: Kitchen sc. [1765].
Engraved plan   51 × 27 cm   (scale 1 inch to 4 miles).
MCL
    Henshall was Brindley's assistant, and resident engineer on the Trent & Mersey Canal. A smaller version of the plan, also drawn by Henshall, is annexed to BENTLEY's *View of the Advantages of Inland Navigations*, 1765 [q.v.]

while a later version to the same scale, including the Staffs & Worcs (i.e. Trent to Severn) Canal, is in [BRINDLEY] *History of Inland Navigations*, Part II, 1766 [q.v.].

144 *[Report relating to the Petition of the Proprietors of the London-Bridge Water-Works].*
Signed: James Brindley. New Chapel, near Burslem, 16 Jan 1767.
p. 2.
[In]
Kite, Mayor. A Common-Council holden ... on Wednesday the 25th Day of February 1767.
    Corporation of London, 1767.
fol   33 cm   7 pp.
CLRO

145 *A Plan of the intended Navigation from Birmingham, in the County of Warwick, to the Canal at Aldersley near Wolverhampton in the County of Stafford.*

Survey'd 1767.   Jnᵒ Wedge del.   Westwood sc.
Engraved plan   23 × 53 cm   (scale 2½ inches to 1 mile).
BM
    The preferred line of two schemes produced by Brindley in June 1767.

146   Second state
As No. 145 but Rob.ᵗ Whitworth del. [and with some modifications to the line near Birmingham].
MCL, BDL (Gough)

147 *A Plan of the intended Navigable Canal from the City of Coventry to the Canal on Fradley Heath, in the County of Stafford.*

Surveyed in 1767. Rob.ᵗ Whitworth del. Tho.ˢ Kitchin sc.
Engraved plan   29 × 70 cm   (scale 1 inch to 1 mile).
BM, BDL (Gough)
    Drawn and probably surveyed by Robert Whitworth, Brindley's principal assistant engineer 1767–1772.

148 *A Plan of the River Salwarp, and of the intended Navigable Canal, from Droitwich to the River Severn in the County of Worcester.*

Survey'd in Oct. 1767. Robt Whitworth del. Westwood sc.

Engraved plan   21 × 35 cm   (scale 3 inches to 1 mile)   with table of lengths and falls.

BDL (Gough)

149 *A Plan of the intended Navigable Canal, from the Coventry Canal near the City of Coventry, to the City of Oxford.*

Survey'd in 1768. Robt Whitworth del.   J. Cole sc.

Engraved plan   26 × 66 cm   (scale 1 inch to 2 miles).

BM, MCL, BDL (Gough)

150 *A Plan of the Navigable Canals now making in the Inland Parts of this Kingdom, for opening a Communication to the Ports of London, Bristol, Liverpool and Hull, with the adjacent Towns and Rivers.*

By James Brindley Engineer. To the most Noble Francis Duke of Bridgewater . . . this Plan is most Humbly Dedicated by . . . James Brindley.   Drawn   by   Robt   Whitworth. Engraved by Thos Jefferys. Apr^l 27th 1769 . . . Published by and for James Brindley.

Engraved plan, hand coloured   51 × 83 cm (scale 1 inch to 4 miles).

BM, MCL, BDL (Gough)

Shows the following canals, with tables of length, rise and fall: Bridgewater, Trent & Mersey, Trent to Severn, Droitwich, Birmingham, Coventry and Oxford.

151 *The Report of James Brindley, upon his being requested to give his opinion of the best plan of making a navigable communication between the friths of Forth and Clyde.*

Dated: [Edinburgh], 13 Sept 1768.

pp. 8–13
                    [In]
Reports by James Brindley Engineer, Thomas Yeoman Engineer, and F.R.S. and John Golborne Engineer, relative to a Navigable Communication betwixt the Friths of Forth and Clyde . . . With Observations.

Edinburgh: Printed by Balfour, Auld, and Smellie. 1768.

4to   25 cm   (iii) + 44 pp.   Folding engraved plan showing east end of the canal with New Cut proposed by Mr Brindley.

NLS, CE, ULG, Elton

See BRINDLEY and WHITWORTH. Report on a proposed canal from Stockton by Darlington to Winston, 1768–1769.

152 *A Plan of the intended Navigable Canal, from Chesterfield to the River Trent near Stockwith.*

Surveyed in 1769. Jno Varley del. Thos Kitchin sc.

Engraved plan   38 × 80 cm   (scale 1 inch to 1 mile).

RGS, MCL, BDL (Gough)

The line proposed by Brindley in August 1769, and finally adopted.

See GRUNDY. Plan of the intended canal from Chesterfield, by Retford, to the River Trent, as proposed by Mr Brindley, with alterations proposed by John Grundy, extracted from Mr Varley's plan. 1770.

153 *A Plan of the Intended Canal in Berkshire from Sunning to Monkey Island.*

Surveyed in 1770. R. Whitworth del. Engraved by Thos Jefferys.

Engraved plan   30 × 40 cm   (scale 1 inch to 1 mile).

BM, CE (Page), BDL (Gough)

Survey, commissioned 9 Jan 1770, carried out by Whitworth under Brindley's direction.

154 *Queries proposed by the Committee of the Common-Council of the City of London, about the intended Canal from Monkey Island to Isleworth, answered.*

Corporation of London, 1770.

Signed: James Brindley. London, 15 June 1770.

fol   32 cm   11 pp.   Folding engraved plan, hand coloured.

A Plan of the River Thames from the Kennets Mouth to London shewing the Intended Canal from Sunning Lock, to Monkey Island and from thence (by Order of the City of London) Surveyed to Isleworth.

By Mr Brindley. R. Whitworth del. Engraved by Thomas Jefferys . . . 1770. Published . . . 27 June 1770.

BM, CLRO, CE, BDL (Gough, plan only)
    City Navigation Committee Minute Book, 16 June 1770: Ordered that 1,000 copies of Mr Brindley's plan together with the Berkshire Survey to Sunning with the Questions . . . and Mr Brindley's Answers thereto, with the Estimates, be printed. Prints from an early state of the plan are in BM and BDL (Gough).

155   Revised scheme
A Plan of the River Thames from the Kennets Mouth . . . to Isleworth [and] *A Plan of the Intended Canal in Berk Shire from Reading to Monkey Island.*
    Surveyed 1770. R̃. Whitworth del. Engraved by Thoˢ Jefferys.
Engraved plan, hand coloured   36 × 98 cm (scale 1 inch to 1 mile).
BM
    Exactly as the plan in No. 154 but with added title and the canal extended to Reading.

156   *The Report of James Brindley, Esq; Engineer, for Improving the Navigation and Drainage at Wisbeach.*
n.p.
Signed: James Brindley. Wisbeach, 22 Aug 1770.
fol   33 cm   (2) pp.   Docket title on p. (4).
CRO, BDL (Gough), Sutro

157   Second edition
Title, etc as No. 156.
n.p.
Single sheet   24 × 18 cm   2 pp.
Sutro

158   [Report] *To the Committee of the Common Council of the City of London* [and] *A Description of the Profile or Section of the River Thames, from Boulter's Lock to Mortlake.*
Corporation of London, 1771.
Signed: James Brindley, 12 Dec 1770.
fol   33 cm   3 + 3 pp.   2 folding engraved plans, hand coloured.
Engraved by Thomas Jefferys.
[1] *A Profile of the River Thames from Boulters Lock*

to Mortlake. Surveyed . . . in 1770 by James Brindley Engineer. R. Whitworth del. Published 14 Jan 1771.
30 × 305 cm   (Scale 2 inches to 1 mile and 1 inch to 12 feet vertical).
[2] *A Plan of the River Thames from Boulters Lock to Mortlake.* Surveyed . . . in 1770 by James Brindley Engineer. R. Whitworth del.
Published 18 Jan 1771.
52 × 106 cm   (scale 2 inches to 1 mile).
BM, CLRO, CE, MCL (plan only), BDL (Gough)
    City Navigation Committee Minute Book, 12 Dec 1770: Ordered that Mr Brindley's report together with the Profile and Plan of the River be printed and that the number printed be one thousand. Stitched as issued in grey wrappers. The surveying was carried out by Whitworth.

See WHITWORTH. Plan of the River Thames from Boulters Lock to Mortlake, Surveyed in 1770 by James Brindley . . . Revised and continued to London Bridge in 1774 by Robert Whitworth.

BRINDLEY, James and WHITWORTH, Robert

159   *The Report of Mess. Brindley and Whitworth, Engineers, concerning the Practicability and Expence of making a Navigable Canal, from Stockton by Darlington to Winston, in the County of Durham.*
Newcastle: Printed by T. Slack, 1770.
Report signed: Rob. Whitworth, 24 Oct 1768. Estimate signed: James Brindley, Rob. Whitworth. New-Chapel, 19 July 1769.
4to   25 cm   20 pp.   Folding engraved plan, the lines entered in colours.
A Plan of the River Tees, and of the intended Navigable Canal from Stockton by Darlington to Winston in the Bishoprick of Durham.
    Surveyed by Robᵗ Whitworth 1768. Tho. Kitchin sc.
BM (plan only), SML, CE (Page, plan only)

[BRINDLEY, James]

The following work, compiled by an anonymous editor and constituting the first published account of canals in England, is traditionally catalogued under [Brindley].

160   *The History of Inland Navigations. Particularly those of the Duke of Bridgwater, in Lancashire and Cheshire; and the intended one promoted by Earl Gower and other Persons of Distinction in Staffordshire, Cheshire, and Derbyshire.*

London: Printed by T. Lowndes . . . 1766. Price Two Shillings and Six-pence.

Signed: The Editor. Manchester, 24 Feb 1766.

8vo   21 cm   (vi) + 88 pp.   2 folding engraved plans. (J. Prockter sc.)

[1] *A Plan of the Duke of Bridgewaters Navigable Canal already made With the Extention proposed from Altrincham to Liverpool.* [with inset view of Barton aqueduct].

[2] *A Plan of a Navigable Canal, intended for a Communication between the Ports of Liverpool and Hull.*

BM, NLS, SML, ULG

161   First edition, Second Part

The History of Inland Navigations. Particularly those of the Duke of Bridgewater . . . and the intended one . . . in Staffordshire, Cheshire, and Derbyshire.

Part the Second. Containing the different Essays which have been lately wrote, some to establish, others to prevent, a Navigable Canal being made from Witton Bridge, to Knutsford, Macclesfield, Stockport, and Manchester.

London: Printed for T. Lowndes . . . 1766. Price Two Shillings and Six-pence.

8vo   21 cm   (iv) + 104 pp.   Large folding engraved plan, the lines entered in colours.

A Plan of the Navigable Canals, intended to be made for opening a Communication between the Interior parts of the Kingdom & the ports of Bristol, Liverpool and Hull.

J. Prockter sc.

ULG

162   Parts One and Two bound together.

BM, ULG, BDL (Gough)

163   Part One, second edition

The History of Inland Navigations. Particularly those of the Duke of Bridgwater . . . and the intended one . . . in Staffordshire, Cheshire, and Derbyshire.

The Second Edition, with Additions.

London: Printed for T. Lowndes . . . 1769. Price 2s. 6d. each Part.

8vo   21 cm   (vi) + 111 pp.   3 folding engraved plans. (J. Prockter sc.)

[1] and [2] as in first edition.

[3] *A Plan of the Intended Navigable Canal from Liverpool to Leeds.*

MCL, CE (Vaughan), ULG

164   Part One, 2nd edn, and Part Two (1766) bound together.

ULG

165   Part One, third edition

The History of Inland Navigations, particularly that of the Duke of Bridgwater, illustrated with Geographical Plans . . . the Whole shewing the Utility and Importance of Inland Navigations.

The Third Edition, with Additions.

London: Printed for T. Lowndes . . . 1779. Price 2s. 6d.

Signed: The Editor. Manchester, 24 Feb 1776.

8vo   22 cm   iii + (1) + 107 pp.   2 folding engraved plans (J. Prockter sc.)

[1] *A Plan of the Duke of Bridgewaters Navigable Canal.* [with inset view of Barton aqueduct].

[2] *A Plan of a Navigable Canal, made for a Communication between the Ports of Liverpool and Hull.*

BM, CE, ULG

On p. iii are given the length, rise and fall of fourteen canals as shown in the map by Hugh HENSHALL, 1779 [q.v.].

BRISTOL DOCKS

166   *A Further Explanation of the Dam and Works proposed to be erected across the River Avon, at Rownham-Meads, for the Improvement of the Harbour of Bristol, as laid down in an engraved plan and section designed by Mr Jessop, Engineer.*

Bristol: Printed by J. Rudhall, 1793.
8vo   19 cm   20 pp.
CE, Bristol Library, Bristol R.O.

167 *Explanation of the Plan proposed for the Improvement of the Harbour of Bristol; and also a Scheme and Proposals for Raising the Sum necessary to carry the Plan into Effect.*
Bristol: Printed by J. Rudhall. [May] 1802.
4to   25 cm   (i) + 18 pp.   Folding engraved plan (for details *see* JESSOP).
NLS (Rennie), Bristol Library, BDL
   Bristol Dock Co. Minute Book, 24 April 1802: Engraving of plan to be completed and 1,000 copies made of plate and Explanation.

168 *Explanation of the Plan for Improving the Harbour of Bristol, submitted to the public by the Corporation and Society of Merchants, of the City, in the month of December 1802.*
Bristol: Printed by J. Rudhall. 1802.
fol   31 cm   (i) + 10 pp.   Folding engraved plan (the new scheme; *see* JESSOP).
Bristol R.O.

169 *Explanation of the Design for Improving the Harbour of Bristol.*
[London]: Luke Hansard, Printer. [1803]
fol   33 cm   2 pp.
Bristol Library
   Issued, separately, to accompany the revised plan of 1803; *see* JESSOP.

BRITTON, John [1771–1857] F.S.A.

170 *Lecture on the Road-Ways of England, pointing out the peculiarly Advantageous Situation of Bristol for the Commerce of the West; with Remarks on the benefits likely to arise from a Rail-Road between that Port and London.*
   By John Britton, Esq. F.S.A. &c.
Bristol: Printed by Gutch and Martin.
Dated: 19 Oct 1833.
8vo   22 cm   14 pp.
BM, CE

See BOURNE, 1839. Drawings of the London and Birmingham Railway . . . with an Historical and Descriptive Account by John Britton.

BROOKS, William Alexander [1802–1877] M.Inst.C.E.

171 *Proposed Asylum Harbour and Naval Station at Redcar, on the Coast of Yorkshire . . . to be called Port William.*
   Projected by W. A. Brooks, Esq. Civil Engineer.
London: Printed by J. Moyes . . . 1834.
Signed: W. A. Brooks. Stockton-on-Tees, Dec 1832 (report) and 14 Jan, 1 March and 19 March 1834 (letters).
8vo   22 cm   54 pp.   Large folding litho plan of the coast line, by Brooks, with two views drawn by L. Haghe (Day & Haghe lithog.).
BM, CE
   The report also published separately in 1832 (CE).

BROWN, George [1747–1816]

172 *General Report of the most necessary and useful Lines of Road, betwixt the East and West Sea, through the Four Northern Counties of Scotland.*
   By Mr George Brown.
Signed: Geo. Brown, 26 July 1803.
                    [In]
[First] *Report of the Commissioners for Highland Roads and Bridges.*
House of Commons, 1804. pp. 23–25.

See TELFORD. Map of intended Roads and Bridges in the Highlands of Scotland [1804]. This includes thirty roads surveyed by George Brown 1790–1799.

BROWN, *Sir* Samuel [1774–1852] R.N., F.R.S.E.

173 *Plan and Elevation of the Patent Iron Bar Bridge over the River Tweed near Berwick.*
> Designed and Erected by Cap.ᵗ Sam.ˡ Brown. Engraved by M. Dubourg.
> London: Published by J. Taylor . . . 1823.
> Engraved elevation and plan with aquatint landscape   25 × 60 cm   (scale 1 inch to 25 feet).
> SML, CE
>> Price 10s.

174 *Specification of a Bridge of Suspension over the River South Esk, at Montrose* [with a letter from] *Captain Samuel Brown, R.N., to . . . the Commissioners of Montrose Bridge.*
> n.p.
> Signed: Sam. Brown. Brighton, 8 July 1823.
> fol   30 cm   3 pp.   Folding engraved plan. (W. H. Lizars sc.)
> *Design for a Bridge of Suspension over the River Esk at Montrose.*
> By Cap.ᵗ S. Brown R.N.
> NLS, CE
>> Probably issued with George BUCHANAN's report, published at Montrose 1824 [q.v.].

See CHAPMAN. Report on proposed suspension bridge over the Tyne, with a design by Capt. S. Brown, 1825.

175 *The Stockton and Darlington Railway Suspension Bridge erected over the River Tees near Stockton.*
> By Sam.ˡ Brown Esq. R.N. . . . This Engraving is by permission most respectfully Inscribed to the Stockton and Darlington Railway C.ᵒ by Ja.ˢ Dixon their assistant Engineer. Drawn by J. Dixon. Engraved by W. Miller [c. 1831].
> Engraved elevation and plan, with landscape and a train on the bridge   38 × 53 cm   (scale 1 inch to 20 feet).
> SML, CE
>> The Stockton & Darlington extension over the Tees to Middlesbrough was opened in 1831, the probable date of this engraving. The SML copy is hand coloured.

176 *Description of a Bronze or Cast-Iron Columnal Light-House . . . designed for the Wolf Rock, situated between the Island of Scilly and Lands-End; or the Skerry Vore, on the west coast of Scotland.*
> By Samuel Brown, Commander R.N., K.H., F.R.S.E.
> Edinburgh: The Edinburgh Printing Company . . . 1836.
> Dated: Netherbyres House, Berwickshire, 1836.
> 8vo   24 cm   28 pp.   3 litho plates.
> NLS, CE

177 *Private Report of the Present State of Westminster Bridge, with Plans for its Security and Improvement.*
> By Saml. Brown, Commander R.N., K.H.
> Brighton: Printed by Edward Hill Creasy . . . and John Baker . . . 1836.
> Signed: Saml. Brown. Netherbyres House, 1 Nov 1836.
> 8vo   21 cm   22 pp.
> CE

178 *Plans, Specifications, and Estimates, accompanied by Drawings, for forming a Western Deep Water Entrance to the present Floating Docks of Leith, and adapting them for the reception of the largest Ships and Steam Vessels at all times of tide.*
> By Samuel Brown, Commander R.N., K.H.
> London: Gray, Son, and Fell . . . 1837.
> Dated: London, 1 Jan 1837.
> 8vo   21 cm   18 pp.   2 folding hand-coloured litho plans.
> [1] *Plans for Various Harbours Proposed to be constructed* [*at Leith*] *1837.* (Turner & Co).
> [2] *Plans (and Sections) for forming a Western Deep Water Entrance to the . . . Docks of Leith . . .* by Samuel Brown. (C. Hullmandel).
> NLS

BRUFF, Peter Schuyler [1812–1900] M.Inst.C.E.

179 *A Treatise on Engineering Field-Work; containing Practical Land Surveying for Railways, &c. with the theory, principles and Practice of Levelling* [etc].
By Peter Bruff, Surveyor.
London: Simpkin, Marshall, & Co . . . 1838.
8vo   23 cm   vi + (ii) + 162 pp.   4 folding engraved plates (C. Cobley) + 4 litho plates (Ravenscroft).
BM, CE, IC

180 Second edition, Part 1
A Treatise on Engineering Field-Work; comprising the practice of surveying, levelling, laying out works, and other field operations connected with engineering.
By Peter Bruff, C.E.
Second edition, corrected and enlarged.
London: Simpkin, Marshall, & Co . . . 1840.
8vo   23 cm   viii + 176 pp.   2 folding engraved plates + 2 plates of field book notes.
BM, NLS, SML, CE
  Second edition, Part 2, on *Levelling* was published in 1842.

BRUNEL, Isambard Kingdom [1806–1859] F.R.S., M.Soc.C.E., M.Inst.C.E.

181 *Great Western Railway between London and Bristol.*
  I. K. Brunel Esq.ʳ F.R.S. Engineer.
Litho map, lines entered in colour   29 × 44 cm (scale 1 inch to 12 miles).
[In]
Great Western Railway between Bristol and London.
  1833. [Prospectus].
n.p.
fol   44 cm   (4) pp.
BM

182 *Great Western Railway between Bristol and London.*
  I. K. Brunel, F.R.S. Engineer. B. Baker lith. Printed by Maguire & Co.
Litho map, lines entered in colour   30 × 47 cm (scale 1 inch to 16 miles).
[In]
Great Western Railway Prospectus.
  1834.
n.p.
fol   45 cm   (4) pp.
BM, CE

183 Another edition
Great Western Railway between Bristol and London.
  I. K. Brunel, F.R.S. Engineer. Baker lith. Printed by Day & Haghe.
Litho map, lines entered in colour   29 × 40 cm (scale 1 inch to 16 miles).
Printed on verso: *Plan of the Line of the Great Western Railway.* 1835.
BM, CE

184 Deposited plan
Plan (and Sections) of the Proposed Great Western Railway with Branches to Bradford and Trowbridge.
  1835. I. K. Brunel Esq.ʳ F.R.S. Engineer.
Oblong folio   47 × 68 cm   Litho, uncoloured, comprising title + 23 sheets of plans + 26 sheets of sections   (scale 4 inches to 1 mile and 1 inch to 60 feet vertical).
HLRO, CE
  The CE copy has the sheets folded and bound in large quarto volume.

185 *Bristol and Exeter Railway.*
  I. K. Brunel Esq.ʳ F.R.S. Engineer
Litho map, lines entered in colour   44 × 55 cm (scale 1 inch to 10 miles).
[In]
Bristol and Exeter Railway, Statement.
[London]: Printed by Richard Taylor . . . [1835].
fol   46 cm   (4) pp.
BM
  The survey was superintended by William Gravatt.

186 *Report (and Second Report) by I. K. Brunel, Esq. to the Directors of the Great Western Railway.*
Westminster: J. Bigg & Son, Printers.
Dated: London, 13 Dec 1838 [and] 27 Dec 1838.
8vo   22 cm   34 pp [and] 22 pp.
BM, SML, CE, ULG, Elton
  Issued with reports by Nicholas WOOD and by John HAWKSHAW (q.v.).

BRUNEL, *Sir* Marc Isambard [1767–1849] F.R.S., M.Soc.C.E., M.Inst.C.E.

187 *Plan of a Bridge on the Principle of Suspension representing the catenary combination & peculiar arrangement adapted in the construction of several Bridges destined for the Island of Bourbon in the East Indian Seas.*
  By Mᶜ Jᵈ Brunel, Esq. C.E. F.R.S. Wᵐ Read sc. Sold by J. Taylor [1824].
Engraved plan and elevation   38 × 54 cm (scale 1 inch to 16 feet).
SML, CE (Rennie)

188 *A New Plan of Tunnelling, calculated for opening a Roadway under the Thames.*
  By M. J. Brunel, Esq. C.E. F.R.S.
London: Printed by Richard Taylor . . . [1823].
8vo   25 cm   4 pp.   5 folding plates.
No. 1 *Transverse Section of the body of the Tunnel; shewing the double Archway.*
  W. Sheldrick Lithog.
No. 2 *Elevation of the Framing, shewing the Cells, in three of which the Workmen are represented in operation.*
  W. Sheldrick Lithog.
No. 3 *Longitudinal Section of the body of the Tunnel, with the Framing preceding it.*
  W. Sheldrick Lithog.
[4] Plan.
[5] *Tunnel across the River Thames.*
  Drawn by J. Pinchback. Engraved by W. Lowry.

SML, CE, ULG, Elton
  The text and plates 2 and 5 reprinted in *Phil. Mag.* Vol. 62 No. 304 (August 1823) pp. 139–142.

189 Second edition
Title as No. 188.
London: Printed by Richard Taylor . . . [1824].
8vo   25 cm   4 pp.   5 folding plates.
No. 1 and No. 2 as in first edition, but with I. Brunel jun. [del.]
[3] *Longitudinal Section of the Tunnel.*
  Printed by C. Hullmandel.
[4] and [5] as in first edition.
BM, SML, CE
  Written in 1824 after the share issue had been subscribed and giving a summary of the river-bed borings. Plate 3 shows a revised design of the tunnelling shield.

190 *Report of Mr Brunel, the Engineer of the Thames Tunnel Company. To the Directors . . . of the Company.*
Signed: M. J. Brunel, 29 May 1828.
pp. 11–13.
[In]
The Thames Tunnel. Report of the Court of Directors and of M. J. Brunel, Esq . . . upon the State of the Works . . . with Resolutions passed at a Special General Assembly.
London: H. Teape and Son, Printers . . . [1828].
8vo   21 cm   15 + (1) pp.
BM, SML, ULG, Elton

191 Reprint
Title as No. 190.
pp. 26–27.
[In]
Documents relating to the Thames Tunnel.
London: Printed by Arthur Taylor . . . 1829.
8vo   22 cm   31 + (1) pp.
BM, CE
  The title on printed wrappers.

192 *A Letter to the Proprietors of the Thames Tunnel.*
  By M. I. Brunel, Esq.
London: Taylor, Printers.
Signed: M. I. Brunel. Bridge Street, Blackfriars, 27 Aug 1829.

8vo   25 cm   (i) + 15 pp.
SML, CE

193   *The Tunnel under the Thames. An Exposition of Facts and Circumstances relating to the Tunnel under the Thames; its object, its progress, and its completion; as the same was most respectfully Submitted to the King by Mr Brunel, when he was honoured with an audience . . . at St. James's Palace, on the 24th of May, 1833.*
London: Printed by A. J. Valpy.
8vo   24 cm   20 pp.   Frontispiece aquatint and
    folding litho plate.
A View of the Western Archway of the Thames Tunnel.
    Robert Cruikshank fec.   The Tunnel sketched
    by M. Dixie.
The Thames Tunnel. Section of the Tunnel as it has been executed to the extent of 600 Feet . . . [and] Section of the Driftway . . . abandoned in 1808 . . . [and geological strata beneath the river bed].
    C. Ingrey lithog.
BM, SML, CE, Elton
    The CE copy printed on 4to paper with separate title page.

194   *An Explanation of the Works of the Tunnel under the Thames from Rotherhithe to Wapping.*
London: W. Warrington, Engraver and Printer
    . . . 1836.
Dated: January 1836.
Oblong   8vo   21 × 26 cm   24 pp.   Frontispiece vignette + 9 engraved plates (including an isometric view of the tunnelling shield drawn by R. Beamish).
BM, SML, ULG
    The ULG copy inscribed 'Presented to Jos.ʰ Hume Esq M.P. by the author M. I. Brunel 8 April 1837'. Another issue, March 1838 (SML, CE).

195   As No. 194 but small format.
Oblong   12mo   11 × 14 cm   24 pp.   9 engraved plates.
SML, B
    Later issues in 1837, 1838 and 1839 include a folded engraved *Plan shewing the Progress of the Thames Tunnel* (all in SML).

196   *Particulars of some Experiments on the mode of binding Brick Construction.*
    Made by M. J. Brunel, Esq. C.E. F.R.S.
Dated: 14 March 1836.
4to   27 cm   (4) pp.   2 engraved plates.
CE
    Reprint from *Trans. Inst. British Architects* Vol. 1 (1836) pp. 61–64.

BRUNTON, William [1777–1851] M.Inst.C.E.

197   [Report] *To the Trustees of the Swansea Harbour.*
Signed: W. Brunton, 2 July 1831.
pp. (3)–14.
    [In]
Reports on the Formation of a Floating Harbour at Swansea with reference to plans submitted to the Trustees.
Swansea: Printed . . . by W. C. Murray and D.
    Rees. 1831.
8vo   20 cm   78 pp.
Nat. Lib. of Wales, SML

See GILES and BRUNTON. Report on proposed
    railway from Basingstoke to Bath, 1834.

198   *Description of a practical and economic method of Excavating Ground and Forming Embankments for Railways, with practical observations on the Construction of Railways.*
    By W. Brunton, Civil Engineer.
London: John Weale . . . 1836.
Signed: W. Brunton. Charlotte Row, Mansion
    House, 8 March 1834. Now [at] Cwm Avon
    Tin-Plate, Copper and Coal Works, Glamorgan.
8vo   22 cm   30 pp.   2 folding litho plates
    (Waterlow & Morland Lith.).
BM, Elton

BUCHANAN, George [c. 1790–1852] F.R.S.E.

199 *Report on the Present State of the Wooden Bridge at Montrose, and the practicability of erecting a Suspended Bridge of Iron in its stead.*
By George Buchanan, Civil Engineer.
Montrose: Printed for the Commissioners, by D. Hill. 1824.
Signed: Geo. Buchanan. Edinburgh, 15 Feb 1823.
fol   31 cm   (i) + 18 pp.   3 folding engraved plans. (W. H. Lizars sc.)
[1] *Plan of the Basin of Montrose and of the River South Esk.*
[2] *Plan and Elevation of the present Bridge at Montrose.*
[3] *Views of the Proposed Iron Bridge of Suspension over the River South Esk at Montrose.*
NLS, CE
Probably issued with Samuel BROWN's Specification, 1823 [q.v.].

200 *Report on the erecting of Low-Water Landing-Places on the coast between Kinghorn & Pettycur.*
By George Buchanan, Civil Engineer.
Edinburgh: Printed by R. Wallace & Co . . . 1827.
Signed: Geo. Buchanan. Edinburgh, 24 Feb 1827.
8vo   23 cm   14 pp.
CE

201 *Views of the Opening of the Glasgow and Garnkirk Railway. By D. O. Hill, Esq. Member of the Scottish Academy . . . Also, an Account of that and other Railways in Lanarkshire. Drawn up by George Buchanan, Esq., Civil Engineer.*
Edinburgh: Printed by Ballantyne and Co. Published by Alexander Hill . . . 1832.
Oblong folio   44 × 59 cm
(iv) + 11 + (1) pp.   7 litho illustrations in text (J. R. Findlater del., J. Miller lith.)   Map (*Sketch of the Lanarkshire Railways shewing the Public Works in connection therewith*) + 4 litho plates (W. Day lith.).

NLS, Elton
Buchanan's *Account of the Lanarkshire Railways* is on pp. (1)–(12) with the map. The four views, dated 2 Jan 1831, are 'from Nature & on Stone' by David Octavius Hill.

See FINDLATER. Report on proposed railway between Dundee and Perth, 1835.

See CUBITT. Plan of the harbour and docks at Leith, 1839.

BUCK, George Watson [1789–1854] M.Inst.C.E.

202 *A Practical and Theoretical Essay on Oblique Bridges.*
By George Watson Buck.
London: John Weale . . . 1839.
Dated: Ardwick, Manchester, June 1839.
4to   28 cm   vii + 43 pp.   12 folding engraved plates (C. Henfrey del., S. Bellin sc.).
BM, SML, CE, RIBA, IC, BDL

BULL, William

203 *Map or Plan of the Proposed Improvements in the Calder & Hebble Navigation, in the West Riding of the County of York.*
Surveyed by James Day under the Direction of Wᵐ Bull, Civil Engineer. Halifax, Nov. 1833.
Franks & Johnson, engravers.
Engraved plan   33 × 72 cm   (scale 2 inches to 1 mile).
BM

BURDETT, Peter Perez [d. 1793]

See BURDETT and BECK.

BURDETT, P. P. and BECK, R.

204 *A Plan of an Intended Navigable Canal from Coln to Liverpool.*
By P. P. Burdett & R. Beck 1769.   Billinge sc.
Engraved plan   36 × 19 cm   (scale 1 inch to 5 miles).
[In]
A Cursory View, of a Proposed Canal: from Kendal, to the Duke of Bridgewater's Canal ... by the several Towns of Milnthrop, Lancaster, Garstang, Kirkham, Preston, Chorley, Wigan and Leigh.
n.p.
8vo   21 cm   56 pp.   Folding engraved plan (as above).
CE

BURGES, Alfred [c. 1801–1886] M.Soc.C.E., M.Inst.C.E.

See   WALKER   and   BURGES.   Reports 1830–1836.

205 *Accounts of the Old Bridge at Stratford-le-Bow, in Essex.*
Communicated to the Society of Antiquaries, by Alfred Burges, Esq.
London: Printed by J. B. Nichols and Son ... 1837.
Signed: Alfred Burges. Great George Street, 7 May 1836.
4to   27 cm   21 pp.   Plan in text. Engraved plate.

Bow Bridge. Plan & Elevation of the South Side.
Meas^d & Drawn by A. Burges 1834. Js. Basire sc.
CE
Offprint, with new title-page, from *Archaeologia* Vol. 27 (1837).

BURGOYNE, *General Sir* John Fox [1782–1871] R.E., F.R.S.

See DRUMMOND *et al.* Reports of the Commissioners ... [on] Railways in Ireland, 1837 and 1838.

See BURGOYNE *et al.* Reports of the Commissioners for the Improvement of the River Shannon, 1837 and 1839.

BURGOYNE, *Sir* John, JONES, *Sir* Harry, GRIFFITH, *Sir* Richard, CUBITT, *Sir* William and RHODES, Thomas

206 *Second Report of the Commissioners for the Improvement of the River Shannon.*
Signed: J.F. Burgoyne, Harry D. Jones, Richard Griffith, W. Cubitt, Tho. Rhodes. Dublin, 5 Dec 1837.
With 75 folding litho plans.
[In]
Second Report of the Commissioners ...
Dublin, 1837. pp. (5)25 and pls. 1–75.

207 *Fourth Report of the Commissioners for the Improvement of the River Shannon.*
Signed: J.F. Burgoyne, Harry D. Jones, Richard Griffith, W. Cubitt, Tho. Rhodes. Dublin, 21 Feb 1839.

With 121 folding litho plans.

[In]

Fourth Report of the Commissioners . . .
  Dublin, 1839. pp. (3)–136 and pls. 1–121.

BURLEIGH, Mark and THOMPSON, Isaac

208 *A Plan of the Mouth of the River Wear, Harbour & part of the Town of Sunderland & Towns adjacent [and] A Plan of the River Wear from Newbridge to Sunderland Barr as it appeared at Low Water.*
  By Burleigh & Thompson 1737. M. Burleigh del. John Tinney sc.
  Engraved plan in 4 sheets, altogether 49 × 213 cm   (scale 1 inch to 600 feet).
  BM, Tyne & Wear R.O., MCL, BDL (Gough)

BURRELL, Andrewes

209 | *A Briefe Relation Discovering Plainely the true Causes why the great Levell of Fenns . . . being Three hundred and seven thousand Acres of Low-Lands, have been drowned . . . And as briefly how they may be drained, and preserved from Inundation.*
  By Andrewes Burrell, Gent.
  London: Printed for Francis Constable. 1642.
  4to   18 cm   (vi) + 22 pp.
  BM, BDL (Gough)

210 *Exceptions against Sir Cornelius Virmudens Discourse for the Draining of the great Fennes, &c. Which in January 1638 he presented to the King for his Designe. Wherein His Majesty was mis-informed and abused, in regard it wanteth all the essentiall parts of a Designe.*
  By Andrewes Burrell, Gent.
  London: Printed by T. H. and sold by Robert Constable . . . 1642.
  4to   18 cm   (iv) + 19 pp.
  BM, BDL (Gough)

BURY, Thomas Talbot [1811–1877] Assoc.Inst.C.E., F.R.I.B.A

211 *Coloured Views on the Liverpool and Manchester Railway . . . From Drawings made on the Spot by Mr T. T. Bury. With Descriptive Particulars, serving as a Guide to Travellers on the Railway.*
  London: Published by R. Ackermann . . . 1831.
  4to   35 cm   (i) + 8 pp   13 hand-coloured aquatint plates (H. Pyall and S. G. Hughes sc.)
  BM, CE, Elton, P, B
    Published in two parts: Pls 1–7 (*Six Coloured Views . . . with a Plate of the Coaches . . .*), Pls. 8–13 and text (general title as above). Some copies have both title pages bound in. All plates dated 1831, eight of them exist in two or more states. Price, complete, £1 4s.

212 Second edition
  Collation as No. 211, title page dated 1832.
  SML, B
    Published 1833. Alterations in seven of the plates. Pl. 10 dated 1832, the others 1833. The SML copy has a new setting of the text with a note referring to two large folding plates [*Travelling on the Liverpool and Manchester Railway*. Drawn by I. Shaw, S. G. Hughes sc.] which 'may be had of the publisher separately or together with this Description'.

213 Second edition, second issue
  As No. 212 but dated 1833.
  P, B
    Like No. 212, can be found with or without additional plates.

CARMICHAEL, John Wilson [1799–1868]

Drawings of the Newcastle & Carlisle Railway by J. W. Carmichael. *See* BLACKMORE, 1836.

CARY, John [c. 1754–1835]

214 *Inland Navigation; or Select Plans of the several Navigable Canals, throughout Great Britain: accompanied with abstracts of the different Acts of Parliament relative to them; likewise the Width, Depth, Length and number of Locks on each: with the principal Articles of Carriage &c.*
London: Printed for J. Cary . . . 1795 [–1808].
4to   35 cm   (i) + 132 pp.   Engraved title-page + 16 folding engraved maps   (25 × 45 cm, scale 1 inch to 2 miles).
Issued in four parts, each with 4 maps and accompanying text: dated 1 Nov 1795, 20 Dec 1796, 1 July 1798 and 2 May 1808.
BM, NLS, CE (Page), ULG (lacking final part)

CASEBOURNE, Thomas [1797–1864] M.Inst.C.E.

See TELFORD. Maps of the roads between Weedon and Lichfield, and between Liverpool and Talk on the Hill, surveyed by Thomas Casebourne, 1828.

CHAMBERS, Abraham Henry [1763–1853]

215 *Observations on the Formation, State and Condition of Turnpike Roads and other Highways with Suggestions for their Permanent Improvement on Scientific Principles* [etc].
By A. H. Chambers, Esq.
London: Printed for the Author (by Charles Atwell . . . 1820.
Dated: Bryanstone Square, 1 June 1820.
8vo   22 cm   28 pp.
BM, CE, ULG

CHAPMAN, William [1749–1832] M.R.I.A., M.Soc.C.E.

216 *Observations on the Advantages of bringing the Grand Canal round by the Circular Road into the River Liffey.*
By William Chapman, Engineer.
Dublin: Printed by P. Byrne . . . 1785.
Signed: William Chapman. Dublin, 28 May 1785.
8vo   20 cm   (i) + 29 pp.   Folding engraved plan.
A Plan for Communicating the Grand Canal with the Liffey and Improving the City of Dublin.
By Wilᵐ Chapman, Engineer. G. Gonne sc.
CE, ULG, CUL

217 *Report on the Means of Perfecting the Navigation of the River Barrow from St. Mullin's to Athy.*
By William Chapman, Engineer.
Dublin: Printed by W. Sleater . . . 1789.
Signed: William Chapman. Old-Town, Naas, 9 Oct 1789.
8vo   20 cm   23 pp.
CE, Elton

218 *Estimates of the Expences of completing the Navigation of the River Barrow, from St. Mullin's to Athy.*
By William Chapman, Engineer.
Dublin: Printed by W. Sleater . . . 1789.
Signed: William Chapman. Old-Town, Naas, 9 Oct 1789.
8vo   20 cm   33 pp.
BM, CE

219 *Report on the Navigation of the River Shannon, from Lough Allen to Killaloe, with Estimates.*
By William Chapman, Engineer.
Limerick: Printed by Order of the Shannon Navigation Company . . . 1791.
Signed: William Chapman. Limerick, 3 Oct 1791.
8vo   20 cm   19 pp.
BM, CE

220 *Report on the Improvement of the Harbour of Arklow, and the Practicability of a Navigation from thence by the Vales of the . . . Ovoca.*
By William Chapman, Engineer.
Dublin: Printed by J. Chambers . . . 1792.
Signed: William Chapman. Dublin, 6 Dec 1791.
8vo   19 cm   15 pp.   Folding engraved plan.
Plan of the Harbour of Arklow.
   G. Gonne sc.
BM, CE, Elton

221 *Report of William Chapman, Engineer, on the means of making Woodford River Navigable, from Lough-Erne to Woodford-Lough.*
Limerick: Printed by A. Watson . . . 1793.
Signed: William Chapman. Limerick, 30 April 1793.
8vo   19 cm   15 pp.
BM, CE

222 *Report on the Measures to be attended to in the Survey of a Line of Navigation, from Newcastle upon Tyne to the Irish Channel; with an Estimate of the probable Annual Revenue.*
By William Chapman, M.R.I.A. Engineer.
Newcastle: Printed by Hall and Elliot, 1795. Price One Shilling.
Signed: William Chapman. Newcastle, 5 Jan 1795.
8vo   21 cm   44 pp.
NLS, CE, ULG (Rastrick)

223 *Report on the Proposed Navigation between the East and West Seas, so far as extends from Newcastle to Haydon-Bridge, with observations on the separate advantages of the North and South Sides of the River Tyne.*
By William Chapman, Engineer, M.R.I.A.
Newcastle: Printed for J. Whitfield . . . 1795.
Signed: William Chapman. Newcastle, 26 June 1795.
8vo   21 cm   20 pp.
BM, NLS, CE, ULG (Rastrick)

224 *Second Part of a Report on the Proposed Navigation between the East and West Seas; viz. from Haydon-Bridge to Maryport.*
By William Chapman, Engineer, M.R.I.A.
Newcastle: Printed for J. Whitfield . . . 1795.
Signed: William Chapman. Newcastle, 10 July 1795.
8vo   21 cm   41 + (3) pp.
BM, NLS, CE, ULG (Rastrick)
   Errata in First Part listed on p. (44).

225 *Third and last part of a Report on the Proposed Navigation between the East and West Seas, viz. on the Advantages, and Disadvantages of Carrying the Navigation on the South Side of the River Tyne, in the different Courses that it is capable of.*
By William Chapman, Engineer, M.R.I.A.
Newcastle: Printed for J. Whitfield . . . 1795.
Signed: William Chapman. Newcastle, 10 Aug 1795.
8vo   21 cm   30 pp.
BM, NLS, CE
   Publisher's advertisement on p. (32) for First and Second Parts: price 6d. each.

226 *Report on the Proposed Line of Navigation, between Newcastle and Maryport, by William Jessop, Engineer. With Abstracts of the Estimates of this Line. And also of that from Stella to Hexham.*
By Wm. Jessop and Wm. Chapman, Engineers.
Newcastle: Printed for J. Whitfield . . . 1795. Price Three-Pence.
Signed: W. Jessop. Maryport, 26 Oct 1795 (and) W. Jessop, W. Chapman. Maryport, 26 Oct 1795.
8vo   21 cm   11 pp.
CE
   Jessop's report pp. 3–9; Estimates pp. 10–11.

227 *Mr Chapman's Postscript to Mr Jessop's Report.*
n.p.
Signed: William Chapman. Newcastle, 15 Dec 1795.
8vo   21 cm   4 pp.
CE

228 Collected second edition
Report on the Measures to be attended to in the Survey of a Line of Navigation from Newcastle upon Tyne to the Irish

Channel. *By William Chapman, Engineer. M.R.I.A. To which are added, all the Reports subsequent to the Survey, and the Estimates of Messrs Jessop and Chapman.*

Newcastle: Printed for J. Whitfield . . . 1796.

8vo   21 cm   13, 11, 16 + (2), 17, 6, 2 pp.

CE (Vaughan), BDL

Reports Nos. 222–227 reprinted in smaller type with separate pagination and issued in one volume, in blue wrappers. Title-page reset but (except the first, as above) still dated 1795.

229   *Plan of the Proposed Canal between Newcastle & Maryport and of the Adjacent Country.*

By William Chapman Engineer M.R.I.A. 1795. A. Hunter sc.

Engraved plan   48 × 119 cm   (scale 1 inch to 2 miles).

BM

230   *Plan of the Proposed Navigation from Newcastle upon Tyne, to Haydon Bridge.*

By William Chapman, Engineer, 1796. J. Cary sc.

Engraved plan   39 × 110 cm   (scale 1½ inches to 1 mile).

BM

231   *The Report of Mr Chapman, respecting the Drainage of the Low Grounds lying below the Wolds, on the West Side of the River Hull, and in Frodingham Carrs, and at Lisset, &c.*

Hull: Printed by Thomas Lee and Co . . . 1796.

Signed: William Chapman. Newcastle, 8 Oct 1796.

12mo   17 cm   32 pp. and table.

BM, CE, Hull University Library

The Beverley & Barmston drainage scheme.

232   *Observations on Mr John Sutcliffe's Report on a Proposed Line of Canal from Stella to Hexham. With an Appendix containing Remarks on his Second Report.*

By William Chapman.

Newcastle: Printed by and for Joseph Whitfield.

Dated: Newcastle, 27 Jan 1797.

8vo   20 cm   18 pp.

CE (Vaughan), BDL

233   *Report of William Chapman, Engineer, on the Drainage and Navigation of Keyingham Level, in Holderness.*

Newcastle: Printed by Edward Walker, 1797.

Signed: William Chapman. Newcastle, 30 June 1797.

4to   23 cm   30 pp.   Folding engraved plan.

Plan of the Drainage of the Levels of Keyingham, Burstwick &c. with the Improvements Projected.

By William Chapman Engineer 1797.

HRO, CE, Hull University Library, Sutro

234   *Observations on the Various Systems of Canal Navigation, with Inferences Practical and Mathematical; in which Mr Fulton's Plan of Wheel-Boats, and the Utility . . . of Small Canals are particularly investigated, including an account of the Canals and Inclined Planes of China.*

By William Chapman, Member of the Society of Civil Engineers in London, and M.R.I.A.

London: Published by I. and J. Taylor . . . 1797.

Dated: Newcastle-upon-Tyne, 1797.

4to   28 cm   (viii) + 104 pp.   4 engraved plates.

BM, SML, NLS, CE, RIBA, ULG, BDL, CUL

235   *Report of William Chapman, Engineer, on the Means of Draining the Low Grounds in the Vales of Derwent and Hertford, in the North and East Ridings of the County of York.*

Newcastle: Printed by E. Walker . . . 1800.

Signed: William Chapman. Newcastle 16 Jan 1800.

8vo   22 cm   18 pp. [First issue] or 22 pp. including *Postscript* dated 13 Feb 1800 [Second issue].   Folding engraved plan.

Plan of part of the Rivers Derwent and Hertford, shewing the Alterations in their Course through the flooded Grounds.

Proposed by Willm. Chapman. Jan 1800. R. Beilby sc.

First issue: BM

Second issue: CE, Hull University Library

The Muston & Yedingham Drainage scheme.

236   *Sundry Papers and Reports, relative to the Defence of the Estate of Cherry Cobb Sands, against the Humber.*

Newcastle: Printed by Edward Walker . . . [1800].

8vo   22 cm   23 pp.   Woodcut plan.

Plan of Cherry Cobb Marsh.

CE, Sutro

Includes (pp. 5–8) the report by Joseph HODSKINSON, 10 Feb 1796 [q.v.] and a report by Chapman (pp. 12–21) dated Newcastle, 5 Nov 1799.

237   Another edition

Title as No. 236.

Newcastle: Printed by Edward Walker . . . [1801].

8vo   22 cm   30 pp.   Woodcut plan (as above).

CE

Includes (pp. 8–10) extracts from HODSKINSON's report, (pp. 16–26) Chapman's report of Nov 1799 and (pp. 27–29) a further report by Chapman dated Newcastle, 27 Dec 1800.

238   *Sundry Papers and Reports, relative, first, to the Defence of the Estate of Cherry Cobb Sands against the Humber: secondly, the Drainage of Keyingham Marshes: and, thirdly, the Eventual Improvement and Accretion of the Fore-Shore, opposite Foul Holme Sands.*

London: Printed by C. Clarke . . . 1801.

8vo   21 cm   46 + (1) pp.   Woodcut plan (as above) and folding hand-coloured engraved plan.

A Plan of Cherry Cobb Sands: Paul Holme and Foul Holme Sands, and Keyingham Marshes.

BM, HRO, CE, BDL

Includes (pp. 8–10) extracts from HODSKINSON's report, (pp. 16–29) Chapman's reports of Nov 1799 and Dec 1800, (pp. 31–38) his report on Keyingham Marshes dated Newcastle, 28 Feb 1801 and (pp. 39–43) his undated report on Foul Holme Sand.

239   *Facts and Remarks relative to the Witham and Welland: or, A Series of Observations on their Past and Present State; on the Means of Improving the Channel of the Witham, and the Port of Boston* [etc].

By William Chapman.

Boston: Printed by J. Hellaby . . . 1800.

Dated: Boston, 10 June 1800.

8vo   23 cm   vii + (1) + 75 pp.

BM, LCL, CE, ULG

240   *Mr Chapman's Report on the Proposed Navigation to Knaresbro'.*

Signed: William Chapman. London, 2 July 1800. pp. 9–16.

[Printed with]

Ralph Burton's report, 25 April 1800.

[In]

Report of the Committee appointed to procure the Levels, Surveys, and Estimates of the Proposed Navigation to Knaresbro', in the West Riding of the County of York.

York: Printed by T. Wilson and R. Spence . . . 1802.

8vo   22 cm   20 pp.

CE (Telford)

241   [Report] *To the Proprietors of Lands within the Keyingham Level Drainage.*

Hull: Printed by W. Rawson.

Signed: William Chapman. Hedon, 24 July 1800.

fol   30 cm   3 pp.

CE

242   *Report on the Proposed Branch Navigations from the River Hull, with Estimates of their Expence.*

By William Chapman.

Newcastle: Printed by E. Walker . . . 1800.

Signed: Wm. Chapman. Newcastle, 7 Aug 1800.

8vo   21 cm   22 pp.

CE

243   *Report on the Harbour of Scarborough, and on the Means Necessary for its Improvement.*

By William Chapman, Engineer.

Scarborough: Printed by G. Broadrick. 1800.

Signed: William Chapman. Hull, 20 Sept 1800.

4to   22 cm   32 pp.   Folding engraved plan.

A Plan of Scarborough Harbour.

Sep.¹ 1800. Hampton Prince & Cattle.

BM, CE

244   *Observations on the Improvement of Boston Haven, humbly submitted to the consideration of . . . the Commissioners of the Drainage dependent on this Haven and to the Corporation of Boston.*

By William Chapman.
Boston: Printed and Sold by J. Hellaby. 1800.
8vo   22 cm   8 pp.
BM, CE

245   *Part the Second. Observations on the Improvement of Boston Haven* [etc].
   By William Chapman.
Boston: Printed and Sold by J. Hellaby, 1801.
8vo   22 cm   43 pp.
CE

246   *Mr Chapman's Report to the Committee of Proprietors of the Beverley & Barmston Drainage.*
Hull: Printed by W. Rawson.
Signed: W. Chapman. York-Hotel, Blackfriars, 22 May 1801.
12mo   17 cm   9 pp.
CE
   Subsequent progress reports on the Beverley & Barmston Drainage works were issued in 1803, 1804, 1806 and 1810 (all in CE).

247   [Report] *To Subscribers for Improving the Port of Stockton.*
Signed: William Chapman. York, 20 Dec 1804.
pp. 14–27   with 2 woodcut plans.
No. 1 *Plan of the River Tees between Stockton and Portrack.*
No. 2 *Plan of the Proposed New Cut.*
   W. Chapman, del.   W. Green, sc.
[In]
Report of the Committee appointed to Enquire into the Expediency of Making a Cut across the Neck of Land between Stockton and Portrack. Also Mr Chapman's Report and Estimates, with Plans.
Stockton: Printed by Christopher and Jennett. 1805.
8vo   21 cm   27 pp.   2 plans (as above).
CE

248   *Mr Chapman's Report on the Means of Obtaining a Safe and Commodious Communication from Carlisle to the Sea.*
Carlisle: Printed by W. Hodgson . . . 1807.
Signed: William Chapman. Carlisle, 27 June 1807.

8vo   21 cm   18 pp. [First issue] or 21 pp. including *Appendix* dated 22 Aug 1807 [Second issue].
First issue: BM, ULG
Second issue: CE

249   *Mr Chapman's Further Report; or, Observations made on Mr Telford's Report respecting the intended Cumberland Canal.*
Signed: Wm. Chapman. Newcastle, 24 Feb 1808.
pp. 11–16.
[Printed with]
TELFORD's report, 6 Feb 1808 [q.v.].
[In]
Mr Telford's Report on the Intended Cumberland Canal; and Mr Chapman's Further Report or Observations thereon.
Carlisle: Printed by W. Hodgson and Co. 1808.
8vo   21 cm   (i) + 16 pp.
BM, CE

250   *A Treatise on the Progressive Endeavours to improve the Manufacture and Duration of Cordage.*
   By W. Chapman, Esq. M.R.I.A. Member of the Society of Civil Engineers in London.
London: Printed for W. H. Wyatt . . . 1807.
4to   25 cm   vi + 54 pp.   3 folding engraved plates.
BM, NLS

251   *Report on the Drainage of the Marshes, from Ancroft Fen to Wainfleet Haven; with an Estimate of its Expence.*
Louth: Printed at the Office of John Jackson . . . 1810.
Signed: William Chapman. Newcastle, 28 April 1810.
12mo   17 cm   12 pp.
CE

252   *Mr Chapman's Report on the Measures Necessary to the Final Completion of Scarborough Harbour, to which is added, a Supplementary Report.*
York: Printed by W. Blanchard and Son. 1811.
Signed: William Chapman. Newcastle, 9 Aug 1810 (and) 16 Aug 1811.
4to   20 cm   16 pp.   Folding engraved plan.

A Plan of Scarborough Harbour.
  Sep.ʳ 1811. Cattle & Barber sc.
CE
  Scarborough Harbour Commissioners Minute Book, 29 Aug 1811: Mr Chapman's reports of 9 Aug 1810 & 16 Aug 1811 to be printed, with the plan; 250 copies to be printed.

253 *Mr Chapman's First Report on the Laneham Drainage.*
Retford: Printed by J. Taylor . . . 1813.
Signed: William Chapman. London, 21 March 1813.
4to   21 cm   (i) + 20 pp.
CE

254 *Mr Chapman's Second Report and Supplement on the Laneham Drainage.*
Retford: Printed by J. Taylor . . . 1813.
Signed: William Chapman. London, 27 April 1813 (and) Retford, 5 May 1813.
4to   21 cm   31 + (1) pp.
CE

255 *Report of William Chapman, Civil Engineer, on various projected Lines of Navigation from Sheffield.*
Sheffield: Printed by James Montgomery. 1813.
Signed: William Chapman. Newcastle, Sept 1813.
4to   23 cm   35 + (1) pp.   Folding engraved plan.
Sketch of the Country referred to in Mr Chapman's Report.
  1813. Drawn by W. & J. Fairbank, Sheffield. Neele sc.
CE

256 *Report of William Chapman, Civil Engineer, on the Proposed Canal, from Castle Orchards, Sheffield, to the River Dun below Tinsley.*
Sheffield: Printed by James Montgomery . . . 1814.
Signed: William Chapman. Newcastle, 6 Oct 1814.
8vo   22 cm   17 + (1) pp.   Folding engraved plan.
A Plan of the Intended Canal from . . . Sheffield into the Township of Tinsley . . . in the West Riding of the

County of York. Describing the Country in or through which the same Canal is intended to be . . . carried.
  By William Chapman Civil Engineer. Surveyed by W. & J. Fairbank 1814. T. Harris sc.
CE

See DANCE, CHAPMAN *et al.* Report on London Bridge, 1814.

257 *Report on the Harbour of New Shoreham.*
  By Wm. Chapman, Civil Engineer.
Brighton: Printed by W. Fleet . . . 1815.
Signed: William Chapman. Shoreham, 10 July 1815.
8vo   22 cm   25 + (2) pp.   Folding engraved plan.
A Plan of the Harbour of New Shoreham.
  Surveyed by Capt. Clegram in January 1815, shewing the Improvements, Projected by William Chapman, Esq.ʳ Civil Engineer, in July 1815. T. Overton sc.
BM, CE

258 *Supplementary Report, on the Efficacy of the Measures proposed for the Improvement of the Harbour of New Shoreham.*
  By W. Chapman.
Brighton: Printed by W. Fleet . . . 1815.
Signed: Wm. Chapman. Newcastle, 19 Nov 1815.
8vo   21 cm   16 pp.
BM, CE

259 *Mr Chapman's Report to the Commissioners of Scarborough Piers.*
Newcastle: Printed by Edw. Walker.
Signed: William Chapman Newcastle, 10 April 1816. ʹ
4to   24 cm   7 pp.
CE
  Subsequent progress reports on the works at Scarborough Harbour were issued with titles similar to the above in 1818–1823 (all in CE).

260 *A Treatise containing the Results of Numerous Experiments on the Preservation of Timber from Premature Decay . . . in Ships and Buildings . . . with Remarks on the Means of Preserving Wooden Jetties and Bridges from Destruction by Worms.*

By William Chapman, M.R.I.A. Civil Engineer.
London: Printed by T. Davison . . . 1817.
Signed: William Chapman. Newcastle, June 1817.
8vo   21 cm   xii + 156 pp.
BM, NLS, BDL, CUL

261   *Mr Chapman's Report on the Proposed Canal Navigation between Carlisle and Solway Frith.*
Carlisle: Printed by Charles Thurnam . . . March 1818.
Signed: William Chapman. Newcastle, 5 Feb 1818.
8vo   21 cm   48 pp.
B

262   Second issue
Title and text as No. 261 but *Second Edition.*
Carlisle: Printed by C. Thurnam . . . May 1818.
8vo   21 cm   48 pp.
CE, ULG

See SPECIFICATION for the new west pier of Scarborough Harbour, December 1818.

See MOUNTAGUE, RENNIE *et al*. Report on London Bridge, 1821.

263   *Mr Chapman's Report to the Chairman and Commissioners of Shoreham Harbour.*
Dated: London, 12 April 1821.
pp. 3–7.
[In]
Observations on Shoreham Harbour.
Newcastle: Printed by Edward Walker . . . 1822.
8vo   21 cm   7 pp.
CE

264   *The Report of William Chapman on the Means of Improving and Enlarging the Harbour of Whitehaven.*
Signed: William Chapman, December 1821.
p. 14   with litho plan.
Design for the Extension of Whitehaven Harbour.
By Wm. Chapman Esq. Civil Engineer 1821.
[In]
Plans suggested at different periods for the Improvement of Whitehaven Harbour. See WHITEHAVEN HARBOUR, 1836.

265   *Case of Scarborough Harbour.*
Newcastle: Printed by Edw. Walker.
Signed: William Chapman. Newcastle, Dec 1821.
fol   42 cm   3 pp.   Docket title on p. (4).
CE

266   *A Plan of the Harbour with part of the Town of Scarborough.*
Surveyed by Thos O. Blackett, under the Direction of Willm Chapman Esqr Civil Engineer. Lambert sc.
Engraved plan   23 × 36 cm   (scale 1 inch to 300 feet).
CE
The plan can be dated to 1821–22.

See SPECIFICATION of a proposed jetty at Scarborough, September 1823.

267   *Address to the Subscribers to the Canal from Carlisle to Fisher's Cross.*
Newcastle: Printed by Edw. Walker.
Signed: William Chapman. Newcastle, 31 Jan 1823.
8vo   22 cm   16 pp.
BM, CE

268   *Mr Chapman's Report on the Works, Improvement, Revenue, and Expenditure of Scarborough Harbour.*
Scarborough: J. Ainsworth, Printer.
Signed: William Chapman, 13 Aug 1824.
4to   23 cm   3 pp.   Docket title on p. (4).
CE
The first of a new series of progress reports, titled as above, printed at Scarborough in 4to or fol and issued annually 1824–30 (all in CE).

269   *Observations on the Most Advisable Measures to be adopted in Forming a Communication for the Transit of Merchandise and the produce of Land, to or from Newcastle and Carlisle . . . in a Letter from Mr Chapman to Sir James Graham.*
Newcastle: Printed by Edward Walker . . . 1824.
Signed: William Chapman. Newcastle, 10 May 1824.
8vo   21 cm   8 pp. [First issue] or 10 pp. including *Additional Supplement* dated 21 July 1824 [Second issue].

First issue: BM
Second issue: NLS, CE, ULG

270 *A Report on the Cost and Separate Advantages of a Ship Canal and of a Rail-Way, from Newcastle to Carlisle.*
By William Chapman, Esq. Civil Engineer.
Newcastle: Printed by Edward Walker . . . 1824.
Signed: William Chapman. Newcastle, 27 Oct 1824.
8vo   22 cm   21 pp.
BM

271   Second issue
Title and text as No. 270 but *Second Edition*.
8vo   22 cm   21 pp.
BM, NLS, CE (Telford), ULG
Issued in yellow printed wrappers.

272   *Report relative to the Improvement of the Harbour of Leith.*
By William Chapman, Esq. Civil Engineer.
n.p.
Signed: William Chapman. Edinburgh, 1 Oct 1824.
4to   26 cm   (i) + 9 pp.   Folding litho plan.
Chart of the Entrance to Leith Harbour & Docks shewing the Wear constructed in 1818 and the New Pier proposed to be erected
according to a Plan by William Chapman, Esq. Civil Engineer 1824. Robertson & Ballantine lithog.
NLS, ULG

273   *Supplementary Report relative to the Further Improvement of the Harbour of Leith.*
By William Chapman, Esq. Civil Engineer.
n.p. [1825].
Dated: Newcastle, 8 Dec 1824.
4to   27 cm   (i) + 8 pp.   Folding litho plan (as in previous report).
NLS
Includes corrected estimates dated 8 Jan 1825.

274   *Manchester and Dee Ship Canal.*
Report of William Chapman, Esq. Civil Engineer.

Manchester: Printed by T. Sowler . . . 1825.
Signed: William Chapman. Manchester, 30 June 1825.
4to   27 cm   (i) + iv + 20 p.   3 folding engraved plates.
[1] *A Plan shewing the Line of the intended Manchester & Dee Ship Canal.* G. Bradshaw sc.
[2] *Plan of the Docks &c at Dawpool for the intended . . . Canal.*
[3] *A Chart of the entrance into the River Dee.* By Twyford & Wilson Surveyor & Engineer 1825. G. Bradshaw sc.
CE

275   *Report on the projected patent wrought iron Suspension Bridge, across the River Tyne, at North and South Shields.*
By William Chapman, Esquire, Civil Engineer.
South Shields: G. W. Barnes, Printer.
Signed: William Chapman. Newcastle, 12 July 1825 (and Postscript dated 5 Aug 1825).
fol   32 cm   10 pp.   2 folding engraved plans.
[1] *Plan of the Suspension Bridge proposed to be Erected over the River Tyne between North & South Shields.* Designed by Cap.ᵗ S. Brown, R.N.   I. Green del.   M. Lambert sc.
[2] *Plan.* John Bell, Surveyor.   M. Lambert sc.
CE, IC

276   *Report on the Improbability of the Formation of a Useful Harbour at Lake Lothing, and on the Means of Improving the Navigation from Norwich to Yarmouth, and thence to the Sea* [etc].
By William Chapman, Esq. Civil Engineer.
London: Printed by Thomas Davison.
Dated: Newcastle, January 1828.
8vo   21 cm   15 pp.
BM

277   *Mr Chapman's Report on the Antient, the Intermediate, and Present State of the Harbour of Scarborough, and on the Measures Requisite for its Final Improvements.*
Newcastle: Printed by Edward Walker . . . 1829.
Signed: William Chapman. Newcastle, 4 Aug 1829.
8vo   20 cm   24 pp.   Folding engraved plan.
Plan of Scarborough Harbour in 1829.

With the Improvements proposed by Mr Chapman in his printed Report of August 4th. From Surveys by T. O. Blackett and John Barry . . . Lambert sc.

CE

278 *Supplementary Report on Scarborough Harbour.*
Newcastle: E. Walker, Printer.
Signed: W. Chapman, 2 Oct 1829.
8vo   20 cm   8 pp.
CE

279 *Additional Supplement to Mr Chapman's Report, of August 1829, on the . . . Harbour of Scarborough.*
Newcastle: Printed by Edward Walker . . . 1831.
Signed: William Chapman. Newcastle, April 1831.
8vo   21 cm   8 pp.   Folding engraved plan.
Plan of Scarborough Harbour, 1831.
With the Improvements proposed by Mr Chapman in his printed report of April . . . Lambert sc.
CE

280 *A Description of the Port of Seaham, in Explanation of the Plan of the Harbour, and a Chart of the Coast, shewing the Projected Extent of the Works.*
By William Chapman, Esq. Civil Engineer.
Newcastle: Printed by Edward Walker . . . 1820.
Signed: William Chapman. Newcastle, Aug 1830.
4to   26 cm   8 pp.   Folding engraved plan.
Plan of Seaham Harbour as Projected . . . by William Chapman . . . and a Chart of the Coast surveyed under his Direction. With the Eastern Part of the Town
according to the Plan of I^{no} Dobson, Architect. June 1830.
BM, CE

281 *Report on the Rise, Progress, Present State, and Projected Extension of the Harbour of Seaham.*
By Wm. Chapman, M.R.I.A., Civil Engineer.
Newcastle: Printed by Charles Henry Cook . . . 1832.
8vo   22 cm   11 pp.   Chart and folding engraved plan.
Plan of the Coast from Blyth to Flamborough Head.

Plan of the Harbour of Seaham and of its Future Extensions
as projected by William Chapman.
CE

CLARE, Martin [d. 1751] F.R.S.

282 *The Motion of Fluids, Natural and Artificial; in particular that of Air and Water, in a familiar Manner, proposed and proved, by evident and conclusive Experiments with many useful Remarks.*
By M. Clare, A.M.
London: Printed for Edward Symon . . . 1735.
Signed: Martin Clare. Academy in Soho Square, 1 May 1735.
8vo   21 cm   (xvi) + 323 + (23) pp.   9 engraved plates (J. Cole sc.).
BM, RS, BDL

283 Second edition
Title as No. 282 with *The Second Edition, Corrected and Improved.*
By M. Clare, A.M. & F.R.S.
London: Printed for Edward Symon . . . 1737.
Signed: Martin Clare. Soho Square, 25 March 1737.
8vo   21 cm   (xvi) + 369 + (24) pp.   9 engraved plates (as in first edition).
BM, RS, BDL

284 Third edition
Title as No. 282 with *The Third Edition, Corrected and Improved.*
London: Printed for A. Ward . . . 1747.
Signed: Martin Clare. Soho Square, 30 Dec 1746.
8vo   20 cm   (xvi) + 375 + (25) pp.   9 engraved plates.
BM, BDL

CLARK, William Tierney [1783–1852] F.R.S., M.Soc.C.E., M.Inst.C.E.

285 *Report of Mr W. Tierney Clark, Engineer, on the Practicability and Advantage of a Navigable Canal, from the Sea at Anderby Haven to the Town of Alford.*
London: Printed by W. M. Thiselton . . . 1825.
Signed: W. Tierney Clark. Hammersmith, 18 Aug 1825.
12mo   11 pp.
LCL

286 *Report (and Second Report) of William Tierney Clark, Esq. relative to New London Bridge.*
Signed: Wm. Tierney Clark. Hammersmith, 25 Oct 1831 [and] 17 Nov 1831.
pp. 8–11.
                    [In]
Copy of the Reports presented to the Corporation of London . . . relative to the Stability of the New London Bridge.
House of Commons, 1832.

287 *Plan of a Proposed Canal from the River Thames, to the Town of Dartford, Kent, with a Branch to Crayford Creek: also a Ferry from the entrance to the Canal to the opposite Shore at Purfleet in Essex.*
Surveyed and Drawn under the direction of W. Tierney Clark Esq. Civil Engineer by W. Hubbard 1835. Printed from Zinc by Chapman & Co.
Litho plan   37 × 55 cm   (scale 1 inch to 660 feet).
BM

CLEGG, Samuel [1781–1861]

288 *Clegg's Patent Atmospheric Railway.*
London: Printed by Richard Kinder . . . 1839.

8vo   23 cm   20 pp.
CE, ULG

See CLEGG and SAMUDA, 1840.

CLEGG, Samuel and SAMUDA, Joseph D'Aguilar

289 *Clegg and Samuda's Atmospheric Railway.*
London: John Weale . . . 1840.
8vo   22 cm   23 pp.   Folding litho plate. (Drawn by J. C. Haddon, printed by J. Grieve).
BM, CE

CLEGRAM, *Captain* William [1784–1863] M.Inst.C.E.

See CHAPMAN. Plan of Shoreham Harbour, surveyed by Clegram, 1815.

290 *A Chart of part of the Coast of Sussex . . . by* Wm. Clegram 1823 [with] *A Plan of Shoreham Harbour.*
Surveyed and Drawn by Wm. Clegram.
Engraved plan   20 × 32 cm   (scale of harbour plan 10 inches to 3 miles).
CE
Printed on verso are 'Sailing Directions for New Shoreham Harbour'. W. Clegram.

CLELAND, James [1770–1840] M.Inst.C.E.

291 *A Description of the Manner of Improving the Green of Glasgow, of Raising Water for the Supply of the Public Buildings of that City, &c.*

By James Cleland.
Glasgow: Printed by R. Chapman . . . 1813.
Signed: James Cleland, Glasgow, 15 May 1813.
8vo   21 cm   131 pp.   Folding engraved plan.
Plan of the Green of Glasgow; exhibiting a Design by James Cleland for sundry improvements thereon.
　　The Survey and Delineation by William Kyle. James Haldane sc.
NLS, ULG

292 *A Scheme for Raising and Distributing Water, and erecting Public Baths, in the City of Glasgow.*
　　By James Cleland.
Glasgow: Printed at the Stanhope Press by R. Chapman . . . 1813.
Signed: James Cleland. Glasgow, 4 Aug 1813.
8vo   20 cm   29 + (1) pp.
NLS

293 *Report relative to the Proposed Road to connect the Inchbelly Bridge Road with the Garscube Road* [etc].
　　By James Cleland, LL.D. Superintendent of Public Works.
Glasgow: Printed by Edward Khull and Son. 1829.
Signed: James Cleland. Office of Public Works, Council Chambers, Glasgow 30 April 1829.
4to   26 cm   (i) + 6 pp.   Folding litho plan.
Plan of a Public Road intended to connect the Inchbelly Bridge and Garscube Highways . . .
　　[by] James Clelland . . . from Surveys . . . by William Kyle.
NLS, ULG, (Rastrick)

CLOWES, Josiah [1735–1794]

294 *A Plan of the Canals authorised to be made in different Parts of England shewing the communications they will open between the principal Navigable Rivers and Ports of the Kingdom.*
　　J. C. del.   R. Murray sc.   Chester.
Engraved map, the lines entered in colours
　　56 × 43 cm   (scale 1 inch to 12 miles).
BM, MCL

The map can be dated *c.* 1772. It includes a table giving length, rise and fall for each of the ten canals shown, and plots the line of the proposed Chester Canal. The attribution to Clowes is tentative.

295 *Plan of an intended Navigable Canal from Hereford to Glocester; with a Collateral Branch to Newent.*
　　Jos[h] Clowes, Engineer. Rich. Hall, Surveyor. 1791.
Engraved plan   38 × 58 cm   (scale 1 inch to 1 mile).
BM, CE (Page), BDL (Gough)

CLYDE, RIVER

296 *Account of the Rise and Progress of the Plan for Improving the Navigation of the River Clyde, from Dumbarton Castle to the City of Glasgow; with . . . Hints for Farther Improvements in this River.*
　　Submitted to the Consideration of Thomas Telford, Esq. Engineer.
Glasgow: Printed by Andrew Young . . . 1824.
8vo   22 cm   35 pp.
NLS

COLE, Charles Nalson [1723–1804]

See DUGDALE. *The History of Imbanking and Draining.* 2nd edn. revised and corrected by C. N. Cole, 1772.

297 *Extracts from the Report of a View of the South Level, part of the Great Level of the Fenns, called Bedford Level; taken in the Summer of the Year 1777.*
　　By Charles Nalson Cole, Esq. at the Desire of the Board; and now by their Order printed.
[London, 1784].
Dated: Fen-Office, 25 March 1784.

8vo   18 cm   xii + 120 pp.
CE, CRO

298   *To the Honourable Corporation of Bedford Level this Reduced Map of that Level is Inscribed by Charles Nalson Cole Esq.! their Register.*
   Executed by their Order under his Direction & Inspection. Engraved by S. Neele.
London: Published by C.N.C. April 9th 1789.
Engraved map, hand coloured   55 × 75 cm
   (scale ¾ inch to 1 mile).
BM, CE, RGS, BDL

COLE, William

299   *The Journal of William Cole, Deputy Surveyor of the South Level, in the Year 1731.*
n.p.
Signed: W. Cole. Ely, 8 May 1732.
fol   34 cm   12 pp.
BDL (Gough)

COLLINGWOOD, Edward

See WATSON and COLLINGWOOD. Report on Rye New Harbour, 1756.

COOKE, Edward William [1811–1880] R.A., F.R.S.

See COOKE and RENNIE. Old and New London Bridges, 1833.

COOKE, Edward William and RENNIE, George

300   *Views of the Old and New London Bridges. Drawn and Etched by Edward William Cooke. With Scientific and Historical Notices of the Two Bridges; Practical Observations on the Tides of the River Thames; and a concise Essay on Bridges, from the Earliest Period . . .*
   By George Rennie, Esq. F.R.S.
London: Published by Brown and Syrett . . . 1833.
fol   43 cm   (iv) + vi + 24 pp.   12 etched
   plates.
BM, NLS, SML, RS, CE (Telford), RIBA
   Price £7 7s. in boards.

COPPIN, Daniel and JACKSON, William

301   *Proposals for the more Effectual Draining all the Levels contiguous to the River Witham, from the City of Lincoln to Chapple-Hill, and likewise, all the Fens and Low-Grounds, which empty themselves at Lodowick-Goat, and at the same Time, to restore the almost lost Navigation upon the said River, to a better State than ever it was.*
   By Daniel Coppin.
n.p. Printed in the Year 1745.
4to   21 cm   15 pp.
BM, ULG
   Includes an Estimate of the Expense of the Works, by William Jackson.

302   *A Map of the Ancient River Witham, with its [proposed] Alterations.*
   By William Jackson. To the . . . Mayor and Corporation of Boston in Lincolnshire. This Plan is humbly Inscribed by their most humble Serv.ts Willm Jackson & D. Coppin.
Engraved plan   42 × 70 cm   (scale nearly 1 inch to 1 mile).
BDL (Gough)

CREASSY, James [?1740–1807]

303 *The Report of James Creassy, respecting the Advantages, Facility and Expence, on opening a Navigable Communication from the Town of New Sleaford, in the County of Lincoln, to the present Navigation of the River Witham.*
n.p.
Signed: James Creassy, 12 March 1774.
fol   32 cm   8 pp.
LAO
  Stitched as issued.

304 *A Plan exhibiting the Course of the River, called Kyme Eau, with its two Branches up to New Sleaford, and the Works proposed to be Executed thereon; to open a Navigation from New Sleaford aforesaid, to the present Navigation of the River Witham.*
  Surveyed in Nov.ʳ 1773. By James Creassy.
Engraved plan   30 × 58 cm   (scale 2 inches to 1 mile).
LAO, MCL

305 *The Report and Opinion of James Creassy, respecting the Drainage of the Middle and South Levels of the Fenns, called the Bedford Level.*
pp. 17–31.
               [In]
Observations on the Means of Better Draining the Middle and South Levels of the Fenns.
  By Two Gentlemen who have taken a View thereof.
London: Printed for T. Evans . . . 1777. Price Two Shillings and Six Pence.
4to   26 cm   vii + 31 pp.   2 folding engraved plans.
BM, CRO, BDL (Gough), Sutro

306 *The Report and Opinion of James Creassy concerning the Embanking the Salt Marshes from the Sea, which lie in or adjoining to the several Parishes of Moulton, Waplode, Holbech, and Gedney, in the County of Lincoln.*
Signed: James Creassy. St. John's Street, London, 23 April 1791.

fol   33 cm   8 pp.
LAO
  Stitched as issued.

307 *Report of James Creassy Engineer, on the Drainage and Improvement of the Keyingham Level, and the Adjoining Country.*
Hull: Printed by William Rawson . . . 1801.
Signed: James Creassy. Burwood Cops near Crawley, Surrey, 8 April 1801.
4to   23 cm   10 pp.   Folding engraved plan, as in CHAPMAN's report of 30 June 1797 [q.v.] with addition of hand-coloured line indicating a proposed catchwater drain.
Hull City Library, CE, Hull University Library, Sutro

CRONK, William

308 *A Plan of the River Stour and Sandwich Haven, from a Survey taken by Mr William Cronk, in June 1775.* [with] *A Section of the River . . . shewing the fall of the surface of the stream at low water neap tide and the bed of the river, as laid down by Messrs Hogben and Cronk . . . and by Mess.ʳˢ Hogben jun.ʳ and Cronk.*
  W. Boys delin.   R. Rogers sc.
Engraved plan   41 × 62 cm   (scale 2 inches to 1 mile, and for the section 1 inch to 1 mile and 1 inch to 8 feet vertical).
BM
  Tables of the levels were published by Hogben senior and junior (taken in 1773), by Cronk (June 1775) and by Henry Hogben and Cronk (October–November 1775).

CROSLEY, William [d. 1796]

See RENNIE. Plans of the proposed Rochdale Canal, surveyed by Crosley, 1791 and 1793.

309 *Line of the proposed Canal, between Whitby and Pickering.*

By W. Crosley Engineer, 1793. W.F. 1794.

Engraved plan   22 × 30 cm   (scale 1 inch to 2 miles).

BM, CE (Page)

CUBITT, *Sir* William [1785–1861] F.R.S., M.R.I.A., M.Soc.C.E., M.Inst.C.E.

310 [*Report to the Committee for Improving the Navigation from Norwich to Yarmouth*].

Signed: W. Cubitt. Ipswich, 18 June 1814.

pp. 10–17.

[In]

The Speech of Alderman Crisp Brown, delivered at . . . Norwich, on Tuesday, the Eighth Day of Sept. 1818, together with Mr Cubitt's Report.

Norwich: Burks and Kinnebrook, Printers.

8vo   23 cm   23 + (1) pp.

BM, CE (Page)

311 *The Second Report of Mr William Cubitt* [on] *the best means of making Norwich a Port, by joining the Rivers Yare & Waveney, and opening a Harbour at Lowestoft.*

Norwich: Matchett and Stevenson, Printers.

Signed: Wm. Cubitt. Ipswich, 17 July 1820.

8vo   22 cm   (iii) + 25 pp.   Folding engraved plan.

A Map of the proposed Navigation for Ships from Norwich to the Sea at Lowestoft,

according to the Survey and Report of W. Cubitt, Civil Engineer, Ipswich. Engraved by Sidʸ Hall.

BDL

312 *Map of the proposed Navigation for Ships, from Norwich to the Sea at Lowestoft,*

according to the Survey and Report of W. Cubitt, Civil Engineer, Ipswich, 1820, and a

re-survey by W. Cubitt . . . and R. Taylor, Surveyor, 1825. Engraved by Sidʸ Hall.

Engraved plan, hand coloured   22 × 35 cm   (scale 1 inch to 2 miles).

[In]

Abstract of the Minutes of Evidence . . . on the Bill for making a Navigable Communication for Ships between Norwich and the Sea.

Norwich: 1826.

8vo   23 cm   viii + 56 pp.   Folding engraved plan, as above.

CE

313 *A Report and Estimate, on the River Waveney, between Beccles Bridge and Oulton Dyke, towards making Beccles a Port* [etc].

By William Cubitt, Civil Engineer.

Beccles: Printed and Published by R. B. Jarman . . . 1829.

Signed: W. Cubitt. London, 1 Dec 1827.

8vo   21 cm   12 pp.   Folding litho plan.

Map of the River Waveney from Beccles to Oulton Dyke with a continuation of the proposed Norwich & Lowestoft Navigation from thence to the Sea.

R. Cartwright, lithog.

ULG, BDL

314 *Ouse Lower Navigation and Drainage. The Report of William Cubitt, Esq; Civil Engineer.*

Lewes: Lee, Printers.

Signed: W. Cubitt. London, 15 Oct 1831.

fol   32 cm   4 pp.

East Sussex R.O.

315 *Description of a Plan for a Central Union Canal, which will lessen the distance and expense of canal navigation between London and Birmingham, and unite the Birmingham–Warwick and Birmingham–Coventry–and Oxford Canals, in one direct line of Inland Navigation between those Places.*

By W. Cubitt, F.R.S. Civil Engineer.

London: Roake and Varty . . . 1832.

8vo   22 cm   7 pp.   Folding engraved plan, lines entered in colours.

A Plan for a Central Union Canal . . .

By W. Cubitt . . . 1832. Engraved by E. Turrell.

CE, BDL

Also issued with the collected edition of BRADSHAW's *Lengths and Levels*, 1833 [q.v.].

316 Second edition
Title as No. 315.
London: Roake and Varty . . . 1833.
Signed: W.C. London, February 1833.
8vo   22 cm   7 pp.   2 folding engraved plans.
No. 1. *A Plan of a Central Union Canal, in two parts for improving the Line between London & Birmingham . . .* by W. Cubitt, F.R.S. Civil Engineer. 1832. Engraved by E. Turrell.
No. 2. [as in first edition].
ULG

317 *A Report on the Financial State of the Birmingham & Liverpool Junction Canal.*
By W. Cubitt, F.R.S., M.R.I.A. &c. Civil Engineer. With an Appendix of Correspondence and Calculations relating thereto.
London: Printed by Roake and Varty . . . 1834.
Signed: W. Cubitt. Parliament Street, London, 20 March 1834.
8vo   22 cm   37 pp.
ULG
   Issued in brown printed wrappers. Includes a letter from Alexander EASTON, dated 20 Feb 1834 [q.v.].

318 *Report on the Harbour and Docks of Leith, and the Projected New Harbour and Docks in its Vicinity.*
By William Cubitt, Esq. Civil Engineer.
Leith: Printed by R. Allardice . . . 1834.
Signed: W. Cubitt. Parliament Street, 11 Sept 1834.
8vo   21 cm   23 pp.   2 folding litho plans.
[1] *Enlarged Plan of the New Western Harbour, Leith, as connected with the Docks.* By Mr Cubitt. Drawn on Stone by R. W. Hume.
[2] *Plan of Leith Harbour & Docks. Shewing a Deep Water Entrance at Newhaven . . .* as Suggested in the accompanying Report. Allardice lithog.
NLS, BDL

319 *Report of William Cubitt, Esq. C.E. to the Commissioners of Public Works, Ireland, on the Improvement of the Port and Harbour of Belfast.*

Belfast: George Harrison . . . 1835.
Signed: W. Cubitt. Parliament Street, London, 30 May 1835.
8vo   22 cm   (i) + 9 pp.   2 folding hand-coloured litho plans.
Plan No. 1 (and No. 2) for the Improvement of Belfast Harbour.
   By W. Cubitt, F.R.S., M.R.I.A. &c. Civil Engineer, London, 1835. McBrair lithog.
CE

320 *Dublin Harbour. To the Commissioners for the Preservation and Improvement of the Port of Dublin.*
   The Report of W. Cubitt, F.R.S., M.I.R.A., &c., Civil Engineer.
n.p.
Signed: W. Cubitt. Dublin, 26 Aug 1835.
8vo   22 cm   8 pp.
CE

321 *Mr Cubitt's Reports. [on the foundations of Westminster Bridge].*
Dated: Parliament Street 1 Aug 1835 [and] London, 20 June 1836.
pp. 33–46.
                    [In]
Reports by Messrs. Telford, Cubitt, and Swinburne, Civil Engineers, as to the State of the Foundations, &c of Westminster Bridge.
   London, 1836. *See* WESTMINSTER BRIDGE.

322 *Mr Cubitt's Report to the Board of Directors, Commercial Blackwall Railway.*
Signed: Wm. Cubitt. Great George Street, Westminster, 8 Dec 1836.
pp. 33–41.
                    [In]
A Brief Statement of the Advantages which will result to the Public from the establishment of a Communication by Railroad between London and Blackwall.
London: Printed by A. Spottiswoode.
8vo   21 cm   41 pp.   Folding litho frontispiece and plan.
Brunswick Steam Wharf, East India Docks. Terminus of the Railway. General Plan shewing the River Thames

with the Line of the Commercial Railway, from London to Blackwall.

F. Mansell, lithog.

SML

323   Second issue
Mr Cubitt's Report . . . To the Board of Directors.
pp. 34–42.

[In]

A Brief Statement of the Advantages . . . of . . . a Railroad between London and Blackwall.

London: 1837. Printed by A. Spottiswoode.

8vo   21 cm   42 pp.   Folding litho plan, as in first issue (no frontispiece).

CE

Cubitt's report also reprinted in *Commercial Railway Company . . . Proceedings at a General Meeting of Proprietors held . . . 24th of January, 1837.* London: Printed by F. Mansell . . . 1837.

324   *Engineer's Reports*
Signed: Wm. Cubitt. London 5 Nov 1836, and for each year, except 1838, to 1843.

[In]

South-Eastern Railway. Proceedings of General Meetings of the Proprietors, with Reports of the Directors and the Engineers.

London: Printed by C. Roworth and Sons.

fol   34 cm   154 pp.

CE

Annual reports, probably issued separately; here collected together in one volume.

See BURGOYNE *et al.* Reports of the Commissioners for the Improvement of the River Shannon, 1837 and 1839.

325   *Report on St. George's Harbour and Railway.*
By William Cubitt, Esq., C.E.

Dated: Great George Street, Westminster, 13 Feb 1838.

pp. 86–88   with engraved plan.

Sketch of St. George's Harbour in Llandidno Bay, N. Wales,

to accompany Mr Cubitt's Report.

[In]

Second Report of the Commissioners [on] Railways in Ireland.

Dublin, 1838. Appendix A.

326   *Leith Harbour and Docks. Mr Cubitt's Report.*
Signed: W. Cubitt. Great George Street, Westminster, 13 May 1839.

pp. 6–13   with folding litho plan.

Plan for the Improvement of the Harbour and Docks of Leith, and the establishment of Low Water Piers or Landing Places.

By W. Cubitt, F.R.S. Civil Engineer.

[Printed with]

Report by James WALKER, 19 June 1839 [q.v.].

[In]

Plans and Reports, by J. Walker, Esq., & W. Cubitt, Esq., for the Improvement of Leith Harbour; and Treasury Minutes, and other Documents connected therewith.

Edinburgh: Printed by A. Murray . . . 1839.

8vo   22 cm   34 pp.   3 folding litho plans.

NLS, CE

CUMMING, T. G.

327   *Illustrations of the Origin and Progress of Rail and Tram Roads, and Steam Carriages or Loco-Motive Engines: also, interesting descriptive particulars of the formation, construction, extent, and mode of working some of the principal rail ways now in use* [etc].

By T. G. Cumming, Surveyor.

Denbigh: Printed for the author, and sold by Baldwin, Cradock, and Joy . . . London, and T. Gee, Denbigh. 1824.

Dated: Denbigh, Nov 1824.

8vo   22 cm   64 pp.   2 engraved plates.

BM, NLS, SML, CE, ULG, BDL

328   *Description of the Iron Bridges of Suspension now erecting over the Strait of Menai, at Bangor, and over the River Conway, in North Wales; with two views: also some account . . . of Captain S. Brown's Iron Bar Bridge over the River Tweed . . . and some calculations of the strength of malleable iron, founded on experiments.*

By T. G. Cumming, Surveyor.
London: Printed for J. Taylor . . . 1824.
8vo   22 cm   55 pp.   2 folding aquatint plates
(T. G. Cumming del., M. Dubourg sc.).
BM. NLS, SML, CE (Rastrick), BDL

329   Second Edition
Description of the Iron Bridges of Suspension erected over
the Strait of Menai . . . over the River Conway . . . and
over the River Thames, at Hammersmith [etc].
By T. G. Cumming, Surveyor. Second edition,
considerably enlarged.
London: Printed for J. Taylor . . . 1828.
8vo   22 cm   xv + 71 pp.   2 folding aquatint
plates (as in first edition) + folding engraved
plate of Hammersmith Bridge.
BM, RIBA, BDL

CUNDY, Nicholas Wilcox
[1778–   ?   ]

330   [*Three*] *Reports on the Grand Ship Canal from*
London to Arundel Bay and Portsmouth with . . . an
Estimate of the Probable Expense [etc].
By N. W. Cundy, Esq. Architect and Civil
Engineer.
London: Sold by Thomas Egerton . . . 1827. Price
10*s*. 6*d*.
Signed: N. W. Cundy. London, 1824; London,
Sept 1825 (and) July 1827.
8vo   23 cm   vi + 58 pp.   3 folding litho plans,
partly hand coloured.
[1]   *Map shewing the Line of the Grand Ship Canal,*
from London to Portsmouth made under the direc-
tions of N. W. Cundy, Esq. A.C.E. 1827.
[2]   *Map shewing the Line . . . from Portsmouth to the*
Junction at Arundel.
[3]   *Section of the Grand Ship Canal . . .* made under
the direction of N. W. Cundy A.C.E. 1825.
BM, CE, ULG
The first two reports were also published
(together) in 1825. London: Robson, Brooks &
Co. (BM, NLS)

331   *Imperial Ship Canal from London to Portsmouth.*
Mr Cundy's Reply to . . . Misrepresentations on his Pro-
jected Line.
London: Sold by Messrs Rivington . . . 1828.
Price 2*s*. 6*d*.
Signed: N.W. Cundy. London, 25 Jan 1828.
8vo   23 cm   21 pp.   Small folding litho plan.
BM, ULG
Issued in blue printed wrappers. Includes a
report by George and John RENNIE dated 12
Oct 1827 [q.v.].

332   *Inland Transit. The Practicability, Utility, and*
Benefit of Railroads; the comparative attraction and
speed of Steam Engines on a Railroad, Navigation and
Turnpike Road . . . also, the Plans, Sections, and Esti-
mates of the Projected Grand Southern and Northern
Railroads.
By N. W. Cundy, Civil Engineer. Second Edition.
London: Published by G. Hebert . . . 1834. Price
7*s*. 6*d*.
Signed: N.W. Cundy. London, December 1833.
8vo   22 cm   iv + (1) + 161 pp.   Folding litho
plate + 2 folding litho plans.
[1]   *Grand Northern Railway from London to York.*
With a Branch to Norwich, &c. Projected by
N. W. Cundy, Esq. Civil Engineer. London
1833, C. Ingrey Lithog.
[2]   *Grand Southern Railway.* Projected & Surveyed
by Nicholas Wilcox Cundy, Esq. Civil
Engineer. London, 1833. Baynes & Harris
lithog.
BM, SML, ULG
The first edition as above but 1833 and without
plan of Grand Southern Railway (BM).

333   *Grand Northern Railway from London to York,*
with a Branch from Cambridge to Norwich, &c.
Projected & Surveyed by Nicholas Wilcox
Cundy, Esq. Civil Engineer. London, 1833.
Baker Lithog.
Litho map   44 × 30 cm   (scale 1 inch to 15
miles).
[In]
[*Prospectus*] *Grand Northern Railroad Company, from*
London to . . . Cambridge, York and Leeds, with
Branches to Nottingham [etc].

Consulting Engineer: James Walker Esq.
Engineer: N.W. Cundy Esq.
[London]: Printer W. M. Knight and Co.
fol   42 cm   (4) pp. including map.
BM, Elton

334   *Plan of the Central Kentish Railway from London to Maidstone, Canterbury & Sandwich with a Branch to Gravesend.*
By N. W. Cundy Esq. Civil Engineer 1836. J. Netherclift Lithog.
Litho map   36 × 48 cm   (scale 1 inch to 4 miles).
[In]
[*Prospectus*] *The Central Kentish Railway and Sandwich Harbour Company.*
Consulting Engineers Messrs Rennie. Engineer N. W. Cundy Esq.
n.p.
fol   44 cm   (3) pp. including map.
BM

CURR, John [*c.* 1756–1823]

335   *The Coal Viewer, and Engine Builder's Practical Companion.*
By John Curr, of Sheffield.
Sheffield: Printed for the Author, by John Marshall. And sold by I. & J. Taylor . . . Joseph Whitfield . . . 1797.
4to   25 cm   96 pp.   5 folding engraved plates (Thos. Harris sc.).
BM, NLS, SML, CE (Telford)
Includes the first printed account of plate rails.

DADFORD, John

336   *Plan of a Canal and Rail Road, for forming a Junction between the Glamorganshire and Neath Canals in the County of Glamorgan.*

By John Dadford 1792.
Engraved plan, the lines entered in colours
48 × 102 cm   (scale 2½ inches to 1 mile).
BM, Elton

DADFORD, Thomas *senior* [d. 1809] M.Soc.C.E.

337   *Plan of the Canals between the Ports of Liverpool, Bristol and Hull, the Towns of Manchester and Birmingham, and the Cities of Coventry and Oxford.*
T. Dadford delin.   Downes sc.   [*c.* 1781].
Engraved plan, the lines entered in colours
27 × 45 cm   (scale 1 inch to 8 miles).
BM

DADFORD, Thomas *junior* [d. 1806]

338   *A Plan of an intended Canal from Kington in the County of Hereford to the River Severn near Stour Port in the County of Worcester.*
By Thos Dadford Junr 1789.
Engraved plan   25 × 93 cm   (scale 1 inch to 1 mile).
BM

339   *Plan of a Canal from Newport to Pontnewynydd with a Branch to Crumlin Bridge, and Rail Roads to Blaen-Afon, Beaufort . . . and Blaendir Iron Works, in the Counties of Monmouth and Brecon.*
By T. Dadford Junr Engineer. 1792.
Engraved plan   43 × 86 cm   (scale 2 inches to 1 mile).
Newport Central Library

340   *Plan of a Canal from the Town of Brecknock to join the Monmouth Shire Canal near the Town of Ponty Pool in the Counties of Brecknock and Monmouth.*
By T. Dadford Junr Engineer. 1793.
With a Plan of the Monmouth Shire Canal [and Rail Roads] as taken in 1792.

Engraved by W. Faden . . . 1793.
Engraved plan, the lines of canals and railways entered in colour   36 × 81 cm   (scale 1 inch to 1 mile).
P, B

341   *Brecknock & Abergavenny Canal. The Engineer's Report.*
n.p.
Signed: Thos. Dadford, jnr. 26 April 1798.
fol   30 cm   3 pp.
B

DALLAWAY, John

342   *A Scheme to make the River Stroudwater Navigable, from Framiload, to Wallbridge near the Town of Stroud, Glostershire.*
Glocester: Printed 1755. Price 6*d*.
Signed: John Dallaway. Brimscomb, 8 Dec 1755.
4to   21 cm   iv + viii + 11 + (1) pp.   Folding engraved plan.
A Plan of the River Stroud Water, with the Towns near the River, shewing by a double line on which side of the River, Locks may be Erected.
   T. Jefferys sc.
ULG
   Includes *An Estimate of the General Expenses of making the River Stroudwater Navigable.* By John Willets, Tho. Yeoman, and others.

DANCE, George *senior* [1695–1768]

343   *A Description of the Centre of the Great Arch of London Bridge.*
   This Plate is engraved from the Original Drawing . . . and publish'd by George Dance [and] Robert Taylor. Geo. Bickham sc. Feb 1st 1760.

Engraved plate   42 × 57 cm   (scale 1 inch to 5 feet).
RS (Smeaton)

DANCE, George [1741–1825] F.R.S., R.A.

See DANCE, JESSOP and WALKER. Plans of the proposed West India docks, 1797 and 1798.

344   *A Survey of the River Thames between London Bridge and Blackfriars Bridge, with the Soundings.*
   By Geo. Dance. Surveyors Office, Guildhall, July 1799. R. Metcalf sc.
Engraved plan   35 × 103 cm   (scale 1 inch to 100 feet).

[with]

345   *Plan and Elevation of London Bridge in its Present State.*
   By Mr Dance. Surveyors Office, Guildhall. July 1799. R. Metcalf sc.
Engraved   plan,   partly   hand   coloured 62 × 252 cm   (scale 1 inch to 10 feet).
[In]
Plans and Drawings referred to in the Second Report from the Select Committee upon the Improvement of the Port of London.
   House of Commons. 1799.

346   *Letter from George Dance, Esq; with Drawings or Designs for the Improvement of London Bridge.*
Signed: Geo. Dance. Gower Street, 2 July 1800.
pp. 77–81   with 7 folding engraved plates.
[In]
Third Report from the Select Committee upon the Improvement of the Port of London.
   House of Commons, 1800.
   The plates (Nos. 13–19) are in a separate volume.

347 Second printing
[In] *Reports from Committees of the House of Commons*
  Vol. 14 (1803) pp. 574–576 and pls. 42–47.

348 *Mr Dance's Answers. To the Select Committee of Bridge-House Lands*
  [on rebuilding London Bridge].
Signed: Geo. Dance. Guildhall, 11 April 1801.
pp. 2–3.
[In]
At a Sub-Committee . . . held at Guildhall, on Wednesday the 19th Day of December 1801.
Corporation of London, 1801.
fol   32 cm   17 pp.
CLRO

349 Second printing
[In] *An Abstract of Proceedings and Evidence relative to London Bridge.*
  Corporation of London. 1829. pp. 56–57. *See* LONDON BRIDGE.

See DANCE, CHAPMAN *et al*. Reports on London Bridge, 1814.

DANCE, George, CHAPMAN, William, *et al*.

350 *Report of George Dance, William Chapman, Daniel Alexander, and James Mountague, respecting the Enlargement and Improvement of the Water-way under London Bridge.*
  Corporation of London. 1814.
Dated: November 1814.
fol   32 cm   (i) + 48 pp.
CLRO, CE

351 Second printing
In *An Abstract of Proceedings and Evidence, relative to London Bridge.*
  Corporation of London, 1819, pp. 68–107. *See* LONDON BRIDGE.

DANCE, George, JESSOP, William and WALKER, Ralph

352 *A Plan of the Proposed Canal and Wet Docks for the West India Trade in the Isle of Dogs.*
  Dance, Jessop & Walker 29 Aug.ᵗ 1797.
  Metcalf sc.
Engraved plan, hand coloured   29 × 52 cm
  (scale 1 inch to 1,200 feet).
[In]
Report of a Committee of West-India Planters and Merchants on the subject of a Bill depending in Parliament for forming Wet Docks, &c at the Port of London.
November 16th 1797.
London: Printed for H. L. Galabin . . . 1797.
8vo   21 cm   49 + (1) pp.   with folding plan
  (as above).
ULG

353 Another issue
A Plan of the proposed Docks in the Isle of Dogs for the West India Trade.
  Surveyors Office, Guildhall. Metcalf sc.
CLRO, PLA (Vaughan)
  Also in *A Comparative View* by Ralph WALKER, 1797 [q.v.].

354 Revised plan
A Plan of the Proposed Canal and Wet Docks for the West India Trade in the Isle of Dogs.
  Dance, Jessop & Walker. Metcalf sc. [1798].
Engraved plan, hand coloured   29 × 52 cm
  (scale 1 inch to 1,200 feet).
[with]
An Explanation of the Plan of the proposed Canal and Wet Docks . . . in the Isle of Dogs.
Signed: Ralph Walker. 26 March 1798.
[and]
Estimate of the Expence of the Alteration intended to be made in the Plan.
Signed: W. Jessop, Ralph Walker. November 1798.
[In]
Minutes of Evidence on the Bill for the Port of London.
City Plan.

House of Commons, 1799. pp. 79–80 and p. 47 and pl. (1).

As compared with the 1797 plan the docks are larger and in a different layout.

DAVIDSON, James [1798–1877] M.Inst.C.E.

355 *A General Report of the State and Progress of the Caledonian Canal Works, from May 1825 to November 1826.*
Signed: James Davidson. Clachnacharry, 9 Nov 1826.

[In]

Twenty-Third Report of the Commissioners for the Caledonian Canal.

House of Commons 1827. pp. 18–21.
Subsequent progress reports by Davidson are in the 24th (1828) and 25th (1829) Commissioners' reports.

DAVIDSON, Matthew [1755–1819]

356 *Extract of a Letter from Mr Matthew Davidson, Superintendent of the Canal at Clachnacharry.*
Dated: 1 May 1812.

[In]

Ninth Report of the Commissioners for the Caledonian Canal.

House of Commons, 1812. p. 22.

DAVIES, Arthur and JEBB

357 *A Plan of the Ellesmere Canal, and its collateral Branches uniting the Rivers Severn, Dee & Mersey and the several Canals connected therewith.*

1796. Davies & Jebb, Oswestry, del. Surveyors to the Ellesmere Canal Company. Neele sc.
Engraved map, the canals entered in colours
44 × 25 cm   (scale 1 inch to 3 miles).
BM, RGS
Also issued, hand coloured, in Report to the General Assembly, 1805. *See* ELLESMERE CANAL.

DAVIES, Benjamin

358 *A Plan of the River Loddon, and Intended Navigable Canal from Basingstoke, in the County of Southampton; to the River Thames near Monkey Island. Proposed in the Year 1769.*
Benj. Davies del.   John Ryland sc.
Engraved plan   35 × 72 cm   (scale 1 inch to 1 mile).
BM, CE (Page)

DAVY, Christopher [c. 1802–1849]

See LONDON BRIDGE. Professional Survey, 1831.

359 *The Architect, Engineer, and Operative Builder's Constructive Manual, or, a Practical and Scientific Treatise on the Construction of Artificial Foundations for Buildings, Railways, &c...*
By Christopher Davy, Arch. & C.E.
London: John Williams . . . 1839.
Dated: Furnival's Inn, November 1838.
8vo   23 cm   (iii) + lxvi + 179 + (2) pp   Engraved frontispiece + 11 litho plates (1 hand coloured).   Woodcuts in text.
BM, RIBA

360 Second edition
Title, date, etc as No. 359 but *Second Edition*.
Dated: Furnival's Inn, 24 Jan 1839.

8vo 23 cm vii + lxvi + 180 pp. Frontispiece
and plates as in first edition.
RIBA, ULG, IC

DAY, James

361 *A Practical Treatise on the Construction and
Formation of Railways, showing the practical application
and expense of Excavating, Haulage, Embanking, and
Permanent Waylaying ... illustrated with Diagrams
and Original Useful Tables.*
By Jas. Day ... Second Edition.
London: John Weale ... 1839.
Dated: Hetton-le-Hole, Durham, July 1839.
8vo 19 cm xii + 210 pp.
BM, NLS, SML, ULG
Actually the first edition, as the author explains
in his Introduction. The 'Excavation Tables'
(pp. 165–204) are advertised by Weale as
being published separately for convenient use
in the field. Another edition of the *Treatise*
appeared in 1848.

DE HAVILLAND, *Major* Thomas Fiott (1775–1866)

362 *Chevalier Dubuat's Principles of Hydraulics;
with their Practical Application to the Indian Method of
Irrigation.*
1st Volume. By Major T. F. De Havilland ...
Acting Chief Engineer at Fort St. George.
Madras: Asylum Press, 1822.
Dated: Madras, 15 Feb 1822.
4to 22 cm xxxiii + 222 pp. 4 engraved
plates.
CE

No copy of the intended second volume, or
reference to its existence, has been found.

DESAGULIERS, John Theophilus [1683–1744] F.R.S.

363 *The Motion of Water, and other Fluids. Being a
Treatise on Hydrostaticks. Written originally in French,
by the late Monsieur Marriotte, Member of the Royal
Academy of Sciences at Paris ... and Translated into
English ...*
By J. T. Desaguliers, M.A. F.R.S. ... By
whom are added several Annotations.
London: Printed for J. Senex ... 1718.
8vo 20 cm xxiii + (1) + 290 pp. 7 folding
engraved plates.
BM, NLS, RS, IC

364 *A Course of Experimental Philosophy.*
By J. T. Desaguliers, LL.D. F.R.S. Vol. I.
London: Printed for John Senex ... 1734.
4to 25 cm (xxiv) + 463 + (9) pp. 32 folding
engraved plates.
BM, NLS, RS, IC, BDL

365 *A Course of Experimental Philosophy.*
By J. T. Desaguliers, LL.D. F.R.S. Vol. II.
London: Printed for W. Innys ... 1744.
4to 25 cm xv + (1) + 568 + (7) pp. 46 fold-
ing engraved plates (J. Mynde sc.).
BM, NLS, SML, RS, IC, BDL
Vols. I and II contain much of engineering
interest including articles by Charles Labelye.

366 *A Course of Experimental Philosophy.*
By J. T. Desaguliers, LL.D. F.R.S. Vol. I. The
Second Edition Corrected.
London: Printed for W. Innys ... 1745.
4to 25 cm xii + 466 + (9) pp. 32 folding
engraved plates.
BM, NLS, SML

367 *A Course of Experimental Philosophy.*
By J. T. Desaguliers, LL.D. F.R.S. The Third
Edition Corrected.
London: Printed for A. Millar ... 1763.
2 vols 4to 25 cm **1**, xi + 466 + (9) pp. **2**,
viii + 568 + (8) pp. 32 + 46 folding engraved
plates.
BM, SML, CE

DESMARETZ, *Captain* John (*fl.* 1724–1759) R.E.

See BRETT and DESMARETZ. Plan and report on Ramsgate Harbour, 1755.

DICKINSON, *Reverend*

368 *Queries relating to the Report of James Brindley Esq; Engineer, for Improving the Navigation and Drainage at Wisbech; and the Report of Mr John Golborne Engineer, concerning the Drainage of the North Level of the Fens, and the Outfall of Wisbech River.*

    Delivered in at the Meeting held at Wisbeach, on Wednesday the 5th Day of September, 1770.

Lynn: Printed by W. Whittingham, 1770.

Dated: 3 Sept 1770.

4to   26 cm   8 pp.

BM, Sutro

    For the authorship of these Queries *see* Richard DUNTHORNE's *Remarks*, 1770.

DODD, Barrodall Robert [*c.* 1780–1837]

369 *Report of Mr B.R. Dodd, Civil Engineer, on the proposed Canal Navigation between Newcastle and Hexham.*

Signed: Barrodall Robert Dodd. Newcastle, 22 Oct 1810.

pp. 31–44.

[In]

Observations on Canals and Railways . . . by the late William Thomas, Esq. Also, second edition, Report of Barrodall Robert Dodd, Esq . . . with Appendix.

Newcastle: Printed by G. Angus . . . 1825.

8vo   22 cm   (i) + 52 pp.

BM, SML, ULG

    The first printing of Dodd's report not found.

370 *Important Improvement of the Great North Road, by a Proposed Bridge over the River Tyne . . . [at Newcastle upon Tyne].*

By Mr B. R. Dodd, Civil Engineer.

Newcastle: Edward Walker, Printer. 1826.

fol   34 cm   3 pp.   Docket title on p. (4).

IC

DODD, Ralph [*c.* 1756–1822]

371 *Reports to the Honourable Commissioners of the River Wear, on the state of the River, Harbour, Piers, &c. of the Port of Sunderland.*

By Mr R. Dodd, Civil Engineer.

Sunderland: Printed . . . by James Graham . . . 1794.

Signed: R. Dodd. Sunderland, 28 July 1794.

4to   18 cm   42 pp.

CE (Vaughan)

372 *A Short Historical Account of the Great Part of the Principal Canals in the Known World.*

By R. Dodd, Civil Engineer.

Newcastle: Printed for, and sold by, W. Charnley and J. Bell . . . 1795.

Dated: Newcastle, 6 March 1795.

8vo   21 cm   iv + 27 pp.

BM, BDL, Elton

    Plagiarized from PHILLIPS's *General History of Inland Navigation* [q.v.].

373 *Report on the First Part of the Line of Inland Navigation from the East to the West Sea by way of Newcastle and Carlisle.*

    As originally projected, and lately surveyed, by R. Dodd, Civil Engineer.

Newcastle: Printed for the Subscribers . . . 1795.

Signed: R. Dodd. Newcastle. 5 June 1795.

8vo   21 cm   (i) + 59 + (1) pp.   Folding engraved plan.

A Plan of the first part of the Canal Navigation . . . by way of Newcastle & Carlisle, projected and surveyed by R. Dodd, Civil Engineer, with the Continuation of the River Navigation into the North Sea.

1795. Engraved by Beilby & Bewick.
BM, CE (Vaughan), BDL

374 *Report on the Various Improvements, Civil and Military, that might be made in the Haven or Harbour of Hartlepool*:
as surveyed at the request of the Corporation by R. Dodd, Engineer.
n.p.
Signed: R. Dodd. Newcastle, 1 Dec 1795.
8vo   21 cm   (i) + 17 pp.
BM, BDL

375 *Report, &c. on the Present State of Tynemouth Harbour, with Projected Improvements*.
By R. Dodd, Engineer.
n.p.
Signed: R. Dodd. Durham, 6 June 1796.
8vo   20 cm   (i) + 14 pp.
CE (Vaughan)

376 *Report on the Line of Inland Navigation, from the City of Durham, to the navigable part of the River Wear*.
Projected and Surveyed by R. Dodd, Engineer.
Durham: Sold by L. Pennington.
Signed: R. Dodd. Durham, September 1796.
8vo   20 cm   (i) + 20 pp.
CE (Vaughan)

377 *Report on the Line of Inland Navigation from Stockton to Winston, by way of Darlington and Staindrop*.
By R. Dodd, Engineer.
Stockton: Printed and sold by Christopher and Jennett. 1796. Price Six-Pence.
Signed: R. Dodd. Durham, December 1796.
8vo   20 cm   (i) + 21 pp.
CE (Vaughan)

378 *A Report and Estimate, on the Projected Dry Tunnel, or Subterraneous Passage under the River Tyne, to communicate with the Two Towns of North and South Shields*.
By R. Dodd, Engineer.
North Shields: To be had of W. Barnes . . . [etc].
Signed: R. Dodd. North Shields, Walker Place, December 1797.

8vo   20 cm   12 pp.
CE (Vaughan)

379 *Reports, with Plans, Section, &c. of the proposed Dry Tunnel, or Passage, from Gravesend, in Kent, to Tilbury, in Essex . . . Also on a Canal from near Gravesend to Stroud*.
By R. Dodd, Engineer.
London: Printed for J. Taylor . . . 1798.
Signed: R. Dodd. February 1798 and Gravesend, 10 Aug 1798.
4to   28 cm   viii + 28 pp.   3 folding engraved plates.   Published by J. Taylor 1 Sept 1798.
1. *Plan & Sections of the proposed Tunnel from Gravesend to Tilbury*.
R. Dodd del.   J. Raffield sc.
2. *A View of Gravesend and Tilbury . . . [with] Section of the River in the line of the Tunnel showing the Strata & the depth of Water*.
R. Dodd del.   J. Raffield sc.
3. *A Map showing the Roads and principal Towns within the immediate Influence of the Tunnel and of the Canal*.
BM, CE, BDL (Gough)

380 *Letters to a Merchant, on the Improvement of the Port of London; demonstrating its practicability without Wet Docks*.
By R. Dodd, Engineer.
[London] 1798.
Signed: R. Dodd. London, 9 Nov and 28 Nov 1798.
8vo   22 cm   18 pp.
BM, CE, ULG

381 *Report on the proposed Canal Navigation, forming a Junction of the Rivers Thames and Medway*.
By R. Dodd, Engineer.
[London] 1799.
Signed: R. Dodd. Gilberts Buildings, Lambeth.
8vo   20 cm   14 pp.
BM, CE (Vaughan), Elton

382 *Plan of the Intended Thames and Medway Junction Canal; on one Level*.
Projected and Surveyed 1799 by R. Dodd, Engineer.

Engraved plan 41 × 45 cm (scale 1 inch to 4 miles).
BM

383 *Introductory Report on the proposed Canal Navigation from Croydon to the River Thames at Rotherhithe.*
By R. Dodd, Engineer.
London: Printed by Order of the Subscribers . . . 1799.
Signed: R. Dodd. 35 Charing-Cross, 26 Nov 1799.
CE (Page)

384 *Plan of Part of Surrey & Kent shewing the general line of the Proposed Croydon Canal.*
By R. Dodd, Engineer. [1799].
Engraved plan 29 × 27 cm (scale 1 inch to 1 mile).
BM

385 Revised scheme
Report on the intended Grand Surry Canal Navigation, with General Estimate, &c.
By R. Dodd, Engineer.
London: Printed by Order of the Subscribers, by S. Gosnell . . . 1800.
Signed: R. Dodd. 35 Whitehall, 14 May 1800.
4to 26 cm 17 pp. Folding engraved plan.
Plan shewing the General line of the intended Grand Surry Canal,
as Proposed and Surveyed by R. Dodd, Engineer, 1800.
BM, PLA (plan only)

386 *Plan of part of the proposed Grand Surry Canal.*
Surveyed by R. Dodd, Engineer. 1800.
Engraved plan, the line entered in colours 41 × 75 cm (scale 2 inches to 1 mile).
BM, BDL

387 *Mr Dodd's Account of London Bridge; with Ideas for a New One*
[and a letter on the proposed design].
Signed: R. Dodd. Whitehall, May 1799 and 8 July 1800.
pp. 47–50 and 51*–54* with 6 folding engraved plates.
[In]

Third Report from the Select Committee upon the Improvement of the Port of London.
1800.
The plates (Nos. 2–7) are in a separate volume.

388 Second printing
In *Reports from Committees of the House of Commons* Vol. 14 (1803) pp. 559–562 and pls. 31–36.

389 *Report on the intended North London Canal Navigation; with General Estimate.*
By R. Dodd, Engineer.
London: Printed by S. Gosnell . . . 1802.
Signed: R. Dodd. Parliament Street, 22 Sept 1802.
4to 26 cm 23 pp. with folding engraved plan (untitled).
BM, CE

390 *Observations on Water; with a Recommendation of a more Convenient and Extensive Supply of Thames Water, to the Metropolis, and its Vicinity.*
By Ralph Dodd, Civil Engineer.
London: Printed by George Cooke . . . 1805.
Signed: Ralph Dodd. London, 28 Dec 1804.
12mo 15 cm (i) + 101 + (19) pp.
BM
Includes on pp. (1)–(19) reports to the subscribers to the South and East London Waterworks, dated 5 June and 2 Nov 1804.

391 *Introductory Report on the intended Bridge across the River Thames, from near Vauxhall to the opposite Shore; with accompanying New Roads.*
Most respectfully addressed to His Royal Highness the Prince of Wales, by . . . Ralph Dodd, Civil Engineer.
London: John Abraham, Printer. [c. 1807].
Signed: Ralph Dodd, 'Change-Alley, Royal Exchange.
fol 32 cm 3 pp.
BM
Dodd's 'Elevation of a Bridge to cross the River Thames near Vauxhall' was exhibited at the Royal Academy in 1807.

392 *Report of Mr Ralph Dodd . . . on . . . London Bridge, with Descriptive Plans for a New One.*
[London] 1820.
Signed: Ralph Dodd, 27 Poultry, London.
4to   27 cm   25pp.
BM, CE (lacking title-page)

DODGSON, George Haydock [1811–1880]

393 *Illustrations of the Scenery on the Line of the Whitby and Pickering Railway . . . From Drawings by G. Dodgson. With a short Description . . . by Henry Belcher.*
London: Longman, Rees . . . 1836.
8vo   24 cm   Vignette title-page, viii + 115 pp.   8 woodcuts in text, 12 steel engraved plates (J. T. Willmore sc. and others).
BM, SML, CE, ULG
  Dodgson, an assistant to George Stephenson, prepared the plans for this railway.

DODSON, William

394 *The Designe for the perfect Draining of the Great Level of the Fens (called Bedford Level) lying in Norfolk, Suffolk, Cambridgeshire, Huntingtonshire, Northamptonshire, Lincolnshire, and the Isle of Ely. As it was delivered to the Honourable Corporation for the Draining of the said Great Level, the 4th of June 1664. As also, several Objections answered since the Delivery of the said Designe* [etc].
  By Colonel William Dodson.
London: Printed by R. Wood . . . 1665.
4to   20 cm   (vi) + 39 + (1) pp.   Folding engraved plan.
A Mapp of the Great Levell of the Fens called Bedford Levell. The pricked Lines represent the New Rivers to be made in and about the Great Levell.
  By William Dodson, Gent.
BM, CE, CRO, ULG, BDL (Gough)

DONALDSON, James [1758–1807]

See TELFORD. Map of intended Roads and Bridges in the Highlands of Scotland, 1805. This includes twelve roads surveyed by James Donaldson 1803–1805.

DOUGLAS, *Sir* Howard [1776–1861] R.E., F.R.S.

395 *An Essay on the Principles and Construction of Military Bridges, and the Passage of Rivers in Military Operations.*
  By Col. Sir Howard Douglas, Bt. F.R.S. Inspector General of the Royal Military College.
London: Printed for T. Egerton . . . 1816.
8vo   21 cm   (iv) + ii + 204 + (6) pp.   13 folding engraved plates (Neele sc.).
BM, NLS, RS, IC, BDL

396 Second edition
Title as No. 395, but *Second Edition, containing much additional matter.*
London: Thomas and William Boone . . . 1832.
8vo   22 cm   (vi) + vi + 417 + 28 pp.   Frontispiece + 12 folding engraved plates.
NLS

DOUGLASS, James

397 *Explanation of a Plan for improving the Port of London.*
[London] 1799.
fol   33 cm   3 pp.   Docket title on p. (4).
CE (Telford)

See TELFORD and DOUGLASS. Plans for iron bridges of three or five spans over the Thames, 1800.

See TELFORD and DOUGLASS. Plans and an account of a proposed single-arch iron bridge of 600 feet span over the Thames, 1800–1802.

DOWNIE, Murdoch

398 *Description of the Navigation of the Murray Firth and Loch-Beauly from the Entrance to Cromarty Harbour to above Kessack; and also of the Lochs or Lakes, Loch-Ness, Loch-Oich, and Loch Locky.*
By Murdoch Downie.
Signed: M. Downie [1803]
[In]
[*First*] *Report of the Commissioners for the Caledonian Canal.*
House of Commons, 1804. pp. 24–38.
Downie's survey was completed in September 1803.

See TELFORD. Plans of the Caledonian Canal by Telford and Downie, 1803 and 1805.

399 *Report upon the Harbour proposed to be constructed in Ardrossan Bay.*
Signed: M. Downie. Saltcoats, 5 Jan 1805.
pp. 13–15.
[In]
TELFORD. *Report by Thomas Telford . . . relative to the Proposed Canal from Glasgow to the West-Coast of Ayr; and the Harbour of Ardrossan*, 1805 [q.v.].

DREWRY, Charles Stewart [1805–1881] Assoc.Inst.C.E.

400 *A Memoir on Suspension Bridges, comprising the History of their Origin and Progress . . . with Descriptions of some of the most important Bridges . . . also an*

Account of Experiments on the Strength of Iron Wires and Iron Bars [etc].
By Charles Stewart Drewry.
London: Printed for Longman, Rees . . . 1832.
Dated: Chancery Lane, September 1832.
8vo   21 cm   xii + (1) + 211 pp.   7 double-page litho plates.
BM, NLS, SML, CE (Telford), BDL

DRUMMOND, *Captain* Thomas [1797–1840] R.E.

See DRUMMOND *et al.* Reports of the Commissioners . . . [on] Railways in Ireland, 1837 and 1838.

DRUMMOND, Thomas, BURGOYNE, *Sir* John, BARLOW, Peter and GRIFFITH, *Sir* Richard

401 *First Report of the Commissioners.*
Signed: T. Drummond, J. F. Burgoyne, Peter Barlow, Richard Griffith. Dublin, 11 March 1837.
[In]
First Report of the Commissioners . . . [on] Railways in Ireland.
Dublin, 1837, pp. 3–10.

402 *Second Report of the Commissioners.*
Signed: T. Drummond, J. F. Burgoyne, Peter Barlow, Richard Griffith. Dublin, 13 July 1838.
With large folding engraved map.
Map of England & Ireland explanatory . . . of the Report of the Railway Commissioners which relates to the communication between London and Dublin, and other parts of Ireland.

Prepared & engraved under the direction of Lieut. Larcom, Royal Eng.^rs May, 1837.
[In]
Second Report of the Commissioners . . . [on] Railways in Ireland.
Dublin, 1838. pp. (1)–122 and pl. 1.

DUGDALE, *Sir* William [1605–1686]

403 *The History of Imbanking and Drayning of divers Fenns and Marshes, both in Foreign Parts, and in this Kingdom; and of the Improvements thereby. Extracted from Records, Manuscripts, and other Authentick Testimonies.*
By William Dugdale, Esquire, Norroy King of Arms.
London: Printed by Alice Warren . . . 1662.
fol   34 cm   (vii) + 424 + (2) pp.   11 folding engraved plans.
BM, ULG, BDL

404 Second edition
The History of Imbanking and Draining of Divers Fens and Marshes . . . By William Dugdale, Esq. Norroy King of Arms. Afterwards Sir William Dugdale, Knt. Garter Principal King at Arms.
The Second Edition, Revised and Corrected, by Charles Nalson Cole, Esq; of the Inner Temple, Barrister at Law, and Register to the Honourable Corporation of Bedford Level.
London: Printed by W. Bowyer and J. Nichols . . . 1772.
fol   40 cm   xii + 469 pp.   11 folding engraved plans (as in first edn.).
BM, CE (Telford), LCL, ULG, BDL (Gough)

DUNN, Matthias

405 *Observations upon the Line of Railroad projected by Mr Landale, from Dundee to the Valley of Strathmore.*

By Matthias Dunn, Colliery Viewer and Mineral Surveyor, Newcastle.
pp. 1–5 (second part). Signed: Matthias Dunn, Dundee, 1 Oct 1825.
[with]
pp. 1–10. Report submitted to the Subscribers . . . By Charles LANDALE . . . 30 Sept 1825 [q.v.].
Dundee: Printed by David Hill . . . 1825.
4to   30 cm   10 + 5 + (1) pp.   Folding litho plan.
CE (Telford), ULG

406 *Prospectus of a Railway from Newcastle to Morpeth, to be called the Northumberland Railway, ultimately to be connected by Branches with Blyth and Shields.*
By M. Dunn, Colliery Viewer and Civil Engineer.
Newcastle: Printed by W., E., & H. Mitchell . . . 1835.
Signed: Matthias Dunn. Newcastle, 18 April 1835.
8vo   22 cm   14 pp.   Folding litho plan.
Map of the Course of the Projected Northumberland Railway from Newcastle to Morpeth by way of Cramlington.
By Mr Matt.^s Dunn . . . Mitchells Lithog.
BM, NLS, CE

DUNTHORNE, Richard [1711–1775]

407 *Remarks on certain Queries delivered in by the Rev. Mr Dickinson, at a Meeting held at Wisbeach, on Wednesday the 5th of September, 1770.*
By Richard Dunthorne.
n.p.
4to   28 cm   4 pp.
CE, BDL (Gough)

408 *So much of the Report of Richard Dunthorne, Superintendent of the Works of the Bedford Level Corpo-*

ration, made at the General Whitsun Meeting, 1771, as relates to Wisbeach Outfall.
n.p.
fol   32 cm   3 pp.
CRO, BDL (Gough), Sutro

409   [Report on the River Stour, in Kent].
Signed: Richard Dunthorne. Cambridge, 8 Sept 1774.
pp. (1)–(2).
[Printed with]
Report by Thomas YEOMAN, 8 Oct 1765 [q.v.]
[and]
Tables of levels by Thomas and Henry HOGBEN, 1773 [q.v.].
n.p.   [Printed 1775].
fol   34 cm   (4) pp.
BM

DYSON, John

410   The Report of Mr John Dyson, Engineer to the Bedford Level Corporation.
Cambridge: Hatfield, Printer.
Signed: John Dyson. Downham Market, 29 Nov 1824.
fol   32 cm   3 pp.   Docket title on p. (4).
CRO

See SPECIFICATION for a Sluice near Welney, 1824.

EASTON, Alexander [1787–1854] M.Inst.C.E.

411   Description of the Outlet at Strone.
By Mr. Alexander Easton.
Signed: Alex.ʳ Easton.
[In]

Seventeenth Report of the Commissioners for the Caledonian Canal.
House of Commons, 1820. p. 20.

See TELFORD. Map of roads from Milford and Pembroke to Caermarthen, surveyed by Alexander Easton 1824 (published 1827).

412   Letter to William Cubitt, Esq. Civil Engineer.
Signed: Alex. Easton. Drayton, 20 Feb 1834.
pp. 22–29.
[In]
CUBITT. Report on the Birmingham & Liverpool Junction Canal.
London, 1834 [q.v.].

EASTON, John [d. 1826] M.Inst.C.E.

413   Report of Mr John Easton, Assistant Engineer under Mr Telford; and resident Inspector of the Improvements and Repairs of the Road from Shrewsbury to London.
Signed: John Easton, 1 June 1822.
[In]
Fourth Report of the Select Committee on the Road from London to Holyhead.
House of Commons, 1822. pp. 84–114.

See SPECIFICATION for Holyhead Road; St. Alban's and South Mimms, 1823.

EASTON, John, of Taunton [1788–1860]

414   The Report of John Easton, on the Practicability of forming a Navigation from the River Parrett at Dunball, about 3 Miles below Bridgwater, to or near the Towns of Somerton, Ilchester and Yeovil, in the County of Somerset, effecting thereby a complete Drainage of the low and wet Lands in King's Sedgmoor . . . and other places on the Line.

Signed: John Easton. Taunton, 1 March 1829.
pp. 3–5   with folding litho plan.
Plan of the Proposed Line of Navigation & Drainage from Dunball . . . to Yeovil.
Taken by Messrs John & Abel Easton and approved by George Rennie Esq. 1829.
[In]
Prospectus of a Plan for Rendering Navigable the King's Sedgmoor and other Principal Drains between the River Parrett and the Town of Yeovil, and effecting a more complete drainage of the Lands.
Somerton: 1829.
fol   33 cm   6 pp.   Folding plan (as above).
Somerset R.O.

EAU BRINK CUT

See MAXWELL, George [attributed to]. Reasons attempting to shew the Necessity of the Proposed Cut from Eau Brink to Lynn, 1793.

415   *A Letter to Sir Thomas Hyde Page in Answer to his Letter lately published on the Subject of the Eau-Brink Cut.*
n.p. 1794.
Signed: Investigator.
8vo   21 cm   15 pp.
BM, CUL
Critical remarks on PAGE's letter of 26 Oct 1793 [q.v.].

416   *An Abstract of the Case and Opinions that have appeared in print For and Against the Eau Brink Cut; affecting Drainage, Navigation, and the Port of Lynn.*
By a member of the Committee.
n.p. 1794.
8vo   21 cm   32 pp.
BM, CUL

417   Second edition
Title as No. 416.
Lynn: Printed by Mr Mugridge, 1824.
8vo   23 cm   iv + 35 pp.
CRO

418   *Report of the Joint Committee of Drainage & Navigation, appointed in pursuance of . . . the Act . . . for altering and enlarging the Powers of the previous Eau Brink Acts.*
Lynn: Mugridge, Printer.
Dated: 19 Feb 1823.
8vo   23 cm   16 pp.
CRO, CE

EDEN, *Sir* Frederick Morton [1766–1809]

419   *Porto-Bello: or, a Plan for the Improvement of the Port and City of London.*
By Sir Frederick Morton Eden, Bart.
London: Printed for B. White . . . 1798.
8vo   22 cm   53 pp.   4 folding engraved plates.
BM, CE, ULG, BDL (Gough)

EDGEWORTH, Richard Lovell [1744–1817] F.R.S., M.R.I.A., Hon.M.Soc.C.E.

420   *The Report of Richard Lovell Edgeworth, upon District No. 7*
[the District of the Inny and Lough Ree].
Signed: Richard Lovell Edgeworth, 28 Oct 1810.
With 1 engraved plate and 5 folding engraved plans.
[In]
Second Report of the Commissioners on the practicability of draining and cultivating the Bogs in Ireland.
House of Commons, 1812. pp. 173–204 and pls. 12–17.

421   *The Report of Mr Richard Lovell Edgeworth, on District No. 15, Bogs on the East of the River Shannon, Counties Leitrim and Longford; and Observations on Mr Roscoe's Improvements of Chat-Moss in Lancashire.*

Signed: Richard Lovell Edgeworth. 1 May 1813.
With 2 hand-coloured folding engraved plans
(J. Basire sc.).

[In]

*Fourth Report of the Commissioners on the practicability
of draining and cultivating the Bogs in Ireland.*
  House of Commons, 1814. pp. 103–106 and
  pls. 9–10.

422 *An Essay on the Construction of Roads and Car-
riages.*
  By Richard Lovell Edgeworth, Esq. F.R.S.
  M.R.I.A.
London: Printed for J. Johnson and Co . . . 1813.
8vo   23 cm   (ii) + ix + 202 + (i) + 194 pp.   4
  folding engraved plates (Thomson sc.).
BM, NLS, SML, ULG, IC
  Issued in grey boards, printed paper label on
  spine. Price 14s.

423 Second edition
An Essay on the Construction of Roads and Carriages.
  By Richard Lovell Edgeworth, Esq. F.R.S.
  M.R.I.A. The Second Edition: with a Report
  of Experiments tried by Order of the Dublin
  Society.
London: Printed for R. Hunter, successor to Mr
  Johnson . . . 1817.
Preface dated: 29 Dec 1816.
8vo   23 cm   iv + 3 + 171 pp.   4   folding
  engraved plates (as in No. 422).
BM, NLS, CE, ULG (Rastrick)
  French translation published in Paris, 1827.

EDINBURGH & GLASGOW UNION CANAL

424 *Observations* [by the Committee of Subscribers to the Union Canal].
[Edinburgh, 1814]. Dated: 24 Jan 1814.
4to   27 cm   16 pp.   Folding engraved plan
  (W. & D. Lizars sc.).
*General Plan of the Communications that will be Formed
by the Proposed Edinburgh & Glasgow Union Canal
with the other Navigations.*
NLS, CE, ULG

425 *Observations by the Union Canal Committee, on
the Objections made by the Inhabitants of Leith to this
Undertaking.*
Edinburgh: Oliver & Boyd, Printers. [1817].
8vo   23 cm   31 + 7 pp.
CE
  Includes two reports by TELFORD dated 5
  April 1815 and 29 Jan 1817 [q.v.].

EDINBURGH, Water Supply

426 *Reduced Plan of the proposed Aqueduct from
Crawley Spring, Glencorse Burn and Black Springs to the
City of Edinburgh. September 1818.*
Engraved plan   36 × 51 cm   (scale 2 inches to
  1 mile).

[In]

*An Act for more effectually supplying the City of Edin-
burgh and Places adjacent with Water.*
[London]: Spottiswoode and Robertson . . . 1819.
fol   62 + (2) pp.   Docket title on p. (64).
  Folding engraved plan (as above).
CE (Telford, from James Jardine)
  Jardine was engineer for the scheme, under
  Telford's direction.

EDWARDS, George

427 *The Report of Mr George Edwards.*
Signed: George Edwards. Lowestoft, 6 Jan 1838.
pp. 7–16   with folding hand-coloured litho plan.
Proposed Lock near the Vinegar Yard.
  Sloman, lithog.

[Printed with]

Report by WALKER and BURGES, 26 Dec 1836
  [q.v.].

[In]

Reports to the Commissioners of the Haven of Great Yarmouth respecting the Proposed Lock upon the River Wensum, at Norwich.

By Messrs Walker and Burges, and Mr George Edwards, Civil Engineers.
Yarmouth: Printed by Charles Sloman . . . 1838.
8vo   22 cm   16 pp.   3 folding plans.
CE

EDWARDS, Langley [d. 1774] M.Soc.C.E.

See GRUNDY, EDWARDS and SMEATON. Report on the River Witham, 1761.

428   *A Plan of the Low Fen Lands and Marshes lying between the Rivers Witham, Kyme Eau, and the Glen, in the County of Lincoln. With the Several Brooks, Becks & Rivulets by which the Waters descend into the Levels, and also the Drains by which they are conveyd to their Outfalls into Boston Haven & Fosdyke Wash.*

By Langley Edwards.
Engraved plan   23 × 31 cm   (scale 1 inch to 2 miles).
MCL, LCL, RS (Smeaton), RGS
Black Sluice Orders Book, 4 July 1764: Langley Edwards to make a Survey and plan of the Level. 27 Nov 1764: the plan prepared by Mr Edwards is approved.

429   *The Report of Langley Edwards Engineer, for Amending the Outfall of the Nene, securing the Middle and North Levels of the Fens, and other Countries adjoining to them from Inundations by Breaches of the Banks of Moreton's Leam, and improving the Navigation of the Port of Wisbeach; and for better Draining the Lands above-mentioned.*

n.p.
Signed: Langley Edwards, 16 April 1771.
4to   28 cm   8 pp.
CE, BDL (Gough), Sutro

EDWARDS, Richard

430   *Observations on the Decay of the Outfalls or Loss of the Channels of divers weak Rivers, particularly of the River Neen, otherwise Wisbeach River, and Shire-Drain [etc].*

By Richard Edwards.
London: Printed in the Year 1749.
8vo   20 cm   15 pp.   4 engraved plans.
BDL (Gough), Sutro

EFFORD, W.

431   *A Scheme for the Better Supplying this Metropolis with Sweet and Wholesome Water from the River Coln.*
Most humbly offered to the Consideration of . . . Parliament, the Nobility, Gentry, and Inhabitants of the West End of the Town. By W. Efford, Gent.
London: Printed for the Author by C. Say . . . 1764.
4to   21 cm   16 pp.   Folding engraved plan (untitled). Efford del.   Kitchin sc.
BM, BDL (Gough)
Derived from FORD's scheme [q.v.].

ELLESMERE CANAL

432   *Report to the General Assembly of the Ellesmere Canal Proprietors . . . on the 27th Day of November 1805. To which is annexed The Oration delivered at Pontcysylte Aqueduct, on its First Opening, November 26, 1805.*
Shrewsbury: Printed by J. and W. Eddowes. 1806.
4to   28 cm   42 + 36 pp.   Folding hand-coloured engraved plan.
CE, ULG, Elton

The plan is that by DAVIES and JEBB, 1796 [q.v.].

ELLIS, *Lieut.* Francis Wilson, R.N.

433 *A Brief Historical Report on Southwold Harbour, submitted at a Special Meeting of the Commissioners.*
 By Lieut. F. W. Ellis, R.N.
Southwold: Printed by Fredk. Skill . . . 1839.
Dated: Southwold, 27 June 1839.
8vo   21 cm   (iii) + 23 pp.
BM, CE

ELLISON, Richard [d. 1743]

See ELLISON and PALMER. Survey of the Rivers Swale and Ouze, 1735.

ELLISON, Richard and PALMER, William

434 *A Survey of the Rivers Swale & Ouze from Richmond to York; in order to improve the Navigation thereof, by Richard Ellison & Willm. Palmer. Taken 1735.*
 Clark sc.
Engraved map   23 × 63 cm   (scale 1 inch to 1½ miles).
BM, BDL (Gough)

ELMES, James [1782–1862]

435 *Essay on Foundations.*
 By James Elmes.
Dated: 29 April 1808.
Signed: James Elmes. Tavistock Place.
pp. 169–189 with engraved plate.
[In]
Essays of the London Architectural Society.
London: Sold by J. Taylor . . . 1808.
8vo   25 cm   xii + (iii) + 189 pp.   4 engraved plates.
RIBA

436 *London Bridge, from its original formation of wood to the present time: with a particular account of the Progress and Completion of the New London Bridge.*
 By J. Elmes, Esq. Architect, Surveyor of the Port of London.
London: C. Wood and Son . . . 1831.
8vo   23 cm   (iii) + 24 pp.   Engraved plate.
London Bridge, 1830. Engraved by J. Shury from a drawing by J. Rennie, Esq. F.R.S.
BM

437 *A Scientific, Historical, and Commercial Survey of the Port of London.*
 By James Elmes, Architect & Civil Engineer. Surveyor of the Port of London.
London: John Weale . . . 1838.
fol   55 cm   vi + (i) + 70 pp.   22 engraved plates including title, text vignette and frontispiece.
Pl. 1. *Chart of the Harbour and Port of London.*
 Surveyed and Drawn by James Elmes . . . 1837.
J. & C. Walker sc.
BM, ULG

438   Another issue
As No. 437 but without prelims.
Forming Division IV in SIMMS' *Public Works of Great Britain.* London, 1838 [q.v.].

ELSTOBB, William *senior* [died *c.* 1765]

439 *Some Thoughts on Mr Rosewell's and other Schemes, now proposed for amending Lynn Channel and Harbour, in a Letter to the Merchants, Owners, and Masters of Ships belonging to the said Place.*
    By William Elstobb.
n.p. Printed in the Year 1742.
4to   24 cm   8 pp.
BM, CRO, ULG, BDL (Gough)

ELSTOBB, William [died *c.* 1782]

440 *The Pernicious Consequences of Replacing Denver-Dam and Sluices, &c. Consider'd in a Letter to Mr John Leaford. Wherein his Arguments in a Pamphlet entitled, Observations, &c. are fairly and candidly examined.*
    By William Elstobb, Jun.
Cambridge: Printed by Joseph Bentham, 1745.
8vo   22 cm   (iii) + 20 pp.
BM, CE (Page), BDL (Gough), CUL, Sutro

441 *A Book of References to the Map of Sutton and Mepall Levels, done from a Survey taken in 1750, By Order of the Gentlemen Commissioners . . . for the better Draining and Improving the said Levels.*
    By William Elstobb, jun.
Cambridge: Printed by S. and J. James.
8vo   19 cm   72 pp.
CE, BDL (Gough), CUL
    Includes on p. 2 an advertisement by William Elstobb, jun. of King's Lynn, Surveyor and teacher of geography, astronomy, surveying, algebra and other parts of the Mathematics.

442 *A Map of Sutton and Mepall Levels . . . and Byal Fen; in the Isle of Ely and County of Cambridge. From an Actual Survey taken in the Year 1750.*
    By Will.ᵐ Elstobb Jun.
Engraved plan   (scale 8 inches to 1 mile).

Original not seen. Reduced photocopies in BM and CUL.

443 *Remarks on a Pamphlet intitled An exact Survey of the River Ouse, from Brandon Creek to Denver Sluice, by Mr James Robinson of Ely; in a Letter to the Author.*
    By William Elstobb, Jun.ʳ Land Surveyor.
Cambridge: Printed by J. Bentham . . . 1754.
Signed: William Elstobb jnr. St. Ives, 1 June 1753.
8vo   20 cm   24 pp.   Folding engraved plate [untitled] showing cross-sections of the river.
CRO, CUL

444 *A Chain and Scale of Levels along Wisbeach River and Channel from Peterborough Bridge down to the Eye at Sea.*
    Taken in 1767 by William Elstobb. Vere sc.
Engraved section   26 × 83 cm   (Scale 1 inch to 1 mile and 1 inch to 4 feet vertical).
CRO, Sutro
    Also annexed to SMEATON's and YEOMAN's reports on the North Level, 1768 and 1769 [q.v.].

445 Sections of the River Ouse
Four engraved sections. Scale 1 inch to 1 mile and 1 inch to 8 feet vertical, and for (d) 1 inch to 30 feet and 1 inch to 12 feet vertical.
(a) *Longitudinal Section of the Bottom of the Hundred Feet and Ouse Rivers from near Denver Sluice up to Over Cote.*
    By Will.ᵐ Elstobb 1776.   24 × 62 cm.
(b) *Longitudinal Section of the Bottom of the River Ouse, from the Old Bar Beacon below Lynn up to Denver Sluice.*
    By Will.ᵐ Elstobb.   25 × 70 cm.
(c) *Longitudinal Section of the River Ouse from Denver Sluice up to Clay-Hithe.*
    25 × 72 cm.
(d) *Lateral Sections of the River Ouse and of the Hundred Feet River [with] Longitudinal Section of Sandy's Cutt.*
    26 × 70 cm.
BM, CE, RS (Smeaton) ((b) only)

446 *Observations on an Address to the Public, dated April 20, 1775, superscribed Bedford Level, and sign'd Charles Nalson Cole, Register; . . . and on A Plan and Draught of a Bill intended to be presented to Parliament . . . for preserving the Drainage of the Middle and South Levels.*

By William Elstobb, Land Surveyor and Engineer.

Lynn: Printed and Sold by W. Whittingham . . . 1776.

8vo   21 cm   (i) + 104 pp.

BM, BDL (Gough), CUL

447   *Remarks on the Report of Mr John Golborne, Engineer, dated December 2ᵈ 1777, on a View . . . of the Middle and South Levels, and their Outfall to Sea; with a Plan for the Effectual Draining of the said Levels. Addressed to the Public . . . the Honourable Corporation of Bedford Level; the Merchants, Traders, and Others, interested in the Navigation of the River Ouse.*

By William Elstobb, Land Surveyor and Engineer.

Lynn: Printed and sold by W. Whittingham . . . 1778.

8vo   22 cm   (iv) + 140 pp.

BDL (Gough), CUL

448   *The Report of William Elstobb, Land Surveyor and Engineer, on the State of the Navigation between Clayhithe and Denver Sluice, September 28, 1778. To which will be added, an Appendix, containing some Facts and Observations relative to the Proceedings of the Bedford Level Corporation.*

Cambridge: Printed by F. Hodson . . . 1779.

Signed: William Elstobb. London, 28 Sept 1778.

4to   28 cm   (iii) + 28 + (4) pp.   Folding engraved plan.

Longitudinal Section of the River from Clayhithe to Denver Sluice,

by William Elstobb. September 1778.   I. Turpin del.   Vere sc.

BM, ULG, BDL (Gough), CUL

449   *A Map of the Harbour & Haven of the Port of Wells in the County of Norfolk; with the Imbank'd Marshes on the East Side of the Town, & the Unimbank'd out Marshes lying North.*

Done from an Actual Survey in the Year 1779 by W. Elstobb & J. Turpin.

Engraved plan   (scale 8 inches to 1 mile).

Sutro

450   *An Historical Account of the Great Level of the Fens, called Bedford Level, and other Fens, Marshes and Low-Lands in this Kingdom.*

By the late W. Elstobb, Engineer.

Lynn: Printed by W. Whittingham 1793.

8vo   23 cm   viii + (xii) + 276 pp.   Folding engraved map.

A Map of the Great Level of the Fens.

BM, CE, LCL, ULG, BDL (Gough), CUL

EMERSON, William [1701–1782]

451   *The Principles of Mechanics; Explaining and Demonstrating the general Laws of Motion . . . Mechanic Powers . . . Centers of Gravity, Strength and Stress of Timber, Hydrostatics, and Construction of Machines.*

London: Printed for W. Innys and J. Richardson . . . 1754.

Signed: W. Emerson.

8vo   21 cm   xi + (iv) + 312 pp.   32 folding engraved plates.

BM

452   Second edition

Title as No. 451 but *Second Edition, Corrected and very much enlarged.*

London: Printed for J. Richardson . . . 1758.

Signed: W. Emerson.

4to   24 cm   viii + 284 pp.   43 folding engraved plates (as in first edition, on larger paper, with 11 additional plates).

BM, SML, BDL

453   Third edition

Title as No. 451 but *Third Edition, Corrected.*

London: Printed for G. Robinson . . . 1773.

Signed: W. Emerson.

4to   24 cm   x + (ii) + 284 + (3) pp.   43 folding engraved plates.
BM, NLS

Fourth, fifth and sixth editions were published
virtually unchanged in 1794, 1800 and 1811.
Copies in BM, CE (Telford), RS, etc.

ERSKINE, Robert

454   *A Plan of the Intended Canal from Uxbridge to
Marybone According to an Actual Survey; Taken in 1776
by Robert Erskine.*
Engraved by T. Jeffreys.
Engraved plan   24 × 74 cm   (scale 2 inches to
1 mile).
BM, RS (Smeaton), BDL (Gough)

455   *A Dissertation on Rivers and Tides. Intended to
demonstrate in general the Effect of Bridges, Cuttings,
removing Shoals and Imbankments; and to investigate in
particular the Consequences of such Works on the River
Thames.*
Respectfully Addressed to the Right Honour-
able William Beckford, Esq; Lord Mayor of
London, Conservator of the River Thames. By
Robert Erskine, Engineer.
London: Printed for the Author, and sold by J.
Wilkie . . . 1770.
8vo   22 cm   24 pp.
BM, CE

456   Second edition
Title as No. 455 but . . . Respectfully addressed
and recommended to the Consideration of the
Nobility, Gentry, and Public at large; Particularly the Proprietors of Lands on the Banks of
the Thames, between Cricklade and Gravesend
[etc].
London: Printed for the Author, and sold by G.
Wilkie . . . 1780.
8vo   21 cm   32 pp.
BM, ULG, BDL (Gough)

ETHERIDGE, William [d. 1776]

457   *A Perspective View of the Engine, made use of for
Sawing off under Water, the Piles which help'd to support
the Centers, for turning the Arches of Westminster
Bridge.*
Most humbly Inscrib'd to the . . . Commis-
sioners . . . by the Inventor: Willm Etheridge,
Carpenter.   C Labelye del.   J. June sc.
[1745]
Engraved plate   45 × 35 cm.
RS (Smeaton), Soane Museum

458   *A Plan of the Bridge from Walton upon Thames
in Surrey, to the Opposite Shore in the Parish of Shepper-
ton, in Middlesex.*
This Design, is most Humbly Presented . . . by
. . . William Etheridge. W. Etheridge del.
[*c*. 1750].
Engraved   plan,   elevation   and   section
59 × 89 cm   (scale 12 inches to 100 feet).
RS (Smeaton)
This timber-arch bridge of 130 feet span was
designed in 1747 and opened in 1750.

EVIDENCE, MINUTES OF

The following entries (Nos. 459–493) form a
selection of printed minutes of evidence taken at
parliamentary committees on civil engineering
schemes. Principal witnesses are listed in order of
presentation.

459   *Bedford Level Petition, presented to the House of
Commons, the 10th of February 1777* [and] *Report of
the Committee to whom the Petition was referred; made to
the House, the 21st of March, 1777.*
[London] 1777.
8vo   19 cm   (i) + 167 pp.
CE
Thomas Hogard, John Wing, C. N. Cole,
James Golborne, Philip Cawthorne, John
Robinson.

460 *Report from the Committee appointed to enquire into the Progress made towards the Amendment and Improvement of the Navigation of the Thames.*
[House of Commons] 1793.
fol   33 cm   108 pp.

Charles Truss, Henry Allnut, John Clarke, John Treacher, Josiah Clowes, Robert Whitworth, Robert Mylne, William Vanderstegen. Report and evidence, without statistical appendices, reprinted in 8vo (58 pp. London, 1793). CE (Page).

461   Second printing
[In] *Reports from Committees of the House of Commons* Vol. 14 (1803) pp. 230–266. Includes appendices.

462   *Minutes of Evidence taken in the House of Commons, on the Second Reading of the Bill for improving the Drainage of the Middle and South Levels, Part of the Great Level of the Fens . . . and for altering and improving the Navigation of the River Ouze, from . . . Eau Brink . . . to the Harbour of King's Lynn.*
[House of Commons] 1794.
fol   33 cm   568 pp.

William Walton, Joseph Hodskinson, Francis Noble, George Maxwell. Evidence given on 17 days but minutes for the 7th and 9th days (pp. 283–340 and 407–442) are missing from all extant copies.

463   *Report from the Committee appointed to take the Residue of Evidence . . . in Support of the Bill for improving the Drainage of the Middle and South Levels . . . and . . . the River Ouze [etc].*
[House of Commons] 1795.
fol   33 cm   265 pp.

John Rennie, John Watté, William Jessop, James Golborne, C. N. Cole.

464   *Report from the Committee appointed to enquire into the best Mode of providing sufficient Accomodation for the increased Trade and Shipping of the Port of London.*
[House of Commons] 1796.
fol   33 cm   xl + (4) + 216 + (245) pp.   19 folding engraved plans.

Joseph Huddart, Daniel Alexander, John Rennie, Joseph Hodskinson, William Jessop, Ralph Walker, John Foulds, George Dance, Samuel Wyatt, Jonathan Pickernell, William Vaughan.

465   Second printing
[In] *Reports from Committees of the House of Commons* Vol. 14 (1803) pp. 267–443 and pls. 1–13.

466   *Minutes of Evidence taken at the Committee on the Bill for making Wet Docks . . . within the Port of London: Merchants Plan.*
[House of Commons] 1799.
fol   33 cm   (i) + 374 pp.   Folding engraved plan.

Daniel Alexander, John Rennie, William Vaughan.

467   *Minutes of Evidence taken at the Committee on the Bill for rendering more commodious, and for better regulating the Port of London: City Plan.*
[House of Commons] 1799.
fol   33 cm   80 pp.   Folding engraved plan.

William Jessop, Ralph Walker.

468   *Minutes of Evidence taken before the Committee to whom the Bill for supplying the Inhabitants of . . . Camberwell . . . and Parts of . . . Lambeth, and several other Places in the Counties of Surrey and Kent, with Water, was Committed.*
[House of Commons] 1805.
fol   33 cm   38 pp.

J. W. Rowe, John Timperley, Ralph Dodd.

469   *Report from the Select Committee on the Embankments in Catwater.*
[House of Commons] 1806.
fol   33 cm   18 pp.

Joseph Whidbey, John Rennie, Daniel Alexander, Joseph Huddart, Robert Mylne, William Jessop.

470   *Report from the Select Committee on the Highways of the Kingdom: together with the Minutes of Evidence taken before them.*

House of Commons, 1819.
fol   33 cm   58 pp.

J. L. McAdam, James McAdam, Benjamin Farey, John Farey, James Walker, Thomas Telford.

471   *Report from the Select Committee of the House of Commons appointed in Session 1821, to inquire into the State of London Bridge.*
pp. 49–129.

[In]

Thorp, Mayor. A Common Council holden .. on Saturday, the 2d Day of June, 1821.
Corporation of London, 1821.
fol   32 cm   149 pp.
CLRO, CE, ULG

Francis Giles, James Mountague, John Rennie, William Chapman, James Walker, Stephen Leach, Ralph Walker, Hugh McIntosh.

472   *Report from Select Committee on Mr McAdam's Petition, relating to his improved System of constructing and repairing the public Roads of the Kingdom.*
House of Commons, 1823.
fol   33 cm   100 pp.

J. L. McAdam, James McAdam, Benjamin Wingrove.

473   *Proceedings of the Committee of the House of Commons on the Liverpool and Manchester Railroad Bill. Session 1825.*
London: Printed by Thomas Davison.
fol   32 cm   xi + (i) + 772 pp.   15 litho plates, mostly folding and hand coloured.
BM, SML, CE, ULG, Elton

J. U. Rastrick, George Stephenson, Nicholas Wood, William Cubitt, Francis Giles, Alexander Comrie, H. R. Palmer, George Leather.

474   *Minutes of Evidence taken before the Lords Committees to whom was referred the Bill intituled 'An Act for making . . . a Railway or Tramroad from . . . Liverpool to . . . Manchester, with certain Branches therefrom'.*
House of Lords, 1826.
fol   33 cm   317 pp.

George Rennie, Charles Vignoles, Josias Jessop, Alexander Nimmo, H. R. Palmer, Peter Barlow.

475   *Abstract of the Minutes of Evidence, taken before a Committee of the House of Commons . . . on the Bill for making a Navigable Communication for Ships . . . between the City of Norwich and the Sea at or near Lowestoft.*
Norwich: Published by S. Wilkin . . . 1826.
8vo   23 cm   viii + 56 pp   Folding engraved plan.
CE

William Cubitt, John Timperley, Thomas Telford, Alexander Nimmo, James Walker, William Chapman, Benjamin Bevan, John Macneill.

476   *Minutes of Evidence taken before the Committee of the House of Commons . . . on the Aberdeen Harbour Bill; with the Speeches of Counsel.*
Aberdeen: Printed by D. Chalmers & Co . . . 1828.
8vo   21 cm   xlii + 286 pp.
CE (Telford)

John Gibb, Thomas Telford, Robert Stevenson.

477   *Report from the Select Committee on the Whetstone and St. Albans Turnpike Trusts*
[with Minutes of Evidence].
House of Commons, 1828.
fol   33 cm   59 pp.

Sir Henry Parnell, James McAdam, John Macneill.

478   *Copy of the Evidence, taken before a Committee of the House of Commons, on the Newcastle & Carlisle Railway Bill. Taken from the short hand notes of Mr Gurney.*
Newcastle: Printed by Wm. Boag . . . 1829.
8vo   22 cm   (i) + 229 pp.   2 litho views + 4 folding litho plans.
ULG

Benjamin Thompson, T. O. Blackett, Joshua Richardson, Robert Stephenson, John Dixon, Joseph Locke. Report by George LEATHER [q.v.].

479 *Report of the Select Committee on the Holyhead and Liverpool Roads*
[with Minutes of Evidence].
House of Commons, 1830.
fol   33 cm   52 pp.
Thomas Telford, John Macneill, John Provis, Sir Henry Parnell.

480 *Minutes of Evidence taken before the Lords Committees to whom was referred the Bill, intituled 'An Act for making a Railway from London to Birmingham'.*
House of Lords, 1832.
fol   33 cm   195 pp.
Robert Stephenson, Peter Lecount, Frank Forster, George Hennet, T. L. Gooch, J. U. Rastrick, H. R. Palmer, Joseph Locke.

481 *Extracts from the Minutes of Evidence given before the Committee of the Lords on the London and Birmingham Railway Bill.*
London: Smith & Ebbs, Printers . . . 1832.
8vo   22 cm   viii + 65 pp.
ULG
Stephenson, Palmer, Rastrick, Lecount. There is another edition, as above but 73 pp.   ULG (Rastrick).

482 *Minutes of Evidence taken before the Lords Committees to whom the Bill, intituled An Act for making a Railway from London to Southampton, was committed.*
House of Lords, 1834.
fol   33 cm   481 pp.
Francis Giles, Joseph Locke, William Gravatt, Samuel Jones, George Stephenson, George Hennet, Robert Stephenson, Joseph Gibbs, H. R. Palmer.

483 *Extracts from the Evidence given on the London and Southampton Railway Bill, as printed by order of the House of Lords.*
[London, 1834].
8vo   21 cm   68 pp.
ULG (Rastrick)
Witnesses as listed above.

484 *Extracts from the Minutes of Evidence given before the Committee of the House of Commons on the Great Western Railway Bill.*

Bristol: Printed by Gutch and Martin . . . 1834.
8vo   21 cm   iv + 51 + (1) pp.
BM, SML, CE, ULG (Rastrick)
I. K. Brunel, George Stephenson, James Walker, H. R. Palmer, W. H. Townsend, Joseph Locke.

485 *Great Western Railway Bill. Minutes of Evidence taken before the Lords Committees to whom the Bill intituled 'An Act for making a Railway from Bristol [to] London . . .' was committed.*
House of Lords, 1835.
fol   33 cm   2 vols   640 pp. and pp. 641–1393.
I. K. Brunel, John Hammond, J. O. Weise, Joseph Locke, George Stephenson, H. R. Palmer, G. W. Buck, Robert Stephenson.

486 *In the House of Commons. Minutes of Evidence taken before the Committee on the London and Dover (South-Eastern) Railway Bill . . . Copy from Mr Gurney's Short-Hand Notes.*
London: Printed by James & Luke G. Hansard & Sons . . . 1836.
fol   33 cm   13 parts.
CE
H. R. Palmer, W. A. Provis, P. W. Barlow, William Gravatt, F. W. Simms, William Froude, John Macneill, J. U. Rastrick, William Cubitt, H. H. Price.

487 *Report from the Select Committee on Turnpike Trusts and Tolls; together with Minutes of Evidence.*
House of Commons, 1836.
fol   33 cm   viii + 161 pp.
Benjamin Wingrove, Thomas Penson, John Provis, Sir James McAdam, John Macneill.

488 *In the House of Commons. Minutes of Evidence taken before the Committee on the London and Brighton Railway Bills; Engineering Evidence.*
[By] Robert Stephenson, Esq. George P. Bidder, Esq. Sir John Rennie, and Joseph Locke, Esq.
London: Printed & Sold by James & Luke G. Hansard & Sons . . . [1837].
8vo   22 cm   (ii) + 454 pp.
BM, CE

In addition there are two massive sets of Minutes of Evidence (1836, Commons and Lords) on the original schemes and two further sets (both 1837 from the Commons) on Stephenson's Direct Line and the line (as built) proposed by Rennie and Rastrick.

489 *Glasgow, Paisley, Kilmarnock, and Ayr Railway Bill. Minutes of Evidence taken before the Lords Committees to whom the Bill . . . was committed.*
House of Lords, 1837.
fol   33 cm   275 pp.
   John Miller, John Gibb, George Stephenson.

490 *Edinburgh and Glasgow Railway Bill. Minutes of Evidence taken before the Select Committee of the House of Lords to whom the Bill . . . was committed.*
House of Lords, 1838.
fol   33 cm   421 pp.
   John Miller, A. O. Riddell, John Gibb, John Macneill, Charles Vignoles.

491 *Report from the Select Committee on Railroad Communication; together with Minutes of Evidence, Appendix and Index.*
House of Commons, 1838.
fol   33 cm   viii + 164 pp.
   Joseph Locke, Robert Stephenson, John Macneill.

492 *Second Report from the Select Committee on Railways; together with the Minutes of Evidence, Appendix and Index.*
House of Commons, 1839.
fol   33 cm   xviii + 546 + 55 pp.   2 folding hand-coloured litho plans.
   William Cubitt, George Landmann, Joseph Gibbs, Thomas Grainger, J. U. Rastrick, Robert Stephenson.

493 *Evidence and Proceedings, in the Committee of the House of Commons, in regard to the Aberdeen Harbour Bill.*
Aberdeen: Printed . . . by Geo. Cornwall. 1839.
8vo   21 cm   lxxxiii + 335 pp.
CE
   John Gibb, James Walker.

EYES, John [d. 1773]

See STEERS and EYES. Plan of the River Calder, 1741.

494 *A Plan of that part of the River Calder that lies between Sowerby Bridge and Halifax Brooksmouth in the County of York.*
   Surv⁴ in Jan 1758 by John Eyes. J. Evans sc.
Engraved plan   25 × 33 cm   (scale 4 inches to 1 mile).
MCL, Sutro

495 *A Plan of the River Calder from Wakefield to Ealand, and thence Continued to Salter hebble Bridge, in the County of York. Survey'd in 1740 & 1741 & that part from Brooksmouth to Sowerby Bridge in Jan⁷ 1758.*
   By John Eyes. Thoˢ Bowen sc. 1758.
Engraved plan   24 × 74 cm   (scale $1\frac{3}{4}$ inches to 1 mile).
MCL, Sutro

FAIRBAIRN, *Sir* William [1789–1874] F.R.S., M.Soc.C.E., M.Inst.C.E.

496 *Remarks on Canal Navigation, illustrative of the advantages of the use of Steam, as a moving power on canals. With an Appendix, containing a Series of Experiments.*
   By William Fairbairn, Engineer.
London: Longman, Rees . . . 1831.
8vo   22 cm   93 pp.   5 folding litho plates.
BM, SML, CE (Telford), ULG

497 *Reservoirs on the River Bann, in the County of Down, Ireland, for more effectually supplying the Mills with Water.*
Manchester: Printed by Robert Robinson . . . 1836.
Signed: Wm. Fairbairn. Manchester, 30 Jan 1836.

4to   26 cm   23 pp.   4 folding hand-coloured litho plans.

[1] *Plan of the River Bann . . . shewing the site of the proposed Reservoirs and Feeders*. Wm. Fairbairn, Engineer.   J. Fred. Bateman, Surveyor. 1835.   Day & Haghe lithog.

[2]–[4] *Plans (and a Section) of Reservoirs . . . for the purpose of supplying water to the River Bann in dry seasons*. William Fairbairn, Engineer. J. Fred. Bateman, Surveyor. 1835. T. Physick, lith.

CE, ULG

498   *An Experimental Inquiry into the Strength and other Properties of Cast Iron, from Various Parts of the United Kingdom.*

By William Fairbairn.

Manchester: Printed by Francis Looney . . . 1838.

Dated: 7 March 1837.

8vo   21 cm   105 pp.

CE

Offprint, with new title page, from *Manchester Phil. Soc. Memoirs* Vol. 6 pp. 171–273.

499   *On the Strength and other Properties of Cast Iron obtained from the Hot and Cold Blast.*

By W. Fairbairn, Esq.

London: Printed by Richard and John E. Taylor . . . 1838.

8vo   23 cm   (i) + 377–415 pp.

CE

Offprint, with new title page, from British Association 7th Report, for 1838.

FAIRBANK, Josiah [1777–1844]

See FAIRBANK, W. and J. Plan of the River Dun, 1803.

See CHAPMAN. Plan of the intended Sheffield Canal, surveyed by W. & J. Fairbank, 1814.

FAIRBANK, William [1730–1801]

500   *A Plan of the intended Dearn and Dove Canal, and of that proposed by the Opposition thereto:*
   drawn by W^m. Fairbank, 1793.

Engraved plan, the lines entered in colours
   18 × 38 cm   (scale 1¼ inches to 1 mile).

BM

501   *A Plan of the Intended Navigable Canal from Stainforth Cut into the River Trent at Keadby,*
   by W^m. Fairbank, 1793.

Engraved plan   33 × 64 cm   (scale 2 inches to 1 mile).

BM, CE (Page), BDL, Sutro

FAIRBANK, William [c. 1771–1846]

502   *A Plan of the Course of the River Dun, from Sheffield to its confluence with the Ouse, shewing all the Cuts made, & those intended to be made, to facilitate the Navigation thereof.*
   Drawn by W^m. Fairbank, 1801.

W.F. 1801.

Engraved plan   26 × 91 cm   (scale 1 inch to 1 mile).

BM

See FAIRBANK, W. and J. Plan of the River Dun, 1803.

See CHAPMAN. Plan of the intended Sheffield Canal, surveyed by W. and J. Fairbank, 1814.

FAIRBANK, William and Josiah

503   *A Plan of the Alterations and Improvements intended to be made in the Navigation of the River Dun.*
   W. & J. Fairbank, Surveyors, 1803.

Engraved plan, the lines entered in colours
21 × 39 cm   (scale 4 inches to 1 mile).
BM

See CHAPMAN. Plan of the intended Sheffield Canal, surveyed by W. & J. Fairbank, 1814.

FEATHERSTONE, Joseph

504   *A Map of Deeping Fen in the County of Lincoln: the several Lands and Commons, Surveyed by Vincent Grant, about the Year 1670. The several Rivers & Drains are carried down to their Outfalls.*
By Jos. Featherstone. 1763.
Engraved plan   40 × 52 cm   (scale 1 inch to 1 mile).
BM, MCL, LCL, RGS, BDL (Gough), Sutro

FINDLATER, James R.

505   *Report relative to the formation of a Railway between the Towns of Dundee and Perth, passing through the Districts of the Carse of Gowrie.*
By James R. Findlater, Civil Engineer; and Supplementary Report by George Buchanan, Esq. Civil Engineer.
Dundee: Printed by D. Hill . . . 1835.
Signed: James R. Findlater. Dundee, Oct 1835.
8vo   21 cm   24 pp.   Folding litho plan.
Sketch shewing the Proposed Lines of Railway between Dundee and Perth.
By James R. Findlater. Forrester & Nichol Lithog.
ULG
Buchanan's brief report, dated Edinburgh 4 Nov 1835, is on p. 24.

FLINT, James [d. 1833]

See THOM. Shaws scheme, survey by Flint, 1827.

506   *Report on the Proposed New Road from Perth to Elgin, Grantown, Nairn, and Inverness.*
By James Flint, Civil Engineer.
Edinburgh: Printed by W. Burness. 1832.
4to   28 cm   (i) + 16 + 7 pp.   Folding litho plan, line entered in colour.
Sketch, shewing the line (and section) of a proposed New Road from Blair Gowrie to Elgin and Granton.
NLS

FORD, *Sir* Edward [1605–1670]

507   *A Designe for bringing a Navigable River from Rickmansworth in Hartfordshire, to St. Gyles in the Fields; the Benefits of it declared, and the Objections against it answered.*
London: Printed for John Clarke. 1641.
4to   18 cm   (i) + 10 pp.   Woodcut on title-page and engraved plate.
The Map of Colne River.
BM, ULG, BDL (Gough)
An enlargement by Ford of his earlier [unpublished] water supply scheme.

508   Second edition
A Design for bringing a Navigable River from Rickmansworth in Hartfordshire to St. Giles's in the Fields; the Benefits of it Declared, and the Objections Answered.
London: Printed for John Clarke, 1641. With an Answer to the Whole, Printed in the same Year, and both Reprinted, 1720.
4to   21 cm   24 pp.   Engraved plate.
A Map of the Colne with all its branches and Mills and the particular place whence the Navigation is to be taken.
BM, CE, ULG, BDL (Gough)
Includes (pp. 13–24) *An Answer to Mr. Forde's Book* by Sir Walter ROBERTS [q.v.]. Some copies lack the plate.

FORTH & CLYDE CANAL

509 *Considerations upon the intended Navigable Communication between the Friths of Forth and Clyde.*
In a Letter to the Lord Provost of Edinburgh . . . from a Member of the Convention.
[Edinburgh, 1767].
Dated: Edinburgh, 11 April 1767.
4to   25 cm   21 pp.
BM, NLS, ULG, BDL (Gough)

510 *Copies of three Letters upon the proposed Navigable Communication between the Friths of Forth and Clyde.*
Published in the Edinburgh News-Papers in April 1767.
[Edinburgh, 1767].
4to   25 cm   16 pp.
NLS, ULG

511 *Thoughts on the intended Navigable Communication between the Friths of Forth and Clyde.*
In a Letter to his Grace the Duke of Queensberry, from a Citizen of Edinburgh.
[Edinburgh, 1768].
Dated: Edinburgh, 20 Jan 1768.
4to   25 cm   27 pp.
NLS, ULG, BDL (Gough)

512 *Considerations on the intended Navigation betwixt Forth and Clyde, occasioned by the proceedings of the Proprietors at their last meeting, and by the Reports of Messrs Brindley, Yeoman, and Golborne, together with the Review by Mr Smeaton.*
n.p.   [1768].
4to   24 cm   11 pp.
NLS

513 *State of Facts and Observations, relative to the Affairs of the Forth and Clyde Navigation.*
Submitted, by Direction of the . . . Governor [and] Council, to the Consideration of the Whole Body of Proprietors.
Edinburgh: Murray & Co. Printers . . . 1816.
fol   32 cm   (i) + 29 + 22 pp.   Folding hand-coloured engraved plan.

Plan for the Improvement of the Forth and Clyde Navigation at Grangemouth.
1814. W. & D. Lizars sc.
CE
Includes a report by Hugh BAIRD dated 8 Sept 1814 [q.v.].

FORTREY, Samuel *junior* [*c.* 1650–1688]

514 *The History or Narrative of the Great Level of the Fenns, called Bedford Level, with a Large Map of the said Level, as Drained, Surveyed, & Described by Sir Jonas Moore, Knight, His late Majesties Surveyor-General of his Ordnance.*
London: Printed for Moses Pitt . . . 1685.
8vo   16 cm   (vi) + 81 pp.
BM, CRO, CE, CUL
Attributed to Fortrey. For the map, which was issued separately, *see* MOORE, 1684.

FULLER, John *et al.*

515 *A Report touching Lynn Navigation.*
Signed: John Fuller (Master Pilot), John Edwards and Samuel Long (Ships Captains), Thomas Badeslade (Surveyor and Professor of Mathematicks), William Stafford (Esquire), and (the Reverend Mr) Peter Bateson.
Lynn Regis, 22 Jan 1723/4.
pp. 5–6.
[Printed with]
Colonel ARMSTRONG's report, 12 May 1724 [q.v.].
[and]
The Complaints of the Merchants, Mariners and Watermen of Lynn about Navigation, [August] 1724.
n.p.
fol   33 cm   8 pp.
CE, BDL (Gough)

FULTON, Hamilton (d. 1834)

See TELFORD. Plan of the proposed Stamford Junction Canal, surveyed by Hamilton Fulton, 1810.

516 *Report on the Mail Roads in North Wales.*
By Hamilton Fulton.
Signed: Hamilton Fulton. Newman Street, 1813.
[In]
Report from Select Committee on Holyhead Roads.
House of Commons, 1815. pp. 15–30.

FULTON, Robert [1765–1815]

517 *A Treatise on the improvement of canal navigation; exhibiting the numerous advantages to be derived from Small Canals ... With a description of the Machinery for facilitating Conveyance by Water ... independent of Locks ... [and] Thoughts on, and Designs for, Aqueducts and Bridges of Iron and Wood.*
By R. Fulton, Civil Engineer.
London: Published by I. and J. Taylor ... 1796.
Signed: Robert Fulton. London, 1 March 1796.
4to 28 cm xvi + 144 pp. 17 engraved plates.
BM, NLS, SML, CE, ULG, BDL
Issued in boards. Price 18*s*.

GARSTIN, *General* John [1756–1820] Hon.M.Soc.C.E.

518 *A Treatise on Rivers and Torrents; with the Method of Regulating their Course and Channels. By Paul Frisi ... To which is added an Essay on Navigable Canals, by the same.*
Translated by Major-General John Garstin.
London: Published by Longman, Hurst ... 1818.
Signed: John Garstin. Manchester Square, 12 April 1818.

4to 28 cm vi + xx + 184 pp. 2 folding engraved plates (Neele & Son sc.).
BM, NLS, ULG, IC, BDL

GEDDES, John

519 *Plan of the Forth & Tay Railway, connecting Perth, Dundee & North of Scotland, with the City of Edinburgh, by Newburgh, Newport & Ferry-Port-on-Craig, and by Dysart, Kirkaldy, Kinghorn and Burntisland.*
By John Geddes, Mining & Civil Engineer, Edinburgh 1836. Leith & Smith Lithog.
Litho plan 44 × 55 cm (scale 1 inch to 1 mile).
NLS

GIBB, John [1776–1850] M.Inst.C.E.

520 *Aberdeen Harbour. Mr Gibb's Narrative.*
Dated: September 1837.
pp. 129–140
[In]
TELFORD. *Life of Thomas Telford.* London, 1838 [q.v.].

521 *Extract from Observations on the Crinan Canal.*
Signed: John Gibb. Aberdeen, 20 Jan 1838.
[In]
Report from the Select Committee on the Caledonian and Crinan Canals.
House of Commons, 1839. pp. 172–174.

GIBBS, Joseph [1798–1864]
M.Inst.C.E.

522 *Report of Joseph Gibbs, Esq. Civil Engineer, upon the Great Northern Railway.*
1835.
[London]: Printed by E. Colyer.
Signed: Joseph Gibbs, July 1835.
fol   31 cm   (i) + 23 pp.
CE
> Issued in green printed wrappers. A small-scale map of the line is in the *Prospectus*.

523   Deposited plan
Plan and Section of an intended Railway from the Town of Croydon . . . to join the London and Greenwich line of Railway near Corbetts Lane in the Parish of Rotherhithe.
> Joseph Gibbs, Engineer. Drawn by W. West. E. Colyer, Lith. [1835].
Litho plan, hand coloured   48 × 91 cm   (scale 4 inches to 1 mile).
HLRO
> The accompanying (MS.) estimate is signed: Joseph Gibbs, 5 March 1835.

524   *Plan of the Proposed London and Brighton Railway, being a Continuation of the Croydon Railway, with Branches to the Southampton Railway, and to Leatherhead.*
> Joseph Gibbs, Engineer. R. Martin, Lithog. [1835].
Litho map   61 × 36 cm   (scale 1 inch to 2½ miles).
BM
> The *Prospectus* is dated 29 Oct 1835.

525   *Report of Mr Gibbs, Civil Engineer, upon the several proposed Lines for a Brighton Railway.*
London: Published by Smith, Elder & Co . . . [1836].
8vo   23 cm   vii + 91 pp.   Plate (section of a tunnel) + 2 folding litho plans, the lines entered in colours.
[1] *Map of the Country between London and Brighton shewing the various Proposed Lines of Railway.* J. R. Jobbins Lithog.
[2] Sections of the Lines.
BM

526   Second edition
As No. 525 but *Second Edition*. [1836].
CE, ULG (Rastrick).

GILBERT, Davies [1767–1839]
F.R.S., Hon.M.Soc.C.E.,
Hon.M.Inst.C.E.

527   *On the Mathematical Theory of Suspension Bridges, with Tables for facilitating their Construction.*
> By Davies Gilbert, Esq. V.P.R.S.
London: Printed by W. Nicol . . . 1826.
Dated: 9 March 1826.
4to   29 cm   (ii) + 17 pp.
CE
> Offprint, with new title page, from *Phil. Trans.* Vol. 116 (1826). Reprinted in W. A. PROVIS, Menai Suspension Bridge, 1828 [q.v.].

GILES, Francis [1787–1847]
M.Soc.C.E., M.Inst.C.E.

See RENNIE, John. Surveys by Netlam and Francis Giles, 1810–1815.

See RENNIE, John. Survey of River Wear, 1819–1822 (published 1826).

See RENNIE, Sir John. Report on Boston Haven, 1822.

528   *Report on the Parliamentary Line of Railway from Newcastle to Carlisle.*
Newcastle: Printed by T. & J. Hodgson . . . 1830.
Signed: Francis Giles. Newcastle, 29 May 1829.
8vo   22 cm   7 pp.
BM

529 *Second Report on the Line of Railway, from New-castle to Carlisle, with Estimate of the cost thereof.*
By Francis Giles, Civil Engineer.
Newcastle: Printed by T. & J. Hodgson . . . 1830.
Signed: Francis Giles, 19 Aug 1829.
8vo   22 cm   32 pp.
BM

530 *Newcastle upon Tyne and Carlisle Railway. Report upon the Comparative Qualities of a Line between Scotswood and Crawcrook Mill, by way of Blaydon, and between those points by way of Lemington.*
Newcastle: Printed by T. & J. Hodgson . . . 1830.
Signed: Francis Giles. Newcastle, 22 June 1830.
8vo   22 cm   12 pp.
BM

See SPECIFICATION for two bridges on the
    Newcastle and Carlisle Railway, 1830.

531 *Estimate of the probable Expense of constructing a line for a double Railway, with its necessary passing places, from the town of Southampton to Vauxhall, London, 77 miles in length.*
Signed: Francis Giles, 30 Jan 1832.
p. (2) with litho plan.
Plan & Section of the intended Southampton & London Railway.
By Francis Giles. 1832.
                        [In]
Southampton and London Railway.
    [Prospectus]
[London]: Baynes & Harris, Printers.
fol   40 cm   (4) pp.
BM, CE
    Issued from the Company's Office, 22 Feb 1832. Two other prospectuses, headed *London and Southampton Railway*, were issued in September and December 1833, both with a plan similar to above.

532   Deposited plan
Plan of the Intended London and Southampton Railway.
By Francis Giles, Civil Engineer. London. May 1834. Lithographed by J. Gardner.
Oblong folio   40 × 70 cm   Title + 6 sheets of litho plans   (scale 4 inches to 1 mile).
BM (HLRO in MS.)

See GILES and BRUNTON. Report on proposed
    railway from Basingstoke to Bath, 1834.

533 *Plan of the intended Portsmouth Junction Railway.*
By Francis Giles, Civil Engineer. 1836. Baynes & Harris, litho.
Litho plan   40 × 25 cm.
                        [In]
The Portsmouth Junction Railway.
    [Prospectus]
[London]: Baynes & Harris, Printers.
fol   44 cm   (4) pp.
BM

534   Deposited plan
Plans and Sections of the intended Alterations and Deviations from the Original Plan of the London and Southampton Railway.
Francis Giles, Civil Engineer. November 1836.
Oblong folio   40 × 70 cm   Title + 3 sheets (scale 4 inches to 1 mile).
BM

535 *Plan of the South Midland Railway in continuation of the Midland Counties and North Midland Railways . . . with a Branch to Stamford.*
Francis Giles, Civil Engineer. 1836. Day & Haghe, Lith.
Litho plan   44 × 27 cm (scale 1 inch to 10 miles).
                        [In]
South Midland Counties Railway.
    [Prospectus]
[London]: T. C. Savill, Printer.
fol   37 cm   (8) pp.
ULG

GILES, Francis and BRUNTON, William

536 *Railway from Basing, in the County of Hants, to Bath.*

[Report to the Provisional Committee].
n.p.
Signed: Francis Giles, William Brunton, Civil
Engineers. London, 8 July 1834.
fol   42 cm   (3) pp.   Litho plan.
*Map of the intended Basing and Bath Railway, shewing
its Connection with London and Southampton.*
By Francis Giles and William Brunton, Civil
Engineers. London, 1834. Baynes & Harris
Lithog.
ULG

GILES, Netlam [?1780–1816]

See RENNIE, John. Surveys by Netlam and
Francis Giles of Seaton and Beer Bay 1810,
London & Cambridge Junction Canal 1811,
Weald of Kent Canal 1811, and Arundel &
Portsmouth Canal 1815.

537   *Plan and Sections of the Reservoirs & Feeders in
the Parishes of Kings Norton & Northfield in the County
of Worcester, as proposed to be made for the use of the
Worcester & Birmingham Canal and Mills on the River
Rea. 1815.*
Netlam Giles.
Engraved map, hand-coloured   29 × 36 cm
(scale 4 inches to 1 mile).
BM

GILLONE, John

See TELFORD and McKERLIE. Plans of roads
in south-west Scotland, surveyed by John Gillone, 1807 and 1808.

GLASGOW, Railway and Tunnel

538   *Glasgow Railway and Tunnel. 1829.*
n.p.
8vo   23 cm
Contains 8 items of varying pagination amount-
ing to 41 pages in all. Folding litho plan.
*Sketch [map] of the Canals, Lanarkshire Railways,
Firth of Forth & River Clyde: showing the intended
Railway from the Harbour of Glasgow.*
NLS
Includes reports by GRAINGER and MIL-
LER, and TELFORD [q.v.].

539   Second edition
Glasgow Railway and Tunnel. 1829.
[London]: J. B. Nichols and Son . . . [1830].
8vo   23 cm   63 + 3 pp.
CE (Telford)

GLASGOW, Water Supply

540   *Report by the Deputation appointed by the Com-
mittees of the Lord Provost, Bailies and Councillors, of
Glasgow, the Trades' House, and the Commissioners of
Police, to conduct the Opposition to the Bill, entitled 'A
Bill for the Better Supplying of the City and Suburbs of
Glasgow with Water'.*
Glasgow: Printed by John Graham . . . 1834.
8vo   21 cm   44 pp.   Folding hand-coloured
litho plan (Kirkwood).
NLS
Includes a report by GRAINGER and MIL-
LER [q.v.].

GLYNN, Joseph [1799–1863] F.R.S., M.Soc.C.E., M.Inst.C.E.

541 *Mr Glynn's Report on Loco-Motive Engines.*
Signed: Joseph Glynn. Butterley, 5 Nov 1832.
pp. (29)–38.
[Printed with]
Mr Jessop's Report on the Railways. 25 Nov 1832. *See* JESSOP, William *junior*.
[In]
Midland Counties Railway. Prospectus of the Projected Railway from Pinxton to Leicester, with Reports of the Estimated Cost of that Undertaking, and on the Application of Loco-Motive Steam Power to Railways Generally.
Alfreton: Printed by George Coates. 1833.
8vo   22 cm   42 pp.   2 folding litho plans.
CE

542 *Draining Land by Steam Power.*
From the Transactions of the Society of Arts, Manufactures, Commerce, &c. Vol. 51 . . .
[by] Joseph Glynn, Esq., Civil Engineer.
London: Printed by J. Moyes . . . [1838].
Dated: Butterley, by Derby, 8 Feb 1836.
8vo   21 cm   22 pp.   1 engraved plate.
CE
Offprint of paper in *Trans. Soc. Arts* Vol. 51 (1838 for 1836–37).

See Appendix to TREDGOLD's *Elementary Principles of Carpentry*. By Smirke, Shaw, Glynn and others, 1840.

GODWIN, George [1815–1888] F.R.S.

543 *Prize Essay upon the Nature and Properties of Concrete, and its application to Construction up to the present period.*
By George Godwin, Jun.
[London: 1836].
4to   29 cm   37 pp.
CE (from the author).

Offprint from *Trans. Inst. British Architects* Vol. 1 (1836). The essay was awarded the medal of the Institute on 18 Jan 1836.

GOLBORNE, James [1746–1819] M.Soc.C.E.

544 *The Report of James Golborne, of the City of Ely, Engineer; in pursuance of several Resolutions passed at a Meeting of the Committee of Landowners, and others, Interested in the Improvement of the Outfall of the River Ouse . . . on Thursday the 16th of June 1791.*
Lynn: Printed by W. Whittingham, 1791.
Signed: James Golborne. Ely, 30 Aug 1791.
4to   23 cm   44 pp.
BM, CE, CRO, CUL, Sutro

See HUDSON, GOLBORNE and MAXWELL. Reports on the outfall of the River Welland, 1792 and 1793.

545 *Report of James Golborne, Engineer, on the Drainage of Keyingham Level, in Holderness.*
Cambridge: Printed by Francis Hodson . . . 1799.
Signed: James Golborne. Ely, 3 Oct 1799.
4to   24 cm   10 pp.   Folding engraved plan.
Plan and Section of part of the North Channel of the Humber from No Mans Friend Clough to Ottringham Haven.
Neele sc.
Hull City Library, CE, Hull University Library

GOLBORNE, John [c. 1724–1783] M.Soc.C.E.

546 [*Reports on the Forth & Clyde Canal*]
Signed: John Golborne. [Edinburgh], 30 Sept 1768 and Chester, 13 Oct 1768.
pp. 33–37.
[In]

Reports by James Brindley Engineer, Thomas Yeoman Engineer and F.R.S. and John Golborne Engineer, relative to a Navigable Communication betwixt the Friths of Forth and Clyde . . . With Observations.
Edinburgh: Printed by Balfour, Auld, and Smellie, 1768.
4to .25 cm (iii) + 44 pp. Folding engraved plan.
NLS, CE, ULG

547 *The Report of John Golborne, Engineer, concerning the Drainage of the North Level of the Fens, and the Outfal of the Wisbeach River.*
n.p.
Signed: John Golborne. Chester, 3 Oct 1769.
4to 24 cm 11 pp. Folding engraved plan.
A Map of the River Nene and Bay from Wisbeach to the Eye, taken in 1767.
By Mʳ Golborne. Vere sc.
BM, CE (lacks plan), CRO, ULG, BDL (Gough), CUL, Sutro
Issued in grey wrappers.

548 *Mr John Golborne's Report.*
Dated: Edinburgh, 14 Nov 1773.
pp. 9–11.
[In]
Reports to the Lords Commissioners of Police, relative to the Navigation of the Rivers Forth, Gudie, and Devon.
Glasgow: Printed by Robert and Andrew Foulis . . . 1773.
4to 21 cm 66 + (1) pp.
NLS, ULG

549 *The Report of Mr John Gilborne, Engineer; on a View taken in Pursuance of an Order of the Bedford Level Corporation, in the Months of June and July last, of the Middle and South Levels, and their Outfalls to Sea; with a Plan for the effectual Draining the said Levels.*
[London, 1777].
Signed: John Golborne. Chester, 2 Dec 1777.
fol 33 cm 22 pp. Docket title on p. (24).
BM, CRO, BDL (Gough), Sutro
Stitched as issued.

550 Second edition
Title, etc. as in No. 549.
n.p. Printed in the Year MDCC LXXVII.

4to 24 cm 24 pp.
P
Perhaps published *c.*1793 in relation to the Eau Brink Cut project.

GOOCH, Thomas Longridge [1808–1882] M.Inst.C.E.

See SPECIFICATION (tender) for works on Manchester & Leeds Railway, 1837.

GOODRICH, Simon [1773–1847] M.Inst.C.E.

551 *Report on the Proposed Enlargement and Reparation of Bridlington Harbour.*
By Simon Goodrick.
York: Printed by Wm. Blanchard. 1814.
Signed: Simon Goodrick. Mile End Lodge, near Portsmouth, 29 Sept 1814.
4to 25 cm (i) + 27 pp. 2 folding engraved plans, both signed: Simon Goodrich 29 Sept 1814.
1. *Plan of Bridlington Harbour . . . shewing a Proposed Enlargement and Reparation of it by two new Piers of Stone.*
2. *Design for new Piers of Stone for Bridlington Harbour.*
BM, Science Museum
The misprint of Goodrich's name persists in the British Library catalogue.

GORDON, Alexander [1802–1868] M.Inst.C.E.

552 *The Fitness of Turnpike Roads and Highways for the most expeditious, safe, convenient and economical Internal Communication.*
  By Alexander Gordon.
London: Roake and Varty . . . 1835.
8vo   22 cm   32 pp.
BM, CE, Elton

553 *Observations addressed to those interested in either Rail-Ways or Turnpike-Roads; showing the comparative expedition, safety, convenience, and . . . economy of these two kinds of road for internal communication.*
  By Alexander Gordon, Civil Engineer.
London: John Weale . . . 1837.
8vo   23 cm   31 pp.
BM, CE, ULG, Elton

GRAHAME, Thomas

554 *A Letter addressed to Nicholas Wood Esq. on that portion of Chapter IX of his Treatise on Railroads, entitled, Comparative Performances of Motive Power on Canals and Railroads.*
  By Thomas Grahame, Esq.
Glasgow: Printed for John Smith & Son . . . 1831.
Signed: Thomas Grahame. Glasgow, 27 March 1831.
8vo   22 cm   40 pp.
BM, NLS, CE (Page), ULG (Rastrick)

555 *Letter to the Traders and Carriers on the Navigations connecting Liverpool and Manchester.*
  By Thomas Grahame Esq.
Glasgow: Printed by John Smith & Son . . . 1833.
Signed: Thomas Grahame. Glasgow, January 1833.
4to   27 cm   viii + 29 pp.
NLS, CE (Page)
  Issued in yellow printed wrappers with vignette lithographs. Price 3s.

556 Second edition
Title as No. 555 but *Second Edition.*
Glasgow: Printed for John Smith & Son . . . 1834.
4to   27 cm   36 pp.
BM

557 *A Treatise on Internal Intercourse and Communication in Civilised States, and particularly in Great Britain.*
  By Thomas Grahame.
London: Printed for Longman, Rees . . . 1834.
Signed: Thomas Grahame. Nantes, 14 April 1834.
8vo   23 cm   xiii + 160 pp.
BM, NLS, CE, ULG

558 *Essays and Letters on subjects conducive to the improvement and extension of Inland Communication and Transport.*
  By Thomas Grahame, Esq.
Westminster: Vacher & Son . . . 1835.
Signed: Thomas Grahame. Nantes, 24 April 1835.
4to   28 cm   61 pp.   Litho plate.
BM, CE, ULG

GRAINGER, Thomas [1794–1852] F.R.S.E., M.Inst.C.E.

559 Deposited plan
Reduced Plan and Section of a proposed Line of Railway from the Monkland Coalfield to the Forth & Clyde Canal near Kirkintullock.
  1823.
Engraved plan, hand coloured   25 × 76 cm
(scale 3 inches to 1 mile).
HLRO
  Endorsed: Thomas Grainger. Glasgow 21 April 1824.

560 *Report to the Subscribers for a Survey of a Railway from the Monkland & Kirkintilloch Railway to that part of the Monkland Coal Field situated North and East of Airdrie.*

[By Thomas Grainger].
Edinburgh. 1825.
8vo 14 pp. Folding plan.
Not seen.

561 Deposited plan
Reduced Plan and Section of a Proposed Railway from Monkland & Kirkintilloch Railway near Bedlay by Garnkirk to Glasgow.
By Thomas Grainger 1825.
Grainger & Miller, Surveyors. Engr^d by Kirkwood & Son.
Engraved plan, hand coloured 25 × 62 cm (scale 3 inches to 1 mile).
HLRO
Endorsed: Thomas Grainger. Glasgow 7 April 1826.

562 Amended scheme
Reduced Plan and Section of the Garnkirk & Glasgow Railway
1826. Grainger & Miller.
Engraved plan, hand coloured 19 × 67 cm (scale 3 inches to 1 mile).
HLRO
Endorsed: Thomas Grainger. Glasgow 19 May 1827.

See GRAINGER and MILLER. Reports and plans 1828–1840.

563 *Observations . . . relative to the Proposed Railway, from the North Shore of the Firth of Forth . . . to the River Tay . . . to be called the Edinburgh, Dundee, and Northern Railway.*
By Thomas Grainger.
Edinburgh: Printed by H. & J. Pillans . . . 1841.
Dated: Edinburgh, 12 Jan 1841.
8vo 23 cm 32 pp.
NLS

GRAINGER, Thomas and MILLER, John

See GRAINGER. Plans of the Glasgow & Garnkirk Railway, 1825 and 1826.

564 *Report relative to the Proposed Railway, to connect the Clydesdale, or Upper Coal Field of Lanarkshire, with the City of Glasgow, and the East and West Country Markets.*
By Thomas Grainger & John Miller, Civil Engineers and Surveyors.
Edinburgh: Printed by J. & C. Muirhead. 1828.
Signed: Grainger & Miller. Edinburgh, 24 Sept 1828.
4to 28 cm 17 pp 2 folding engraved plans, the lines entered in colours. (Kirkwood & Son sc.).
1. *Sketch [map] of Proposed Railway . . . 1828.*
2. *Reduced Plan & Section of the Proposed Railway to connect the Clydesdale or upper Coal field of Lanarkshire with the City of Glasgow . . .*
By Grainger & Miller 1828.
NLS, CE

565 *Report of a Survey undertaken for the purpose of ascertaining the line by which the Best Road from the City of Glasgow to Ayrshire (near Flockside) is to be obtained.*
By Thomas Grainger and John Miller, Civil Engineers and Surveyors.
Edinburgh: Printed by J. & C. Muirhead . . . 1829.
Signed: Thomas Grainger, John Miller. Edinburgh, 31 Dec 1828.
4to 28 cm 35 + (3) pp. Folding hand-coloured engraved plan.
Plan of the Country between Glasgow & Ringswell showing the Alterations Proposed on the Road from the City towards Kilmarnock, with the Sections of the respective lines. 1828.
Engraved by Kirkwood & Son.
NLS, CE, ULG (Rastrick)

See McADAM, 1829. Plan of road between Edinburgh and Newcastle by Jedburgh. Surveyed by Grainger and Miller, and N. Weatherly, 1828.

566 [Report] *To the Promoters of the proposed Improvements on Stirling's Road.*
Signed: Grainger & Miller. Edinburgh, 27 May 1829.
pp. 2–4.
[In]
Statement respecting a Northern Approach to the City of Glasgow, along the line of Stirling's Road.
R. Malcolm, Printer.
4to 26 cm 4 pp.
NLS, ULG (Rastrick)

567 *Report and Estimate of the Probable Expense of the Proposed Extension of the Garnkirk and Glasgow Railway to Port Dundas.*
By Thomas Grainger and John Miller, Civil Engineers and Surveyors.
Edinburgh: Printed by J. & C. Muirhead. 1829.
Signed: Grainger & Miller. Edinburgh, 4 June 1829.
4to 28 cm 8 pp. Folding litho plan.
Plan and Sections of the Proposed Extension of the Garnkirk and Glasgow Railway to Port Dundas.
R. H. Nimmo's Lithog.
CE

568 *Report to the Proprietors of, and Traders on the Canals and Railways terminating on the North Quarter of Glasgow, and Other Gentlemen interested therein and in the Harbour and Streets of Glasgow.*
n.p.
Signed: Thomas Grainger, John Miller. Edinburgh, 20 Aug 1829.
fol 32 cm 4 pp.
NLS
Reprinted in GLASGOW, Railway and Tunnel, 1829 [q.v.].

569 *Observations on the Formation of a Railway Communication between the Cities of Edinburgh & Glasgow, with branches to the Frith of Forth at Leith and the River Clyde at Glasgow.*
By Thomas Grainger & John Miller. Members Inst. Civ. Eng.
Edinburgh: Printed for William Blackwood and J. & C. Muirhead. 1830.
Signed: Thomas Grainger, John Miller. Edinburgh, 13 Oct 1830.

4to 27 cm 10 + 4 pp. Folding litho plan, the lines entered in colour.
Sketch [plan] of a Proposed Line of Railway between the Cities of Edinburgh and Glasgow with branches to the Port of Leith, Benhar Coalfield, and Raw Camp Limeworks.
By Grainger and Miller. October 1830. (scale 1 inch to 2 miles). R. H. Nimmo's Lithog.
NLS

570 *Report by Messrs Grainger & Miller, Civil Engineers.*
Signed: Thomas Grainger, John Miller. Edinburgh, December 1830.
pp. 1–18. With folding engraved plan.
Reduced Plan & Section of the proposed Edinburgh Glasgow & Leith Railway with Branches to Raw Camp, Bathgate, Benhar, and Airdrie.
By Grainger & Miller. 1830. (scale 1 inch to 1 mile). Engraved by Kirkwood & Son.
[Printed with]
Report by George STEPHENSON, 30 Dec 1830 [q.v.].
[In]
Edinburgh, Glasgow & Leith Railway. Reports by Messrs Grainger & Miller of Edinburgh, and Mr George Stephenson of Liverpool. Civil Engineers. January 1831.
[Edinburgh, 1831].
4to 29 cm (iii) + 2 + 5 + 18 + 6 pp. Folding engraved plan (as above).
NLS, CE
Stitched as issued in printed wrappers with engraving of Locomotive Steam Engine with a Train of Railway Carriages. (Lizars sc.)

571 Revised edition
Report by Messrs Grainger and Miller, Civil Engineers.
Signed: Thomas Grainger, John Miller. Edinburgh, 23 Nov 1831.
pp. 17–28 with folding engraved plan.
Reduced Plan & Section of the proposed Edinburgh & Glasgow Railway with Branches to Leith, Raw Camp & Benhar.
By Grainger & Miller 1831. (scale 1 inch to 1 mile). Engraved by Kirkwood & Son.
[Printed with]

Report by George STEPHENSON, 6 June 1831 [q.v.].

[In]

Edinburgh and Glasgow Railway. Reports by Mr George Stephenson, of Liverpool, and Messrs Grainger and Miller, of Edinburgh, Civil Engineers. November 1831.
Edinburgh: Printed by J. & C. Muirhead. 1831.
4to   28 cm   28 pp.   Folding engraved plan (as above)
BM, NLS
  Issued in printed wrappers similar to the January 1831 publication. The reports and plan differ substantially from the earlier versions.

572   *Report relative to a Proposed Railway Communication between the City of Glasgow and the Towns of Paisley and Johnstone,*
  by Thomas Grainger and John Miller. Members Inst. Civ. Eng.
Edinburgh: Printed by J. & C. Muirhead. 1831.
Signed: Thomas Grainger, John Miller. Edinburgh, 1 Jan 1831.
4to   27 cm   19 + (2) pp.   Folding engraved plan.
Plan & Section of the Proposed Lines of Railway from the City of Glasgow to the Towns of Paisley & Johnstone.
  By Grainger & Miller. 1830. Kirkwood sc.
NLS

573   *Report to the Trustees of the Roxburghshire Turnpikes, relative to the Proposed Alterations of the Road from Hundalee Smithy, near Jedburgh . . . to Whitelee Toll-Bar.*
  By Messrs Grainger and Miller, Engineers.
Jedburgh: Printed by Walter Easton, 1831.
4to   28 cm   19 pp.   Folding engraved plan.
Plan of the Proposed Line of Road from Jedburgh to Whitelee Toll Bar on the Great Line of Road between Edinburgh & Newcastle [with] *Section of Carter Fell on the line of the Proposed Tunnel* [and] *Transverse Section of Proposed Tunnel.*
  By Grainger & Miller 1831. Kirkwood sc.
NLS, ULG (lacking plan)

574   *Report by Messrs Grainger and Miller, Civil Engineers, Edinburgh, Relative to the Formation of a Harbour and Dock in Trinity Bay, on the southern shore of the Firth of Forth.*

n.p.
Signed: Thomas Grainger, John Miller. George Street, Edinburgh, 26 Feb 1834.
fol   28 cm   5 + 17 pp.   Folding litho drawing.
Projected Harbour and Docks.
  Drawn on Stone by Jobbins and Cheffins. [with a small-scale plan of Trinity Harbour].
NLS, CE

575   *Report by Thomas Grainger and John Miller, Civil Engineers, relative to the best mode of obtaining a plentiful supply of pure water for the City of Glasgow.*
Signed: Thomas Grainger, John Miller. London, 28 April 1834.
pp. 40–44.

[In]

Report by the Deputation . . . to conduct the Opposition to the Bill, 1834. See GLASGOW, Water Supply.

576   *Report relative to the Proposed Railway from the River Clyde, at Renfrew Ferry, to Paisley.*
  By Thomas Grainger & John Miller, Members Inst. Civ. Eng.
Edinburgh: 1834.
Signed: Thomas Grainger, John Miller. Edinburgh, 6 Dec 1834.
8vo   22 cm   12 pp.
NLS

577   *Plan and Sections of the Proposed Railway for connecting the City of Edinburgh with the Shore of the Frith of Forth near to Newhaven and the Harbour and Docks of Leith.* 1835.
  Grainger & Miller. Engineers. Engraved by R. Scott.
Engraved plan   45 × 54 cm   (scale 1 inch to 600 feet and 1 inch to 200 ft).
NLS

578   *Plan and Section of a Proposed Railway from the City of Edinburgh by the Towns of Musselburgh & Haddington to the Town of Dunbar to be called the Edinburgh, Haddington & Dunbar Railway.* 1836.
  George Stephenson Esq. Consulting Engineer. Grainger & Miller Acting Engineers. Engraved by W. H. Lizars.
Engraved hand-coloured plan   48 × 305 cm (scale 4 inches to 1 mile and 1 inch to 100 feet).

MS. endorsement: Edinburgh 30 Nov 1836. Grainger & Miller Engineers.

GRAY, Thomas [1788–1848]

579 *Observations on a General Iron Rail-Way. Shewing its great superiority over all the present methods of conveyance, and claiming the particular attention of all merchants, manufacturers, farmers, and, indeed, every class of society.*
London: Published by Baldwin, Cradock, and Joy . . . 1820.
8vo   21 cm   22 pp.
CE

580 Second edition
As No. 579 but *Second Edition* . . . 1821.
8vo   21 cm   60 pp.
BM

581 Third edition
As No. 579 but *Third Edition, revised and considerably enlarged* . . . 1822.
8vo   22 cm   xii + 131 pp.   3 engraved plates.
SML, ULG

582 Fourth edition
Observations on a General Iron Rail-Way . . . showing its great superiority, by the general introduction of mechanic power, over all the present methods of conveyance by turnpike road and canals . . . [etc].
Fourth Edition, considerably improved.
London: Published by Baldwin, Cradock, and Joy . . . 1823.
8vo   22 cm   xii + 131 pp.   5 engraved plates (2 folding).
BM, SML, ULG

583 Fifth edition
Observations on a General Iron Rail-Way, or Land Steam-Conveyance; to supersede the necessity of horses in all public vehicles . . . Containing every species of information relative to Rail-Roads and Loco-Motive Engines.
By Thomas Gray, Fifth edition.
London: Published by Baldwin, Cradock, and Joy . . . 1825.
8vo   22 cm   xxiv + 233 pp.   5 engraved plates (as in fourth edition).
BM, CE, ULG

GREEN, James *senior*

584 *A Plan shewing the Lines & the relative Situations of the intended Nottingham Canal, and of the Erewash Canal.*
Surveyed in 1791 by James Green.
Engraved plan
Sutro
Not seen. Reduced photocopy in the British Museum (Natural History), London.

GREEN, James [1781–1849] M.Inst.C.E.

585 *A Report on the Alteration and Improvement of the Turnpike Road, between Exeter and Plymouth, through Chudleigh and Ashburton.*
By James Green, Civil Engineer.
Plymouth: Printed by P. Nettleton and Son.
Signed: James Green. Exeter, 30 June 1819.
8vo   21 cm   29 pp.   with table.
Devon & Exeter Institution

586 Deposited plan
Map of the Turnpike Road, from Plymouth to Exeter, with Improvements, as Proposed by James Green, 1819.
J. Cary sc.
Engraved plan, the lines entered in colours
26 × 98 cm   (scale 1 inch to 1 mile).
HLRO

587 *The Report of James Green, Civil Engineer, on a proposed small Canal, to open a Communication between* the English Channel at Beer, and the Bristol Channel near Bridgewater.

Chard: Printed by J. Toms ... 1822.
Signed: James Green. Exeter, 16 Dec 1822.
fol   32 cm   13 pp.
ULG

588   *Plan of Bridport Harbour Dorsetshire, with Improvements proposed by James Green, Civil Engineer, 1823.*

J. Wyld sc.
Engraved plan   23 × 31 cm   (scale 1 inch to 300 feet).
BM

See TELFORD. Plan of the proposed English and Bristol Channels ship canal, surveyed by James Green, 1824.

589   *Plan (and Section) of the Exeter Canal and of the proposed Extensions and Improvements.*

By James Green Civil Engineer 1828. J. Wyld sc.
Engraved plan   29 × 61 cm   (scale 2 inches to 1 mile).
BM

GREEN, John [1787–1852] M.Inst.C.E.

See SPECIFICATION for masonry of proposed suspension bridge over the River Tyne, 1825.

GREGORY, *Professor* Olinthus Gilbert [1774–1841] Hon.M.Inst.C.E.

590   *A Treatise on Mechanics. Theoretical, Practical, and Descriptive.*

By Olinthus Gregory, of the Royal Military Academy, Woolwich.

London: Printed for George Kearsley ... 1806.
Signed: Olinthus G. Gregory, December 1805.
2 vols.   8vo   21 cm   **1**, xx + 547 pp.   **2**, vii + 514 pp.   *Atlas*, 18 + 37 double-page engraved plates (Mutlow sc.).
BM, RS, BDL

591   Second edition
Title as No. 590 but *Second Edition, with Improvements.*
London: George Kearsley ... 1807.
Elton

592   Third edition
Title as No. 590 but *Third Edition, Corrected and Improved.*
London: Printed for F. C. and J. Rivington ... 1815.
Signed: Olinthus G. Gregory, July 1815.
2 vols.   8vo   21 cm   **1**, xx + 569 pp.   18 plates.   **2**, vii + 565 pp.   40 plates.
BM, NLS, BDL

593   Fourth edition
Title as No. 590 but *Fourth Edition, Corrected and Improved.*
London: Printed for Geo. B. Whittaker ... 1826.
Signed: Olinthus G. Gregory, January 1826.
2 vols.   8vo   22 cm   **1**, xx + 575 pp.   **2**, viii + 586 pp.   *Atlas*, 18 + 45 plates.
SML, CE

594   *Mathematics for Practical Men: being a Common-Place Book of Principles, Theorems, Rules, and Tables, in Various Departments of Pure and Mixed Mathematics, with their most useful applications; especially to the pursuits of Surveyors, Architects, Mechanics, and Civil Engineers.*

By Olinthus Gregory, LL.D. ... Professor of Mathematics in the Royal Military Academy.
London: Printed for Baldwin, Cradock, and Joy. 1825.
Signed: Olinthus Gregory. Woolwich, 1 Oct 1825.
8vo   22 cm   (i) + xii + 411 pp.   3 folding engraved plates.
BM, CE, BDL

595 Second edition

Title as No. 594 but *Second Edition, Corrected and Improved.*

London: Printed for Baldwin, Cradock, and Joy, 1833.

Signed: Olinthus Gregory. 1 July 1833.

8vo   22 cm   xii + 427 pp.   3 folding engraved plates (as in first edition).

NLS

GREGORY, Thomas

596 *Elevation of Bridgwater Bridge.*

Deeble sc.   Published . . . by Tho.ˢ Gregory. 8 July 1797.

Engraved elevation   20 × 44 cm   (scale 1½ inches to 10 feet).

CE (Rennie)

The early cast-iron bridge at Bridgwater, Somerset.

GRIEVE, John

597 [Letter to William Vazie, Esq. and Report]

Signed: John Grieve. Edinburgh, 4 Nov and 14 Nov 1805.

pp. 1–5

[Printed with]

Report by James TAYLOR and William VAZIE, 6 June 1805 [q.v.].

[In]

Reports of Surveys made for ascertaining the Practicability of making a Land-Communication by a Tunnel under the River Forth, at or near Queensferry.

1806.

[Edinburgh]: Mundell, Doig, and Stevenson, printers.

4to   27 cm   (i) + 16 pp.   Engraved map.

BM, CE

598 Reprint

Grieve's Letter and Report are reprinted in *Observations on . . . the Proposed Tunnel under the Forth.* By James MILLAR and William VAZIE. Edinburgh, 1807 [q.v.].

See GRIEVE and McLAREN. Report on railway from Edinburgh to Dalkeith, 1824.

GRIEVE, John and McLAREN, James

599 *Report on the utility of a Bar-Iron Railway, from the City of Edinburgh to Dalkeith, and to the Harbour of Fishberrow.*

pp. 1–11.

Signed: John Grieve, James M'Claren. Dalkeith, 17 July 1824.

[with]

pp. 12–18. *Supplement to the Report . . .* Signed: John Grieve. Sheriffhall, 20 Sept 1824.

Edinburgh: Printed by John Stark. 1824.

4to   28 cm   18 pp.

CE

GRIFFITH, *Sir* Richard [1784–1878] F.R.S.E., M.Inst.C.E.

600 *Report on the practicability of Draining and Improving a part of the Bog of Allen, situated in the county of Kildare.*

By Richard Griffith, jun. Civil Engineer.

Signed: Richard Griffith. Robertstown, June 1810.

pp. 15–55   with 3 folding engraved plans.

[In]

First Report of the Commissioners on the practicability of draining and cultivating the Bogs in Ireland.

House of Commons, 1810.

601 *Report on the practicability of Draining and Improving a part of the Bog of Allen, situated in the King's and Queen's Counties.*
By Richard Griffith, jun. Civil Engineer.
Signed: Rich. Griffith, jun. Portarlington, 9 Feb 1811.
pp. 31–76   with 4 folding engraved plans.
[In]
Second Report of the Commissioners on the practicability of draining and cultivating the Bogs of Ireland.
House of Commons, 1812.

602   [*Two*] *Reports of Mr Richard Griffith, junior, on Bogs in the Counties of Galway and Roscommon.*
Signed Richard Griffith, June 1812 [and] Dublin, April 1813.
pp. 109–162   with 4 folding hand-coloured engraved plans.
[In]
Fourth Report of the Commissioners on the practicability of draining and cultivating the Bogs of Ireland.
House of Commons, 1814.

See DRUMMOND *et al*. Reports of the Commissioners . . . [on] Railways in Ireland, 1837 and 1838.

See BURGOYNE *et al*. Reports of the Commissioners for the Improvement of the River Shannon, 1837 and 1839.

GRIMSHAW, John [*c. 1763–1840*]

603   *A Report of the Repairs given to Wearmouth Bridge, in the Year 1805.*
By John Grimshaw.
Sunderland: Printed by G. Garbutt. 1818.
Dated: Bishopwearmouth, 5 March 1818.
8vo   22 cm   24 pp.
BM, Science Museum, Sunderland Public Library

GRUNDY, John *senior* [*c. 1696–1748*]

604   *Here follow some Propositions, which plainly shew the Necessity of Mathematical and Philosophical Knowledge, in the draining of Lands that lie near to the Sea or low Fen-Lands* [*etc*].
Humbly presented to the Nobility and Gentry, who are Proprietors of such Lands that want draining. By their most humble servant, John Grundy, of Congestone in the County of Leicester, Land-Surveyor and Teacher of the Mathematicks.
n.p.
Dated: 15 April 1734.
fol   39 cm   4 pp.
RS (Smeaton), BDL

605   *Philosophical and Mathematical Reasons, humbly offer'd to the Consideration of the Publick: To prove that the Present Works, executing at Chester, to recover and preserve the Navigation of the River Dee must intirely Destroy the same. With some Remarks on Mr Badeslade's Reasons, &c thereon.*
By their most Humble Servant, John Grundy of Congestone, in the County of Leicester, Land-Surveyor, and Teacher of the Mathematics.
London: Printed for the Author 1736.
4to   23 cm   16 pp.   including (as p. 13) an engraved plate.
BDL (Gough)

606   Reprint
Title and text as No. 605 but without the Preface and engraved plate.
pp. 17–22.
[In]
BADESLADE. *The New Cut Canal.* Chester, 1736 [q.v.].

607   *An Examination and Refutation of Mr Badeslade's New-cut Canal, &c. By Quotations from his own Words, as well as from Observations and Experiments made upon the River Welland, and the Country adjacent, in the Years 1731, 2, 3, 4, 5 and 6. Done for the use of Drainers* [etc].
By John Grundy, of Congestone, in the County

of Leicester, Land-Surveyor, and Teacher of the Mathematics.

London: Printed for J. Roberts . . . 1736. Price, One Shilling and Six-pence.

Signed: John Grundy, 11 May 1736.

4to   25 cm   36 pp.   Small folding engraved plate.

CRO, LCL, BDL (Gough)

The CRO copy stitched as issued.

See GRUNDY, *senior* and *junior*. Plan and report on navigation of the River Witham and draining the adjacent low-lands, 1743–44.

608   *A Further Illustration of Messrs Grundy's Scheme, for Restoring and making Perfect, the Navigation of the River Witham, from Boston to Lincoln; and also for Draining the Low-Lands and Fens contiguous thereto. Mr Coppin's Scheme is here answer'd* [etc].

By John Grundy, Agent and Engineer to the Honourable Adventurers of Deeping Fen.

Stamford: Printed by F. Howgrave . . . 1745. Price Four Pence.

Dated: Spalding, 9 April 1745.

8vo   19 cm   20 pp.

CUL

GRUNDY, John *senior* and *junior*

609   *A Scheme for the Restoring and making perfect the Navigation of the River Witham from Boston to Lincoln, and also for draining the Low-Lands and Fenns contiguous thereto.*

By John Grundy, Sen. and Jun. of Spalding, Engineers.

Printed in the Year 1744.

Signed: J. Grundy, Sen. and Jun. [1743]

8vo   20 cm   (i) + 48 pp.

BM

610   *A Map of the Antient River Witham as reduced from the Original (which was plotted from a Scale of 2½ Inches in a mile) from the Primary Station at Wiberton Roads to the Bradon above Lincoln . . . together with*

several of the Parishes, Low Lands and Fens on each side . . . [and] as many of the Station Points of our Leveling Notes as its Size wou'd bear as they were actually taken upon the Spot by a nice Spirit Level [etc].

By John Grundy Sen.ʳ & Jun.ʳ of Spalding in Lincolnshire, Engineers. Anno Dom. 1743. W. H. Toms sc.

Engraved plan   41 × 70 cm   (scale 0.9 inches to 1 mile).

BM, RS (Smeaton), RGS, LCL, BDL (Gough)

GRUNDY, John [1719–1783] M.Soc.C.E.

611   *Objections to the Bill for making the Branch of the Trent running by Newark, Navigable.* [with] *The Observations of Mr Grundy, Junior, the Surveyor.*

n.p.   [1741].

Signed: John Grundy, jun.ʳ

fol   33 cm   3 pp.   Docket title on p. 4.

BM

612   *A Map of the River Trent as it runs from Farndon Ferry to Holme Meadow in its two Branches by Kellum & Newark.*

Survey'd, Level'd and Delineated by John Grundy, Jun.ʳ Surveyor. 1741.

Engraved plan   42 × 53 cm

BM, RGS, BDL (Gough)

See GRUNDY, *senior* and *junior*. Plan and report on navigation of the River Witham and draining the adjacent low-lands, 1743–1744.

613   *Mr Grundy's Plan* [for improving the River Witham]

Signed: John Grundy, [November 1753].

pp. (1)–15 with folding engraved plan, untitled, signed John Grundy.

[Printed with]

At a Meeting held at Boston in the County of Lincoln, October 19th 1753.

Lincoln: Printed by W. Wood.

4to 23 + 15 pp. Folding engraved plan (as above).

CE, RGS (plan only)

614 *A Scheme for executing a Navigation from Tetney-Haven to Louth; and for Draining the low Grounds and Marshes adjoining thereto.*

By John Grundy.
Signed: John Grundy, Spalding
pp. 1–15.
[Printed with]
The report of John SMEATON, 14 July 1761 [q.v.].
Nottingham: Printed by Samuel Creswell, for E. Parker, Bookseller at Louth, 1761.
8vo 24 pp.
LCL, Bristol County Library
Lough Navigation Minute Book, 3 Sept 1760: Smeaton has sent his (MS.) report and returned Mr Grundy's. 20 July 1761: Reports (revised) ordered to be printed. 17 Sept 1761: Mr Parker paid for printing.

615 *A Plan of the Proposed Navigation, from Tetney Haven through Tetney Common, the Low Grounds of North Coats, Marsh Chapel, Gainthorpe, and Avingham . . . and to Kedington, & so to Louth* [etc].

By John Grundy of Spalding, Lincolnshire. Engineer. J. Larken sc.
Engraved plan 51 × 40 cm (scale 2¼ inches to 1 mile).
BM, RGS, MCL, LCL
Louth Navigation Minute Book, John Grundy's Acct. 24 Sept 1760: Reducing and drawing the Plan for engraving, revising & correcting.

See GRUNDY, EDWARDS and SMEATON. Report on the River Witham, 1761.

616 *A Plan of the River Witham and adjoining Fens and Low Ground from Lincoln to Boston, with the New Works proposed to be Executed thereon for Draining the said Fens and Low Grounds, and restoring the Navigation of this River.*

By J. Grundy, Surveyor and Engineer.
Engraved plan 28 × 42 cm (scale 1 inch to 2 miles).

BM, LCL, MCL, RS (Smeaton), RGS, BDL (Gough), Sutro
Issued in 1762.

617 *The Report of John Grundy, of Spalding, in Lincolnshire, Engineer, concerning the Drainage of Low Grounds and Carrs lying in the Parishes, Townships . . . and Territories of Sutton, Ganstead . . . Benningholme . . . Leven . . . Tickton, Weel, Routh, Meaux, and Waghen, in Holderness, in the East Riding of the County of York.*

n.p.
Signed: John Grundy. Beverley, 30 Dec 1763.
fol 33 cm 7 pp. Docket title on p. (8).
HRO
Stitched as issued. Includes (on p. 7) an extract from Smeaton's report of 12 Feb 1764.

See GRUNDY and SMEATON. Plan of part of Holderness with a scheme for draining the same, 1764.

618 *Report in Consequence of a View of the Works carrying on for the Drainage of Holderness Level.*

By J. Grundy.
n.p.
Signed: John Grundy. Beverley 17 Sept 1765.
fol 28 cm 7 pp.
HRO

619 *Mr Grundy's Observations on the River from Great Driffield to Emmotland, and on the adjoining Country, and his Report, Scheme, and Estimate for making a Navigation between these two Places, and also from Emmotland to Frodingham Bridge, in the County of York.*

n.p.
Signed: John Grundy. Driffield, 18 Dec 1766.
fol 33 cm 5 pp. Docket title on p. (6).
HRO, MCL

620 *A Plan of the River formed by the Junction of the several Becks about Great Driffield and running from thence . . . into the River Hull at Emertland with the Situation of the adjoyning Towne and high and low Grounds extracted from Mr Milborns Survey thereof; and also a Scheme for making a Navigation from Driffield to the said River Hull below Emertland.*

By John Grundy Engineer 1766.
[with] *A Section of the Country made from Mr Milborn's Levels.*

Engraved plan 45 × 62 cm (scale 4 inches to 1 mile)
MCL

621 *The Report of John Grundy, Esq; Engineer, for making the River Swale to Moreton Bridge and Bedale Brook navigable; with an Estimate of the Expences thereof. 1767.*
n.p.
Signed: John Grundy, Engineer. London, 21 Feb 1767.
fol 28 cm 3 pp. Docket title on p. (4).
BM, HRO

622 *A Survey of the River Swale from Moreton Bridge to its Junction with the River Ure, and from thence to Widdington Ings upon the River Ouze. Also the Brook from Bedale to the Swale.*
Engraved by T. Jefferys 1767.
Engraved plan 29 × 104 cm (scale 2 inches to 1 mile).
BM, BDL (Gough)

623 *The Report of John Grundy, Engineer, for making the Brook Cod Beck from Thirsk to the River Swale navigable; with an Estimate of the Expence thereof.*
n.p.
Signed: John Grundy, Engineer. Thirsk, 21 March 1767.
fol 28 cm 6 pp.
HRO

624 *A Survey of the Brook Cod-Beck, from Thirsk to the River Swale; taken by Richard Firth and re Survey'd and corrected by Isaac Milbourn. 1767 [with] A Section of the Country formed from the Levels thereof.*
By John Grundy, Engineer.
Engraved plan 22 × 82 cm (scale 6 inches to 1 mile)
BM

625 *A Plan of the Ings Meadows Marshes and other Low Grounds in Laneham, Rampton, Treswell, South & North Levertons, Hebblesthorpe, Fenton, Sturton, Littleborough, Cottam & West Burton, in the County of Nottingham, which are frequently overflowed by the Trent and other Outward Waters.*

Surveyed by John Grundy, Engineer, and George Kell, Surveyor, 1769.
Engraved plan 35 × 53 cm (scale $3\frac{1}{4}$ inches to 1 mile).
LCL

626 *The Report of John Grundy, Engineer; respecting the Proposed Navigation, from Chesterfield to the River Trent.*
Spalding: Printed by J. Albin. 1770.
Signed: John Grundy, Engineer. Spalding, 22 Aug 1770.
fol 30 cm (i) + 25 + (2) pp.
Derbyshire R.O.

627 *A Plan of the Intended Canal from Chesterfield, by Retford, to the River Trent, as propos'd by Mr Brindley. Also Alterations therein as proposed by John Grundy, Engineer, extracted from Mr Varley's Plan, Views & Observations made on the Premises.*
Engraved by T. Jefferys 1770.
Engraved plan 28 × 75 cm (scale 1 inch to 1 mile).
BM, Sutro

628 *The Report of John Grundy, Esq; respecting the Drainage and Navigation proposed for Walling Fenns, &c. 1772.*
[York: Printed by Anne Ward].
Signed: John Grundy, Engineer. Market-Weighton, 2 Sept 1772.
4to 22 cm 30 pp. Folding engraved plan.
A Section of the Country through which the proposed Navigation and Drainage of Walling Fenns &c is to pass.
By John Grundy, Engineer, 1772. Engrav'd by H. Shepherd.
HRO, Hull University Library
Stitched as issued. Market Weighton Navigation & Drainage, Treasurer's Accounts, 20 Nov 1772: To Mr Henry Shepherd for engraving a Section of the Navigation and Drainage and 300 prints therefrom. *Ibid* subsequent entries: Mrs Ward for printing copies of Mr Grundy's report, and Mr Locke the Bookseller for folding, sewing etc 320 of Mr Grundy's reports.

629 *The Report of John Grundy, Esq; Engineer, respecting the Proposition of making Quays and Wharfs on the West Side of the Haven of the River Hull.*

n.p.

Signed: John Grundy, Engineer. Hull, 14 Oct 1772.

4to   20 cm   14 pp.

HRO, Hull City Archives, Sutro
   Stitched as issued.

630 *The Report of John Grundy, Engineer* [on a proposed quay in Hull harbour].

n.p.

Signed: John Grundy, Engineer. Hull, 22 Dec 1773.

4to   20 cm   4 pp.

Hull City Archives

631 *Observations resulting from Surveys, Levels and Views made on the East Fen, the low Grounds and Fens adjoining thereto, belonging to the Soke of Bolingbroke, East Holland, and the Level Towns; with a Report of the Causes of their present drowned State and Condition: Also Schemes for the Drainage thereof, and Estimates of the Expence of executing those Schemes, &c.*

   By John Grundy of Spalding, Lincolnshire, Engineer. 1774.

n.p.

Signed: John Grundy, Engineer. Spalding, 14 Nov 1774.

4to   22 cm   46 pp.   Folding engraved plan.
A Plan of the East Fen & adjoining Low Grounds in the Level & East Holland Towns.
   By I. Grundy Engineer 1774.   M. Darby sc.
BM, LAO, LCL
   LAO copy stitched as issued.

See HODSKINSON. Report on Wells harbour, with note by John Grundy and Thomas Hogard, 6 July 1782.

GRUNDY, John, EDWARDS, Langley and SMEATON, John

632 *The Report of Mess. John Grundy, Langley Edwards, and John Smeaton, Engineers, concerning the present ruinous State and Condition, of the River Witham, and the Navigation thereof, from the City of Lincoln, thro' Boston, to its Outfall into the Sea; And of the Fen Lands on both Sides of the said River. Together with Proposals and Schemes for Restoring, Improving, and Preserving the said River and Navigation. And also for effecting the Drainage of the said Fen Lands. To which is annexed a Plan, and proper Estimates of the Expences in performing the several Works recommended for those Purposes.*

Lincoln: Printed by W. Wood.

Signed: John Grundy, Langley Edwards, John Smeaton. Sleaford, 23 Nov 1761.

4to   20 cm   26 pp.

BM, LCL, BDL (Gough)
   The engraved plan, not ready in time to be annexed to the report, was issued separately: *see* GRUNDY, 1762.

GRUNDY, John and SMEATON, John

633 *A Plan of an Actual Survey, of part of the Middle & North Bailiwicks of Holderness: Describing the Low Grounds &c which are almost continually Overflow'd with Water. By C. Tate, Surveyor. With a Scheme for Draining the same.*

   By Mess.rs Grundy & Smeaton Engineers 1764. H. Shepherd sc.

Engraved plan   41 × 27 cm   (scale $1\frac{1}{4}$ inches to 1 mile).

BM, RS (Smeaton), MCL

GWILT, Joseph [1784–1863]

634 *A Treatise on the Equilibrium of Arches, in which the Theory is demonstrated upon Familiar mathematical Principles.*
By Joseph Gwilt, Architect.
London: Printed for the Author; and sold by J. Taylor . . . 1811.
Dated: Stamford Street, London, December 1810.
8vo   22 cm   xiv + (2) + 80 pp.   3 engraved plates, woodcuts in text.
BM, RIBA

635   Second edition
As No. 634 but *Second Edition*.
London: Priestley and Weale. 1826.
Dated: Abingdon Street, Westminster, January 1826.
8vo   22 cm   xxi + (1) + 104 pp.   Folding engraved frontispiece (Gwilt's design for London Bridge 1822) + 4 plates, woodcuts in text.
BM, RIBA
Issued in quarter cloth, grey boards, printed paper label on spine.

636   Third edition
As No. 634 but *Third Edition*.
London: John Weale. 1839.
8vo   22 cm   xiii + (2) + 104 pp.   Frontispiece and plates as in second edition.
BM, RIBA

GWYNN, George

637   *Plan of Ramsgate Harbour.*
Surveyed by George Gwynn, April 1815. W.F. 1815.
Engraved plan   56 × 51 cm   (scale 1 inch to 100 feet).
CE (Rennie)

GWYNN, John [1713–1786] R.A.

638   *The Plan and Elevation, of the South Side of a Bridge, to be built over the Severn at Shrewsbury.*
J. Gwynn, Arch. May 1768. E. Rooker sc.
Engraved plan and elevation   25 × 40 cm   (scale 3 inches to 100 feet).
CE (Rennie)

HAGUE, John   M.Inst.C.E.

639   *Morecambe Bay Embankment. [Report] To Sir H. Le Fleming Senhouse, and the Gentlemen forming the Provisional Committee of the Caledonian, West Cumberland, and Furness Railway.*
Signed: John Hague. Cable-street, London.
pp. (3)–10
[Printed with]
Report on the West Cumberland Railway by J. U. RASTRICK, 24 Dec 1835 [q.v.].
[In]
West Cumberland, Furness, and Morecambe Bay Railway. Reports of J. U. Rastrick Esq., and John Hague, Esq.
London: J. T. Norris . . . [1839].
8vo   18 cm   23 pp.   Folding litho map.
BM, ULG (Rastrick)

HAMILTON, George Ernest M.Inst.C.E.

640   *Report to the Commissioners of the Second District of the Turnpike Road from Coleshill, through the City of Lichfield and Town of Stone, to the end of the County of Stafford.*
Shrewsbury: Printed by W. and J. Eddowes . . . [1825].
Signed: George E. Hamilton. Shrewsbury, 3 Nov 1825.
4to   30 cm   22 pp.
CE

641 *Report on the Turnpike Road from Sandon through Leek to Hugbridge, in the County of Stafford.*
Shrewsbury: Printed by W. and J. Eddowes . . . 1827.
Signed: George E. Hamilton, Civil Engineer. Stone, 1 Oct 1827.
4to   29 cm   28 pp.
CE

HARDWICK, Philip [1792–1870] F.R.S., R.A., M.Inst.C.E., M.R.I.B.A.

See TELFORD. Plans of St. Katharine Docks, 1824 and 1828.

HARDWICKE, Philip, *Earl of* [1757–1834] F.R.S.

642 *Observations upon the Eau-Brink Cut. With a Proposal offered to the Consideration of the Friends of the Drainage.*
By the Earl of Hardwicke.
Cambridge: Printed by F. Hodson; and sold by Messrs Merrills: and the Booksellers of Ely, Wisbech, and Lynn.
Dated: Ely, 18 Feb 1793.
4to   24 cm   15 pp.
CRO, CUL, Sutro
   Includes (pp. 11–13) a reprint of John HUDSON's report of 3 Nov 1792 [q.v.].

643 Second edition
Title as No. 642
London: 1794.
8vo   21 cm   28 pp.
BM

HARE, Edward [d. 1816]

See MAXWELL and HARE. Report on the drainage of Deeping . . . Spalding and Pinchbeck Commons, 1800.

See JESSOP, RENNIE *et al.* Report on Deeping Fen, 1800.

HARNESS, *Colonel Sir* Henry [1804–1883] R.E.

644 *Report from Lieutenant Harness, Royal Engineers, explanatory of the principles on which the Population, Traffic and Conveyance Maps have been constructed.*
Signed: H. D. Harness, Lieut. Royal Engineers.
With three large engraved maps.
[1] *Map of Ireland . . . showing by the varieties of shading the comparative Density of the Population.*
[2] *Map of Ireland . . . showing the relative Quantities of Traffic in different Directions.*
[3] *Map of Ireland . . . showing the relative Number of Passengers in different Directions.*
[All three maps] Constructed under the Direction of the Commissioners by Henry D. Harness, Lieut. Royal Engineers. 1837.
Engraved by J. Gardner.
                    [In]
Second Report of the Commissioners . . . [on] Railways in Ireland.
Dublin, 1838.
Appendix pp. 41–49, pls. 3–5.

HARRISON, John

645 *A New Method of making the Banks in the Fens almost Impregnable . . . Also some Observations on the River Cam.*

By John Harrison, Botanist, Nurseryman in Cambridge.

Cambridge: Printed for, and Sold by the Author. [1766].

8vo   22 cm   xi + 61 pp.

CE (Page), CRO, BDL (Gough)

Publication date as given in BDL catalogue.

HART, John

646 *A Practical Treatise on the Construction of Oblique Arches.*

By John Hart, Mason.

London: Longman & Co and John Weale . . . 1836.

Signed: J. H. Birmingham, August 1836.

8vo   24 cm   (i) + vi + 32 pp.   7 engraved plates (J. Hart del.   A. Johnson sc.).

BDL

647 Second edition

Title as No. 646 but *Second Edition*.

London: John Weale . . . 1839.

Signed: J. Hart. London, August 1839.

4to   28 cm   (i) + vi + 40 pp.   11 engraved plates (as in first edition with 4 additional plates, J. Roffe sc.).

BM, CE, RIBA, ISE, ULG

HARTLEY, Jesse [1780–1860]

648 *Plan of the Proposed Alterations and Additions to the Docks and Basins in the Port and Town of Liverpool.*
1825.

Jesse Hartley, Surveyor.

Liverpool, 1825.

Engraved plan   17 × 58 cm   (scale 1 inch to 720 feet).

BM, Mersey Docks & Harbour Office

649 Revised plan

Title as No. 648 but 1828.

Engraved plan   17 × 58 cm   (scale 1 inch to 720 feet).

Mersey Docks & Harbour Office

650 Revised plan

Title as No. 648 but 1830.

Engraved plan   17 × 58 cm   (scale 1 inch to 720 feet).

Mersey Docks & Harbour Office

651 [Report] *To the Trustees of Swansea Harbour.*

Swansea: Murray and Rees, Printers.

Signed: Jesse Hartley. Liverpool, 12 Oct 1831.

8vo   20 cm   8 pp.   Folding litho plan.

Plan for the Improvement of the Harbour of Swansea
By Jesse Hartley.

J. Grove, Lithog.

SML

Printed with a letter by J. H. Vivian of 7 Nov 1831 (pp. 9–31).

652 [Reports] *To the Trustees of the Harbour of Swansea.*

Swansea: Murray and Rees, Printers.

Signed: Jesse Hartley. Liverpool 18 April 1832 (to 26 May 1832).

8vo   20 cm   23 pp.

SML

Four reports, printed with a report by Lieut. H. M. Denham R.N. of 6 June 1832 (pp. 24–27).

See WALKER and HARTLEY. Report on Whitehaven Harbour, 17 May 1836.

653 *The Surveyor's Report to the Dock Committee, on the General State and Progress of the Dock Works, since 24th December 1835, and of the whole of his Expenditure, from the 24th March 1824 to the present time.*

Liverpool: Printed by J. and J. Maudsley. 1836.

Signed: Jesse Hartley. Liverpool, 25 Oct 1836.

8vo   28 cm   20 pp.

Mersey Docks & Harbour Office

HASLEHURST, Joseph

654 *Second Report upon the proposed Grand Commercial Canal, having for its object the Union of the Peak-Forest, Sheffield, Chesterfield & Cromford Canals.*
By Joseph Haslehurst, Civil Engineer.
Chesterfield: Printed by J. Roberts. 1824.
Signed: Joseph Haslehurst. Unstone Colliery near
    Sheffield, 10 Dec 1824.
8vo   21 cm   84 + (5) pp.   Folding hand-coloured engraved plan.
Proposed Grand Commercial Canal.
    Drawn by W. Dickin, Surveyor, under direction of J. Haslehurst. Engraved by J. Fothergill.
CE (Telford)
    The first, preliminary report of June 1824 was printed in *Proposals* for the Grand Commercial Canal.

HASSALL, Charles [1754–1814]

See HASSALL and WILLIAMS. Road from Milford to Gloucester, 1792.

HASSALL, Charles and WILLIAMS, John

655 *The Road from the New Port of Milford to the New Passage of the Severn, and Gloucester;*
    Survey'd in the Year 1790, by C. Hassall of Eastwood, Pembroke-shire and J. Williams of Margam, Glamorgan-shire . . . J. Cary, Engraver & Map-seller . . . London 1 Oct 1792.
8vo   20 cm   title-page + 18 folding engraved plans each 15 × 45 cm   (scale 2 inches to 1 mile).

[Issued with]
South Wales Association for the Improvement of Roads, instituted in the Year 1789.
London: Printed by James Phillips . . . 1792.
8vo   20 cm   (i) + 70 pp.
BM, ULG (text only)
    The text includes Estimates of the Expense of Repairing the Road from Milford to Gloucester . . . 1790.

HASSELL, John [1767–1825]

656 *Tour of the Grand Junction, illustrated in a Series of Engravings; with an Historical and Topographical Description of those parts of the Counties . . . through which the Canal passes.*
By J. Hassell.
London: Printed for J. Hassell . . . 1819.
8vo   22 cm   viii + 148 + (4) pp.   24 hand-coloured aquatint plates, drawn (and engraved) by Hassell, of which 16 depict views of the canal.
BM, SML

HAWKSHAW, *Sir* John [1811–1891] F.R.S., M.Soc.C.E., M.Inst.C.E.

657 *Report of John Hawkshaw Esq. to the Directors of the Great Western Railway.*
n.p.
Dated: Manchester, 4 Oct 1838.
8vo   22 cm   31 pp.
BM, SML, CE, ULG
    Issued with reports by I. K. BRUNEL and Nicholas WOOD [q.v.].

HAWKSMOOR, Nicholas [*c.* 1661–1736]

658 *A Short Historical Account of London-Bridge; with a Proposition for a New Stone-Bridge at Westminster.*
By Nicholas Hawksmoor, Esq.
London: Printed for J. Wilcox . . . 1736.
4to   25 cm   47 pp.   5 folding engraved plates (B. Cole sc.).
BM, CE (Telford), RIBA, ULG, BDL (Gough)
  Includes (pp. 18–21) LABELYE's Calculation of the Fall of the Water at the intended Bridge at Westminster [q.v.].

HENSHALL, Hugh [1734–1816] M.Soc.C.E.

See BRINDLEY. Plan of the intended Trent & Mersey Canal, 1765. Drawn by Henshall. A smaller version, also drawn by Henshall, is annexed to BENTLEY's *View of the Advantages of Inland Navigations*. London, 1765 [q.v.].

659 *A Plan of the Navigable Canals made and now making in England.*
Henshall del.   J. Cary sc.   Published by T. Lowndes 29 May 1779.
Engraved plan, hand coloured   66 × 61 cm (scale 1 inch to 7 miles)
BM
  Issued in conjunction with [BRINDLEY] *The History of Inland Navigations*, 1779 [q.v.].

660 [Report] *To the Committee of the Brecknock and Abergavenny Canal.*
Brecknock: Printed by W. and G. North, 1794.
Signed: H. Henshall, 28 May 1794.
fol   31 cm   3 pp.
B

HILL, David Octavius [1802–1870]

See BUCHANAN. Glasgow & Garnkirk Railway, views by Hill, 1832.

HODGKINSON, *Professor* Eaton [1789–1861] F.R.S., Hon.M.Inst.C.E.

661 *Experimental Researches on the Strength of Pillars of Cast Iron, and other Materials.*
By Eaton Hodgkinson, Esq.
London: Printed by R. and J. E. Taylor . . . 1840.
Dated: 22 April 1840.
4to   30 cm   (i) + [385–456] pp.   3 litho plates (J. Basire lith.)
CE (from the author)
  Offprint, with new title page, from *Phil. Trans.* Vol. 130 (1840).

HODGKINSON, John [1773–1861]

662 *A Plan of an intended Rail Way, or Tram Road, from Sirhowy Furnaces, in the Parish of Bedwellty, in the County of Monmouth, by Tredegar Iron Works . . . to communicate with the Monmouthshire Canal and the River Usk, at or near the Town of Newport . . .*
  Taken from the Line marked out by John Hodgkinson, Engineer, by David Davies, Surveyor. Septem.r 26th 1801.
Engraved plan   41 × 107 cm   (scale 2 inches to 1 mile).
Newport Central Library

HODSKINSON, Joseph
[?1735–1812] M.Soc.C.E.

663 *The Report of Joseph Hodskinson, Engineer, respecting the State of Wells Harbour, in the County of Norfolk. 1782.*
Dated: Arundel-street, London, 5 July 1782.
15 pp.   with folding hand-coloured engraved plan.
A Plan of the Harbour & Haven of the Port of Wells in the County of Norfolk, with the Marshes Adjoining thereto. 1782.
[Printed with]
A Note (p. 16) by John Grundy and Thomas Hogard, 6 July 1782, concurring with Hodskinson's report.
[And]
The Report of Joseph NICKALLS . . . 5 July 1782 [q.v.].
[In]
The Reports of Mess. Hodskinson, Grundy, Hogard, and Nickalls, on the State and Causes of Decay in Wells Harbour.
[London]: 1782.
fol   33 cm   16 + 5 pp.   Docket title on p. (6)
Folding engraved plan (as above).
BM, RS (Smeaton, plan only), ULG, Sutro
A manuscript in the Sutro Library entitled 'The Report of Mess.*rs* John Grundy, Thomas Hogard and Joseph Hodskinson, Engineers, respecting the state of Wells Harbour in the County of Norfolk 1782' is almost identical to the printed report noted above.

664 *The Report of Joseph Hodskinson, Engineer, on the Probable Effect, which a New Cut, now in contemplation, from Eau-Brink to a little above Lynn, will have on the Harbour and Navigation of Lynn; with a Plan for Improving the Present Channel, both above and below the Town.*
Lynn: Printed by R. Marshall, 1792.
Dated: Arundel Street, London, 2 Jan 1792.
4to   22 cm   23 pp.
CRO, Northants R.O., ULG, BDL (Gough), Sutro
Stitched as issued.

665   Second edition
Title as No. 664 but re-set.
London: Printed by E. Hodson . . . [1793].
4to   23 cm   31 pp.
To the report of 2 Jan 1792 is added (pp. 23–31)
Observations by way of Supplement to the preceeding Report.
Signed: J. Hodskinson. Arundel Street, 25 Feb 1793.
BM, CUL

666 *Plan of the Outfall of the River Ouse Explanatory of Mr Hodskinsons Report on the Intended Cut at Eau Brink.* [1792].
Engraved plan   24 × 50 cm   (scale 1 inch to 1 mile).
CE (Page)

667   Later state
Plan of the Outfall of the River Ouse Explanatory of Mr Hodskinsons Report on the Intended Cut at Eau Brink [and inset table] *Difference of the Levels between Eau Brink & the Crutch on the Surveys taken in 1777 and 1791.* [1793].
Engraved plan   24 × 50 cm   (scale 1 inch to 1 mile).
BM, BDL
Includes notes on soundings, and position of the outfall channel in March 1793.

668 *A Plan and Estimates, for Improving and Extending the Navigation of the River Stour from Sandwich to Canterbury, in the County of Kent.*
Canterbury: Printed by Simmons, Kirby and Jones. 1792.
Signed: J. Hodskinson. Arundel Street, 17 May 1792.
8vo   21 cm   16 pp.   Folding engraved plan.
Plan of the Proposed Cut, for extending the Navigation of the River Stour, from Fordwich to Canterbury. 1792.
Barlow sc.
BM

669 *The Report of Mr Hodskinson, Engineer, As to the effects of the proposed navigable canal from Stainforth to Keadby, upon the drainage of the level of Hatfield Chace.*

Doncaster: Printed by W. Sheardown.
Signed: J. Hodskinson. London, 9 Jan 1793.
fol   33 cm   2 pp.   Docket title on p. (4).
Nottingham Univ. Library

670   *Plain and useful Instructions to Farmers; or, An Improved Method of Management of Arable Land; with some Hints upon Drainage, Fences, and the Improvement of Turnpike and Cross Roads.*
   By Joseph Hodskinson.
London: Printed for the Author; and sold by F. and C. Rivington . . . [1794].
Price One Shilling.
Dated: Arundel-street, London, 1 Jan 1794.
8vo   21 cm   38 pp.
BM, ULG

671   *The Report of J. Hodskinson, Civil Engineer, to Edward Constable, Esq. containing Instructions for preserving the Estate.*
Signed: J. Hodskinson. Arundel-Street, 10 Feb 1796.
pp. 5–8   with woodcut plan.
Plan of Cherry Cobb Marsh.
                    [In]
Sundry Papers and Reports, relative to the Defence of the Estate of Cherry Cobb Sands, against the Humber.
Newcastle: Printed by Edward Walker . . . [1800].
8vo   22 cm   23 pp.   Woodcut plan (as above).
CE, Sutro

672   *The Report of Joseph Hodskinson, Engineer, on the Keyingham Drainage.*
n.p.
Signed: J. Hodskinson. Arundel-Street, 10 Feb 1796.
4to   22 cm   8 pp.   Folding engraved plan.
Hull University Library

673   *Explanation of the Plan for an Improved System of Mooring Vessels in the River Thames, from London Bridge to the King's Mooring Chains at Deptford.*
Signed: J. Hodskinson. Arundel-Street, 31 March 1796.
                    [In]
Report from the Committee on the Port of London.
   House of Commons, 1796. Appendix Q (pp. 29–33).

674   Second printing
In *Reports from Committees of the House of Commons* Vol. 14 (1803) pp. 360–363.

HOGBEN, Thomas [1703–1774] and Henry [d. 1822]

675   *Tables [of Levels of the River Stour] according to the Survey made by Messrs Hogben in 1773, taken at Low Water.*
pp. (3)–(4).
                    [Printed with]
Report by Richard DUNTHORNE, 8 Sept 1774 [q.v.].
                    [and]
Report by Thomas YEOMAN, 8 Oct 1765 [q.v.].
n.p.   [Printed 1775].
fol   34 cm   (4) pp.
BM

See CRONK. Plan of River Stour, with longitudinal sections by Hogben *senior* and *junior* (1773) and Henry Hogben and William Cronk (1775).

HOLDERNESS DRAINAGE

676   *Observations on the Drainage of Certain Low Grounds on the east side of the River Hull, in Consequence of a View of the Works the 13th of February, 1786.*
[Beverley, 1786].
Signed: A Friend of the Undertaking.
8vo   21 cm   20 pp.
HRO
   Holderness Drainage Minute Book, 13 May 1786: Paper on 'Observations' etc to be printed and copies sent to the landowners, and to Mr Jessop.

HOLLINSWORTH, James [d. 1828]

677 *Plan of the Town and Harbour of Great Grimsby in the County of Lincoln.*
Surveyed by James Hollinsworth 1801.
Engraved plan   36 × 79 cm   (scale 1 inch to 300 feet).
BM

678 *Plan of the River Arun Navigation from Burpham to New Bridge.*
Surveyed by J. Hollinsworth 1820. Ja.ˢ Wyld del. Lithog.
Litho plan   33 × 58 cm   (scale 2 inches to 1 mile).
West Sussex R.O.

HOLT, Luke [*c. 1723–1804*]

679 *A Plan of the intended Navigable Canal from Cooper Bridge to Huddersfield, in the County of York taken November the 6th 1773.*
Engraved by Faden and Jefferys . . . 1774.
Engraved plan   31 × 58 cm   (scale 6 inches to 1 mile).
MCL
Plan for Sir John Ramsden's Canal, known to be by Holt.

HOMER, Henry [1719–1791]

680 *An Enquiry into the Means of Preserving and Improving the Publick Roads of this Kingdom, with Observations on the probable Consequences of the present Plan.*
By Henry Homer, A.M., Rector of Birdingbury in Warwickshire.

Oxford: Printed for S. Parker . . . 1767. Price One Shilling.
8vo   21 cm   (iv) + 87 + (1) pp.
BM, CE, ULG, BDL

HOPKIN, Evan

681 *An Abstract of the particulars contained in a Perambulatory Survey of above Two Hundred Miles of Turnpike-Road through the Counties of Carmarthen, Brecknock, Monmouth, and Glocester . . . in August and September, 1805.*
By Evan Hopkin, Civil Engineer.
Swansea: Printed by T. Jenkins. 1805.
Signed: Evan Hopkin. Swansea, 30 Nov 1805.
8vo   20 cm   16 pp.
BM, CE

HOPKINS, Roger   M.Inst.C.E.

682 *The Report of Roger Hopkins, Civil Engineer, on the Works proposed by him for the Formation of a Floating Harbour, and otherwise Improving the Navigation of the River; also, for effecting a Communication from Swansea to the opposite shore by a Bridge across the River and the purposed New Channel.*
Signed: Roger Hopkins, Civil Engineer, M.I.C.E. Swansea, 1 July 1831.
pp. (26)–42.
[In]
Reports on the Formation of a Floating Harbour at Swansea, with reference to plans submitted to the Trustees.
Swansea: Printed . . . by W. C. Murray and D. Rees. 1831.
8vo   20 cm   78 pp.
Nat. Library of Wales, SML

HOPKINS, Roger and Sons

683   Deposited plan
Map of a Proposed Railway, from Wadebridge to Went-ford Bridge, with a Branch to Bodmin, and Communications to Ruthern Bridge and Nanstallan, all in the County of Cornwall.
Roger Hopkins & Sons, Civil Engineers.
Bodmin, 30 November 1831.
Drawn and Printed at Hackett's Lithographic Office, Exeter.
Litho plan, lines entered in colour   40 × 52 cm (scale $2\frac{2}{3}$ inches to 1 mile).
HLRO
The accompanying MS. Section of the Line is signed Thomas Hopkins.

684   *Map of the proposed Railway from the proposed Harbour* [at Tremoutha Haven] *to Launceston.*
Roger Hopkins and Sons, Civil Engineers.
Litho map   20 × 46 cm   (scale 1 inch to $1\frac{1}{4}$ miles).
[In]
The Duke of Cornwall's Harbour and Launceston and Victoria Railway. Prospectus. March 1836.
Engineers: Roger Hopkins and Sons, Plymouth. Consulting Engineers: Sir John Rennie and George Rennie
fol   38 cm   7 + (1) pp.   including map (as above) and *Map of the Proposed Harbour.*
BM

HOWELL, John

See TELFORD. Map of a canal from Glasgow to Ardrossan, surveyed by John Howell, 1805.

HUDDART, *Captain* Joseph [1741–1816] F.R.S., M.Soc.C.E.

685   *Copy of the Report of J. Huddart, Esquire, relative to the making of a New Dock at Kingston-upon-Hull.*
n.p.
Signed: J. Huddart. Highbury Terrace, 6 Aug 1793.
fol   38 cm   3 pp.   Docket title on p. (4).
Hull City Archives

686   *Captain Huddart's Report. To Sir Joseph Banks.*
Signed: Joseph Huddart, 2 Sept 1793.
pp. 1–4   with folding engraved plan.
[In]
Captain Huddart's Report upon the Improvement of the Port and Harbour of Boston . . . accompanied by Mr Rennie's Report thereon.
London: Luke Hansard & Sons . . . [1800]
fol   30 cm   10 pp.   Folding engraved plan.
BM, CE

687   *Swansea Harbour. Extracts from the Report of Joseph Huddart, Esq . . . shewing his Plan for the Improvement thereof.*
London: Printed by T. Burton . . . 1804.
Signed: Joseph Huddart. Highbury Terrace, 27 Sept 1794.
8vo   19 cm   19 pp.
BM, SML, CE

688   *Swansea Harbour. Second Report of Joseph Huddart, Esq . . . shewing his Plan for the farther Improvement thereof.*
London: Printed by T. Burton . . . 1804.
Signed: J. Huddart. Highbury Terrace, 11 May 1804.
8vo   19 cm   19 pp.   Folding hand-coloured engraved plan.
Plan for the Improvement of Swansea Harbour, 1804.
BM, SML, CE

689   *On the Improvement in Manufacturing of Cordage, whereby a more equal Distribution of the Strain upon the Yarn is acquired.*

London: Printed by J. Skirven.
Signed: J. Huddart. Highbury Terrace, Feb 1804.
4to 28 cm 16 pp.
CE

690 *Eau Brink New River, or Cut. Deed Poll, stating the Opinion (in the Nature of an Award) of Joseph Huddart, Esq. In Pursuance of the reference to him by Sir Thomas Hyde Page, and Robert Mylne, Esq.*
London: Printed by J. and E. Hodson . . . 1804.
Signed: Joseph Huddart, 10 Sept 1804.
4to 27 cm 11 pp. 2 folding engraved plans.
[1] *Plan of the intended New River or Cut from the River Ouze near Eau Brink, to rejoining the Old Channel of the Ouze near Kings Lynn.* November 1804.
[2] *Sections for the Intended New River or Cut, from the Ouze near Eau Brink to rejoin the River Ouze near Kings Lynn.*
CRO, RS, CE, Sutro

691 *The Report of J. Huddart, Hydrographer, accompanied by a Plan for the Improvement of Whitehaven Harbour.*
Signed: J. Huddart. Allonby, 27 Oct 1804.
p. 6 with litho plan.
Whitehaven Harbour with Capt.ⁿ *Huddart's additions 1804.*
[In]
Plans suggested at different periods for the Improvement of Whitehaven Harbour. See WHITEHAVEN HARBOUR, 1836.

See RENNIE and HUDDART. Reports on Howth Harbour, 1808 and 1809.

692 *Copy of a Letter from Captain Huddart, to the Right Honourable Nicholas Vansittart* [on Howth Harbour].
Signed: Joseph Huddart. Highbury Terrace, 5 Oct 1815.
pp. 41–45
[In]
Fourth Report from the Select Committee on the Roads from Holyhead to London.
House of Commons, 1817.

HUDSON, John [d. 1801]

See HUDSON and DYSON. Report on imbanking and draining in the parish of Timberland, 1784.

See HUDSON and PARKINSON. Report, with estimates, on draining low lands in Metheringham, Dunston, Nocton . . . and Heighington, 1788.

See JESSOP and HUDSON. Report on the Sleaford Navigation, 1791.

693 *A Plan exhibiting the Course of Kyme Eau and the two Branches of Sleaford River; from the Witham to Castle Causeway, above the Town of Sleaford, in the County of Lincoln, and the Works proposed to be executed thereon for making a Navigation from the River Witham to Castle Causeway.*
By John Hudson 27th Feb.ʸ 1792.
Engraved plan 28 × 55 cm (scale 2 inches to 1 mile).
BM, CE (Page), RGS, Sutro

See HUDSON and BONNER. Report on the Horncastle Navigation, 1792.

See HUDSON and MAXWELL. Report on the drainage of South Holland, 1792.

See HUDSON, GOLBORNE and MAXWELL. Reports on the outfall of the River Welland, 1792 and 1793.

694 *The Report of John Hudson, Engineer, on the Probable Effect the Proposed Cut, from Eau-Brink, to Lynn, will have on the Banks and Drainage of Bedford South Level.*
n.p.
Signed: John Hudson. Kenwick Thorpe near Louth, 3 Nov 1792.
fol 33 cm 2 pp. Docket title on p. (4).
BM, Northants R.O., Sutro

695 Reprint
[In] *Observations upon the Eau-Brink Cut. By the Earl of* HARDWICKE, 1793 [q.v.].

696 *The Report and Estimate of Mr John Hudson, on the Means and Expence of Embanking and Draining the Dales in Walcot, Timberland . . . Linwood and Blankney.*
[Lincoln]: Thornhill. Printer.
Signed: John Hudson. West Ashby, 6 Dec 1796.
8vo   24 cm   3 pp.   Docket title on p. (4).
LAO

HUDSON, John and BONNER, William

697 *Messrs Hudson and Bonner's Report, relative to the intended Navigation from Lincoln to Horncastle.*
Printed by Order of the General Commissioners for Drainage by the River Witham.
n.p.
Signed: John Hudson, William Bonner. Lincoln, 9 Feb 1792.
fol   33 cm   (3) pp.
Spalding Gentlemen's Society, Sutro

HUDSON, John and DYSON, John

698 *The Report of John Hudson, and John Dyson, Engineers, on a Survey taken by them, to ascertain the general Practicability of Imbanking and Draining the Fen Lands in . . . the Parish of Timberland, in the County of Lincoln.*
n.p.
Signed: John Hudson, John Dyson. Lincoln, 21 June 1784.
fol   31 cm   3 pp.
LCL

HUDSON, John, GOLBORNE, James and MAXWELL, George

699 *The Report of Messrs Hudson, Golborne, and Maxwell; on the Improvement of the Outfall of the River Welland, in the County of Lincoln.*
Spalding: Printed by J. Albin. 1792.
Signed: John Hudson, James Golborne, George Maxwell. Spalding, 15 Oct 1792.
fol   32 cm   6 pp.
LAO, Sutro

700 *The Further Report of Messrs Hudson, Golborne and Maxwell on the Improvement of the Outfall of the River Welland.*
Spalding: Printed by J. Albin, 1793.
Signed: John Hudson, James Golborne, George Maxwell.
fol   32 cm   5 pp.
LAO, Sutro

HUDSON, John and MAXWELL, George

701 *The Report of John Hudson, and George Maxwell; on the Drainage of South Holland, in the County of Lincoln. 1792.*
Spalding: Printed by J. Albin.
Signed: John Hudson, George Maxwell. Spalding, 18 Aug 1792.
fol   32 cm   7 pp.
LAO
   Stitched as issued.

HUDSON, John and PARKINSON, John

702 *The Report of John Hudson and John Parkinson, on a Survey taken by them pursuant to an Order of the*

Proprietors of Low Lands within the several Parishes, Townships, and Hamlets of Metheringham, Dunston, Nocton, . . . and Heighington, in the County of Lincoln. [and] *An Estimate of the Probable Expences.*
Lincoln: Brooke, Printer.
Signed: John Hudson, John Parkinson. 22 Nov 1788.
fol   33 cm   3 + 3 pp.
LCL

HUGHES, John D'Urban [1807–1874] M.Inst.C.E.

703   *A Practical Inquiry into the Laws of Excavation and Embankment upon Railways; being an attempt to develop the natural causes which affect the progress of such works . . . with an Appendix and Plates, illustrative of their application in Practice.*
By a Resident Assistant Engineer.
London: Saunders and Otley . . . 1840.
Dated: Bredon's Norton, 4 March 1840.
8vo   23 cm   xiv + 173 pp.   2 folding engraved plates (F. Mansell sc.) + folding litho drawing.
Sketch of the Bredon Cutting as it appeared [under construction] *in March 1840.*
W. Clerk, lith.
BM, ULG
> The author can be identified as John Hughes, resident engineer on the Birmingham & Gloucester Railway under Capt. W. S. Moorsom, to whom the work is dedicated.

HUGHES, Thomas

704   *The Practice of Making and Repairing Roads; of constructing footpaths, fences and drains; also, a method of comparing roads, with reference to the power of draught required; with Practical Observations, intended to simplify the mode of estimating earthwork in cuttings and embankments.*

By Thomas Hughes, Esq. Civil Engineer.
London: John Weale . . . 1838.
8vo   23 cm   iv + 108 pp.
BM, IC, BDL
> Issued in brown boards, printed paper label on upper cover.
Price 3/6ᵈ.

HUSTLER, John

705   *An Explanation of the Plan of the Canal from Leeds to Liverpool: exhibiting the extensive inland navigation it would open between . . . the Ports of Liverpool and Hull and all the Principal Towns in the counties of York, Lancaster, Lincoln, & Nottingham.*
Bradford: Printed by George Nicholson. 1788.
Signed: John Hustler.
8vo   20 cm   (i) + 16 pp.
BM

706   *A Plan of the Canal from Leeds to Liverpool which when compleated will open a safe, expeditious and cheap navigation . . . from the Ports of Hull and Liverpool.* [etc].
John Hustler jun. del.   W. Darton & Co. sc.
Engraved map   71 × 103 cm   (scale 1 inch to 3 miles).
BM, BDL
> The plan referred to in the *Explanation*.

See HUSTLER, BIRKBECK and PRIESTLEY. Report on Leeds & Liverpool Canal, 1789.

HUSTLER, John, BIRKBECK, William and PRIESTLEY, Joseph

707   *The Report of John Hustler, William Birkbeck and Jo. Priestley.*
Dated: Liverpool, 9 Oct 1789.
p. 3

[Printed with]
Robert WHITWORTH's Report, 9 Oct 1789
[q.v.].

[In]

Leeds & Liverpool Canal. At a General Assembly
of the Proprietors . . . held at Liverpool 9th and
10th October 1789 [etc].

n.p.

fol   32 cm   3 pp.

BM, Lancashire R.O.

Ordered to be printed 10 Oct 1789.

HUTTON, *Professor* Charles [1737–1823] F.R.S.L. & E., Hon. M.Soc.C.E.

708 *The Principles of Bridges: containing the Mathematical Demonstrations of the Properties of the Arches, the Thickness of Piers, the Force of Water against them, &c. Together with Practical Observations and Directions drawn from the whole.*

By Cha. Hutton, Mathematician.

Newcastle: Printed by J. Saint; and sold by J. Wilkie . . . London; and by Kincaird and Creech, Edinburgh. 1772.

8vo   21 cm   iv + 102 pp.   Woodcuts in text.

BM, RIBA, ULG, BDL

709   Second edition

Title as No. 708 but *The Second Edition, with Corrections and Additions.*

By Charles Hutton, F.R.S.   Professor of Mathematics in the Royal Military Academy.

London: Printed for the Author, by W. Glendinning . . . 1801.

8vo   21 cm   104 pp.   Woodcuts in text.

BM, RS, CE, RIBA, ULG, IC

710   *Answers [to the Questions respecting the Construction of a Cast Iron Bridge, of a Single Arch, 600 Feet in the Span, and 65 Feet Rise]*

By Dr Charles Hutton.   Royal Military Academy, Woolwich.

Signed: Cha.ˢ Hutton. Woolwich, 21 April 1801.

[In]

Report from the Select Committee upon Improvement of the Port of London, 1801. pp. 42–48.

711   Second printing

In *Reports from Committees of the House of Commons* Vol. 14 (1803) pp. 618–621.

712   *Tracts on Mathematical and Philosophical Subjects; comprising, among numerous important articles, the Theory of Bridges* [etc].

By Charles Hutton, LL.D. and F.R.S. &c.

London: Printed for F.C. and J. Rivington . . . 1812.

Signed: Cha. Hutton. London, July 1812.

3 vols   8vo   21 cm   **1**,   x + (2) + 485 pp.   Woodcuts in text. Portrait frontispiece.   **2**, (iii) + 384 pp.   6 folding engraved plates.   **3**, (iii) + 383 pp.   4 folding engraved plates. (Mutlow sc.)

BM, SML, RS, CE (Telford), BDL

INGLIS, James

713   *The Report of James Inglis, Engineer, concerning the Practicability and Expense of making a Road from Glasgow to Elvanfoot . . . with a Map of the country through which it passes.*

Addressed to the Honourable the Trustees on the Douglas-Mill and Lesmahagow Road.

Glasgow: Printed by D. Mackenzie. 1814.

Signed: Ja. Inglis. Glasgow, 15 Aug 1814.

4to   28 cm   15 pp.   Folding engraved plan, the lines of roads entered in colours.

Plan of different lines of Road from Glasgow to Elvanfoot . . . surveyed by Ja. Inglis 1814.

Engraved by R. Gray.

NLS, CE

JAMES, John [c. 1672–1746]

714 *A Short Review of the several Pamphlets, and Schemes, that have been offered to the Publick, in relation to the Building of a Bridge at Westminster. With Remarks on the different Calculations made for the Rise and Fall of Water, which the Peers of a Stone-Bridge may occasion.*

By John James, of Greenwich.

London: Printed by H. Woodfall . . . 1736. Price One Shilling.

Signed: John James. Greenwich, 30 July 1736.

8vo 20 cm 59 pp.

BM, CE, BDL (Gough)

JAMES, William [1771–1837]

715 *Report, or Essay, to illustrate the Advantages of Direct Inland Communication through Kent, Surrey, Sussex, and Hants, to connect the Metropolis with the ports of Shoreham, (Brighton) Rochester, (Chatham) and Portsmouth, by a line of Engine Rail-Road . . . [etc].*

London: Published (for the Author. . .) by J. and A. Arch . . . 1823.

8vo 23 cm 31 pp. Large folding litho map.

Plan of the proposed Line of Engine Rail Road to connect the Ports of London, Shoreham, Rochester and Portsmouth.

As designed by W. James, April 1823. Lithographic drawing & writing by R. Padley.

BM, SML, CE, ULG (Rastrick), Elton

JARDINE, James [1776–1858]
F.R.S.E., M.Soc.C.E., M.Inst.C.E.

716 *Report of James Jardine, Engineer in Edinburgh, respecting Rathillet Mill-Dam. Unto the . . . Lords of Council and Session.*

[Edinburgh, 1814].

Signed: James Jardine. Edinburgh, 22 June 1814.

4to 28 cm 11 pp.

NLS

717 *Report by Mr Jardine [respecting Craigie Mill-Dam].*

Signed: Ja. Jardine. Edinburgh, 8 Feb 1817.

Appendix pp. 17–33 with folding engraved plan.

Plan of Craigie Mill and Brewery.

1816. Kirkwood sc.

[In]

Unto the . . . Lords of Council and Session. The Petition of Sir Thomas Moncrieff.

[Edinburgh, 1818].

4to 14 + 23 pp. Folding engraved plan (as above).

NLS, CE (Telford)

718 *Plan of Callender Park situate in the County of Stirling . . . Shewing the Deviation through the Park from the Parliamentary Line of the Edinburgh & Glasgow Union Canal, proposed by the Canal Company November 1818.*

Drawn under the direction of James Jardine, Civil Engineer, by James Anderson Nov. 1818. Engr.d by R. Scott.

Engraved plan, hand coloured 35 × 54 cm (scale 1 inch to 600 feet).

NLS

See EDINBURGH, Water Supply. Plan of proposed aqueduct, 1818.

719 *Abstract of Mr Jardine's Report regarding the Levels of the proposed New Streets in Edinburgh.*

Dated: Edinburgh, 18 Oct 1824.

pp. 1–5.

[Printed with]

Report by William Burn and Thomas Hamilton (Architects) on the proposed New South and West Approaches, 18 Nov 1824.

[In]

Report relative to the proposed Approaches from the South and West to the Old Town of Edinburgh.

[Edinburgh, 1824].

4to 28 cm (i) + 10 + 7 pp. Folding engraved plan.

Plan of the Proposed Approaches from the South & West to the High Street of Edinburgh.

W. H. Lizars sc.
NLS, ULG (Rastrick)

720 *Plan of the Proposed Edinburgh & Dalkeith Railway and of the Branches to it from Cowpits, Fishberrow Harbour, and Edinburgh.*

Surveyed under the Direction of James Jardine Civil Engineer. 1825. W. H. Lizars sc.

Engraved plan   26 × 35 cm   (scale 2 inches to 1 mile).

NLS

721 *Plan and Section of the Edinburgh and Dalkeith Railway, of the proposed extension of it from Dalhousie Mains to Newton Grange, and of the proposed Branch Railway from Niddrie North Mains Bridge to Leith Harbour.*

Surveyed under the direction of James Jardine, C.E.   Nov 1828.   A. & S. Arrowsmith sc.

Engraved plan   25 × 40 cm   (scale 1¾ inches to 1 mile).

NLS

722 *Report to the Subscribers of the Survey of the Edinburgh and Glasgow Railway.*

By James Jardine, Civil Engineer.

Signed: Ja. Jardine. Edinburgh, December 1826.

pp. 11–14   with large folding engraved plan.

Reduced Plan & Section of the Proposed Railway between Edinburgh & Glasgow and of Branches from it to Leith, Broomielaw [etc].

Surveyed under the direction of James Jardine, Civil Engineer. 1826. Lizars sc.

[Printed with]

Supplementary Report and Estimate of Expence.

Signed: Ja. Jardine. Edinburgh, 16 Oct 1830.

pp. 14–21.

[In]

Report to the Subscribers for a Survey and Plan of a Railway from Edinburgh and Leith to Glasgow, by the Committees appointed for the purpose . . . 1830.

Edinburgh: Printed by John Stark.

4to   28 cm   (i) + 22 pp.   Folding engraved plan (as above).

NLS

723 *Report as to the proposed measure of Supplying the City of Perth and Suburbs with good Water.*

Signed: A. Anderson, J. Jardine.

pp. 1–4.

[Printed with]

Appendix to the report, by the Provost of Perth, 29 Nov 1828.

Perth: R. Morison, Printer.

8vo   20 cm   8 pp.

BM

Report by Dr Anderson and Mr Jardine laid before the Committee 20 Nov 1828.

724 *Plan of Dundee Harbour exhibiting the Proposed Improvements,*

Surveyed under the direction of James Jardine, Civil Engineer, 1829.

Engraved by J. Fenton.

Engraved plan   29 × 60 cm   (scale 1 inch to 400 feet).

Scottish Record Office

JESSOP, Josias [1781–1826] M.Soc.C.E.

725 *A Plan of a proposed Canal to connect the Rivers Arun and Wey, in the Counties of Sussex & Surrey.* 1813.

Engraved plan, the lines entered in colours   21 × 41 cm   (scale 1 inch to 1 mile).

BM

The estimate accompanying the MS. deposited plan of 1812 in HLRO (the original of this engraved plan) is signed: Josias Jessop, Petworth, 27 May 1812.

726 Deposited plan

Plan (and Section) of the Intended Railway from Mansfield in the County of Nottingham to Pinxton Basin, in the County of Derby.

By Wm. Chrishop, Surveyor. W. F. 1817.

Engraved plan, the line entered in colour   29 × 81 cm   (scale 4 inches to 1 mile).

HLRO

The accompanying estimate is signed: Josias Jessop. London, 21 May 1817.

727 [Report] *To the Committee of the Promoters of the intended Rail-way from the Cromford Canal to the Peak Forest Canal.*
[and]
Mr Jessop's second Report to the Committee of the proposed Rail-way from Cromford to the Peak Forest Canal at Whaley Bridge.
Both signed: Josias Jessop. Butterley, 1 Sept 1824 [and] 29 Nov 1824.
[In]
GRAY. *Observations on a General Iron Rail-way.* 5th edition. London, 1825 [q.v.] pp. 92–100 and 101–108.

728 *Report of Josias Jessop, Esq. Civil Engineer, drawn up by order of and addressed to the Committee of Enquiry into the most desirable mode of Improving the Communication between Newcastle and Carlisle.*
Newcastle: E. Walker, Printer.
Signed: Josias Jessop. Butterley Hall, 4 March 1825.
fol   38 cm   2 pp.
CE

JESSOP, William [1745–1814]
M.Soc.C.E.

729 *The Report of W. Jessop, Engineer, relative to a proposed Canal from Haddlesey Dam, to Brier Lane End, in the Township of Newlands, in Consequence of a Survey of the River Air, from Haddlesey downwards . . . made in the Months of September and October 1772.*
n.p.
Signed: W. Jessop, 25 Nov 1772.
fol   31 cm   6 pp.   Docket title on p. (8).
   Folding engraved plan.
Plan of the River Air from Haddlesey Bridge to Armin with the Track of a proposed Canal from Haddlesey to Brier-Lane.
   By W. Jessop, Engineer, 1772.
ULG, BDL (plan only)
   Includes a short report by SMEATON, dated 5 Dec 1772 [q.v.].

See SMEATON and JESSOP. Plan of the Rivers Air and Calder, 1773.

730 *Plan of the proposed Canal from the River Air at Haddlesey to the River Ouse at Selby.*
   W. Jessop, Engineer. [1774]
Engraved plan   21 × 56 cm   (scale 4 inches to 1 mile).
Leeds City Library, MCL
   Jessop gave evidence in Parliament on this scheme in February 1774.

731 *Report of William Jessop, Engineer, on the State and Means of improving and enlarging the Harbour of Workington; from a Survey thereof taken in January 1777.*
n.p.
Signed: W. Jessop. Pontefract, 24 Jan 1777.
fol   33 cm   4 pp.   Folding engraved plan.
Plan of the Harbour of Workington, with the Works proposed for the improvement of the same.
   By W. Jessop, Engineer. 1777. F. Chesham sc.
Carlisle R.O., RS (Smeaton, plan only).

732 *Report of William Jessop, Engineer, on a Survey of the River Trent, in the Months of August and September, 1782, relative to a Scheme for improving its Navigation.*
Nottingham: Burbage and Son, Printers.
Signed: W. Jessop. Fairburn, 3 Nov 1782.
fol   33 cm   14 pp.
PRO

733 *Plan of the River Trent from Gainsborough to Cavendish Bridge from a Survey taken in August and September 1782.*
   By William Jessop. J. Butterworth sc.
[with]
Profile of the River Trent from the Mouth of the Lincoln Navigation to Cavendish Bridge.
Engraved plan   31 × 115 cm   (scale 1 inch to 1 mile and 1 inch to 40 feet).
BM, Sutro

734 *Report of W. Jessop, Engineer, on the present navigable state of the river Severn, and on the means of improving the navigation; from a Survey taken in May 1784.*

Dated: Newark, 10 Aug 1784.
pp. 4–6.

[In]

T. R. Nash. *Supplement to the Collections for the History of Worcestershire.*
London: Printed for John White . . . 1799.
fol   42 cm   (i) + 104 pp.
BM, BDL

735   *The Report of William Jessop, Engineer, on an Enquiry into the Practicability of forming Reservoirs for Water . . . tending to produce a more regular Supply for the Works, in the Town and Environs of Sheffield, in dry Seasons.*
n.p.
Signed: W. Jessop. Newark, 11 Oct 1785.
8vo   20 cm   12 pp.
Sheffield City Library

See WHITWORTH and JESSOP. Plan of intended canal from Thrinkston Bridge to Leicester, 1785.

736   *River Severn. To The Noblemen and Gentlemen Owners of Land on the Banks of the Severn.*
n.p.
Signed: W. Jessop. Newark, 30 Jan 1786.
fol   33 cm   4 pp.
Staffordshire R.O.

737   *Report of William Jessop, Engineer, on the State of the Drainage, of the Low Grounds and Carrs . . . in Holderness, in the East Riding of the County of York; and on the means of compleating the same; and on the present state of the River Hull . . . From a Survey taken in the Month of June, 1786.*
Hull: Printed by J. Ferraby.
Signed: W. Jessop. Newark, 17 July 1786.
4to   22 cm   15 pp.   Docket title on p. (16).
HRO
   Stitched as issued.

738   *Report of W. Jessop, Engineer, on a Survey of the River Hull, from the North-Bridge to near Stone-Ferry, in the Month of October, 1787.*
Hull: Printed by G. Price.
Signed: W. Jessop. Newark, 11 Nov 1787.
fol   39 cm   3 pp.   Docket title on p. (4).

Hull University Library
   There is another report by Jessop on this subject, also in Hull University Library: 4 pp. in the same format, dated Newark 3 July 1788.

739   *Correspondence between the Right Hon. Thomas Pelham, and Mr Jessop, relative to the Improvement of the River Ouse and of the Drainage of Lewes Levels.*
n.p.
fol   33 cm   4 pp.   Folding engraved plan.
Sketch of the Piers at Newhaven with the proposed Extension.
   W. Jessop 1787.
East Sussex R.O.
   Includes two letters from Jessop, dated 23 Aug and 1 Sept 1787, and an Estimate for improving the Ouse from Lewes to Newhaven.

740   *Estimate of the Expence of making Navigable the River Ouse, from Lewes to Barcomb Mill-Pond.*
n.p.
Signed: W. Jessop. Lewes, 26 Oct 1788.
fol   32 cm   4 pp.
BM, ULG

741   *Report of Mr Jessop, Engineer, on a Design for a Canal, from Langley Bridge to Cromford, in the County of Derby, from a Survey taken in the Months of September 1787, and November 1788.*
[Nottingham]: G. Burbage, Printer.
Signed: W. Jessop. Newark, 13 Dec 1788.
fol   32 cm   3 pp.   Docket title on p. (4).
Derbyshire R.O., Sutro

742   *A Plan of the Intended Canal from Cromford to Langley-Bridge, in the Counties of Derby & Nottingham; with collateral Branches to the Parishes of Pinkston and Selston.*
   By E. G. Fletcher 1789.
Engraved plan   26 × 54 cm   (scale 2 inches to 1 mile).
BM, Sutro
   Surveyed under Jessop's direction by Edward G. Fletcher.

743   *Mr Jessop's First (and Second) Report.*
Signed: William Jessop. Oxford, 4 Aug 1789 [and] London, 7 Aug 1789.
pp. 3–19 [and] 20–23.

[Printed with]
Mr MYLNE's Report, 8 May 1791 [q.v.].
[In]
Reports of the Engineers appointed by the Commissioners of the River Thames and Isis to Survey the State of the said Navigation, from Lechlade to Days Lock.
[London]: Printed in 1791.
8vo   22 cm   60 pp.   Folding engraved plan.
BM, CE (Page), ULG, BDL

744   *Plan of the Intended Navigation from Stowmarket to Ipswich.*
By Isᶜ Lenny, Surveyor; & W. Jessop, Engineer. 1790.
Engraved plan, the lines entered in colour 28 × 159 cm   (scale 1 inch to 1½ miles).
BM, Ipswich & East Suffolk R.O.

See JESSOP and STAVELEY. Plan of intended navigation from Loughborough to Leicester and Melton Mowbray, and railways to Swannington, etc. 1790.

745   *The Report of W. Jessop, Engineer, on the Practicability and Expence of making a Navigable Communication between the River Witham and the Town of Horncastle.* [And] *on the means of making a compleat Navigable Communication between the Witham and the Fosdike at Lincoln.*
Horncastle: Printed by J. Weir. 1791.
Signed: W. Jessop. Newark, 30 June 1791.
8vo   20 cm   15 pp.
BM, LCL, Sutro
For a plan of the Horncastle Navigation, largely based on this report, *see* STICKNEY and DICKINSON, 1792.

See JESSOP and HUDSON. Report on the Sleaford Navigation, 1791.

746   *The Report of William Jessop, Engineer on the Practicability of Widening and Deepening the Communication between the River Witham and the Fossdike Canal, without Injury to the low Lands East of Lincoln.*
n.p.
Signed: W. Jessop. London, 29 Feb 1792.
fol   34 cm   3 pp.   Docket title on p. (4).
BDL, Sutro

747   *A Plan of a navigable Canal from Grantham in the County of Lincoln to the River Trent at Ratcliffe in the County of Nottingham.*
Taken in September 1791.
Engraved plan   39 × 59 cm   (scale ¾ inch to 1 mile).
BM, RGS, BDL
This and the following revised plan were both surveyed under Jessop's direction.

748   *A Plan of the intended Navigable Canal from the Town of Grantham in the County of Lincoln, to the River Trent near Nottingham Trent Bridge; and also of a collateral Branch . . . to the Town of Bingham.* 1792.
Engraved plan   39 × 73 cm   (scale ¾ inch to 1 mile).
BM, RGS

749   *Observations on the Use of Reservoirs for Flood Waters.*
Signed: W. Jessop, May 1792.
pp. 39–44.
[In]
General View of the Agriculture of the County of Stafford.
By Mr W. Pitt of Pendeford.
London: Printed by T. Wright. 1794.
4to   25 cm   168 pp.
BM, ULG
Also in the second edition (pp. 118–122). London, 1796. BM, SML

See JESSOP, GOTT and WRIGHT. Plan of a canal from the Calder to Barnby Bridge, 1792.

750   *Grand Junction Canal. Mr Jessop's Report.*
n.p.
Signed: W. Jessop. Northampton, 24 Oct 1792.
fol   33 cm   3 pp.   Docket title on p. (4).
BM

See BARNES. Plan and section of the Grand Junction Canal, 1792.

751   *Mr Jessop's Report on the several Lines of the intended Derby Canals.*
n.p.
Signed: W. Jessop. Newark, 3 Nov 1792.

fol   33 cm   3 pp.   Docket title on p. (4).
Nottingham University Library

See JESSOP and STAVELEY. Plan of intended
   navigation from Melton Mowbray to Oakham,
   1792.

752   *A Plan of Part of the Rivers Avon & Frome viz.
from Rownham Ferry, to Temple Back; and the Quay's
Mouth, to Traitor's Bridge. Also Sections of the same,
together with the Proposed Dam, Locks & Canals for the
Improvement of the Harbour of Bristol.*
   J. Jessop, Engineer. W. White, Surveyor. 1792.
Engraved by W. Faden . . . 1793.
Engraved plan   62 × 186 cm   (scale 1 inch to 6
   chains).
With   this   was   published   a   broadsheet:
   57 × 20 cm.
*Explanation of the Plan, Sea Front, and Section of the
Dam and Bridge to be built across the River at Redcliffe;
and Description of the Locks and Canal to be formed at
Rownham Meads.*
BM, Bristol Library, Bristol R.O. (plan only)
   Bristol Dock Co. records, Aug 1793: William
   Faden paid for engraving and printing 500
   copies of plate and Explanation.

See BRISTOL DOCKS. A Further Explanation
   of the Dam and Works at Rownham Mead,
   1793.

753   [Report] *To the Commissioners of the Witham
Navigation.*
n.p.
Signed: W. Jessop. Newark, 20 June 1793.
fol   32 cm   2 pp.
Scottish Record Office

754   *Copy of the Report, of William Jessop, Esquire,
relative to the making of a New Dock, at Kingston upon
Hull.*
n.p.
Signed: W. Jessop, Newark, 1 July 1793.
fol   36 cm   3 pp.
Hull City Archives

See WHITWORTH and JESSOP. Report on
   Trent navigation, 1793.

755   *Mr Jessop's Report, on his Survey of the Grand
Western Canal . . . from Topsham to Taunton.*
n.p.
Signed: W. Jessop. Bristol, November 1793.
fol   33 cm   10 pp.
Exeter City R.O.

See WHITWORTH and JESSOP. Plan of the
   Ashby de la Zouch Canal, 1794.

756   *Mr Jessop's Report* [on the Present State of
the Works].
Signed: W. Jessop. London, 24 May 1794.
                    [Printed with]
[Begins] *At a Meeting of the Grand Junction Canal
   Committee.* 27 May 1794.
n.p.
fol   33 cm   2 pp.
Northants R.O.

757   *Letter from William Jessop, Esq. to the President
of the Board of Agriculture, on the Subject of Inland
Navigations, and Public Roads.*
Signed: W. Jessop. Chatham Place, London, 19
   March 1795.
[First printed in *Communications to the Board of
   Agriculture* Vol. 1, part 3, pp. 176–182. London,
   1797].
Reprinted in *Georgical Essays* (edited by Alexander
   Hunter). Third edition. Vol. 6, pp. 74–88.
York. Printed by T. Wilson and R. Spence . . .
   1804.
8vo   21 cm   582 pp.
BM, ULG

758   *Report on the Proposed Line of Navigation be-
tween Newcastle and Maryport.*
   By William Jessop, Engineer. [with] *Abstracts of
   Estimates . . .* By Wm. Jessop and Wm. Chap-
   man, Engineers. 26 Oct 1795.
First edition. CE
Second edition. BM, CE, BDL
   For details *see* CHAPMAN, 1795, where these
   items are listed with his other reports on the
   subject.

759   *Estimates for Floating Dock on the Isle of Dogs
(and) in Rotherhithe.*

Signed: W. Jessop, 7 April 1796.
Appendix Dd (pp. 71–72) with folding hand-coloured engraved plate (R. Metcalf).
[In]
Report from the Committee on the Port of London.
House of Commons, 1796.

760 Second printing
In *Reports from Committees of the House of Commons*
Vol. 14 (1803) p. 385 and pl. 4(b).

761 *The Report of Mr Jessop, respecting the Drainage of the Low-Grounds lying on the Wolds, or West-Side of the River Hull, Frodingham Carrs, Lisset, &c.*
Hull: Printed by Thomas Lee and Co . . . 1796.
Signed: W. Jessop. Newark, 14 July 1796.
12mo   16 cm   10 pp.
CE, Beverley County Library.

See DANCE, JESSOP and WALKER. Plans of proposed docks in the Isle of Dogs, 1797 and 1798.

762 [Report] *To the Chairman of the General Committee of the Ellesmere Canal Company.*
Chester: Printed by J. Fletcher.
Signed: Wm Jessop. Shrewsbury, 24 Jan 1800.
fol   33 cm   3 pp.
Shrewsbury County Library

763 *Mr Jessop's Report respecting the practicability and expence of making a harbour for large vessels at Dunleary, to communicate with the new Docks by means of a Canal.*
Signed: W. Jessop. Dublin, 29 April 1800.
pp. 69–76.
[In]
Facts and Arguments respecting the Great Utility of an Extensive Plan of Inland Navigation in Ireland.
By a Friend to National Industry.
Dublin: Printed by William Porter . . . 1800.
8vo   23 cm   77 pp.
National Library of Ireland, CE, BDL, CUL

764 Deposited plan
Plan of part of the Township of Leven, and parts adjacent, describing the Course of the intended Canal, from the River Hull, to Leven Bridge.

As recommended by Mr Jessop, 1800.
Engraved plan, the line entered in colour
39 × 50 cm   (scale 5 inches to 1 mile).
HLRO, Sutro

See WALKER, Ralph. Plan of the West India Docks, 1800.

765 *Report on the Effect of Deepening and Embanking the River Thames.*
Signed: W. Jessop.
pp. 138–141   with folding engraved plate
Section of the River Thames opposite the Steel Yard, shewing its present width, depth & form; and the supposed Improvement by deepening & embanking the same.
W. Jessop. Basire sc.
[In]
Third Report from the Select Committee upon the Improvement of the Port of London.
House of Commons, 1800.
The plate (No. 20) is in a separate volume.

766 Second printing
In *Reports from Committees of the House of Commons*
Vol. 14 (1803) pp. 601–603 and pl. 49.

See JESSOP, RENNIE *et al.* Report on Deeping Fen, 1800.

767 [Report] *To the Chairman of the Committee for the Surry Iron Railway.*
Signed: W. Jessop, 10 Dec 1800.
pp. 8–10.
[In]
Copy of Minutes of the 24th July, 1800, for an Iron Railway from Wandsworth to Croydon with the Report of the Committee and Mr Jessop thereon.
[London]: Brooke, Printer.
fol   32 cm   10 pp.   Docket title on p. (12).
Guildford Muniment Room
Includes (pp. 2–4) Jessop's preliminary report dated 9 Dec 1799.

768 Deposited plan
Plan of the Intended Iron Railway from Pitlake Mead at Croydon to the River Thames at Wandsworth; also a Branch beginning at the South West Corner of Mitcham

Common to Hack Bridge in the Hamlet of Wallington in the . . . County of Surrey.
   Blake sc. [1801]
Engraved plan, the line entered in colour
   36 × 96 cm (scale 5 inches to 1 mile).
HLRO
   The first railway deposited plan. The accompanying estimate is signed W. Jessop, 31 Jan 1801.

See JESSOP and RENNIE. Report on Lancaster Canal, 1801.

769 [Report] *To the Owners of Land interested in the intended Inclosure of the Commons and Wastes in Thorne and Hatfield.*
Doncaster: Sheardown, Printer.
Signed: W. Jessop. Newark, 19 March 1801.
fol   33 cm   2 pp.
Nottingham University Library

770   *Mr Jessop's Answers. To the Select Committee on Bridge-House Lands.*
[on rebuilding London Bridge].
Signed: W. Jessop. London, 11 April 1801.
pp. 4–5.
                [In]
At a Sub-Committee . . . held at Guildhall, on Wednesday the 19th Day of December 1801.
Corporation of London, 1801.
fol   32 cm   17 pp.
CLRO

771   Second printing
In *An Abstract of Proceedings and Evidence relative to London Bridge.* Corporation of London, 1819. pp. 57–58. *See* LONDON BRIDGE.

772   [Report] *To the Chairman of the Committee of the Proprietors of the River Dun Navigation.*
n.p.
Signed: W. Jessop. Sheffield, 2 Aug 1801.
fol   33 cm   4 pp.
Sheffield City Library
   Also in this library there is a short preliminary report by Jessop dated 11 Feb 1801.

773   [Report] *To the Chairman of the Committee for the Extension of the Surrey Iron Railway.*

Signed: W. Jessop. London Coffee-House, 7 Oct 1802.
pp. 8–9.
                [In]
Minutes of the Proceedings for the Extension of the Surrey Iron Railway.
[London]: Printed by M. and S. Brooke.
fol   33 cm   10 pp.   Docket title on p. (12).
NLS (Rennie), Lambeth Minet Library
   Ordered to be printed 14 Oct 1802.

774   Deposited plan
A Plan of the Intended Iron Railway, from Pitlake Mead, in the Parish of Croydon, to the Town of Reigate; and a Collateral Branch, from Merstham, to Godstone, all in the County of Surrey.
   1802. Brook sc.
Engraved plan, the lines entered in colour
   34 × 73 cm (scale 2½ inches to 1 mile).
HLRO
   Surveyed under Jessop's direction by Samuel Jones and George Wildgoose. The accompanying estimate is signed W. Jessop.

775   *Design for Improving the Harbour of Bristol.*
   By William Jessop 1802.
Engraved by Cook & Johnson.
Engraved plan   22 × 45 cm (scale 1 inch to 6 chains).
                [In]
Explanation of the Plan proposed for the Improvement of the Harbour of Bristol. [May] 1802. *See* BRISTOL DOCKS.
NLS, Bristol Library, BDL
   Surveyed by William White.

776   New scheme
Design for Improving the Harbour of Bristol.
   Wm. White del. 1802. Cook & Johnson sc.
Engraved plan   30 × 69 cm (scale 1 inch to 10 chains).
                [In]
Explanation of the Plan for Improving the Harbour of Bristol, submitted . . . in the month of December 1802.
   See BRISTOL DOCKS.
Bristol R.O.
   Jessop's MS report on this scheme is dated 14 Aug 1802.

777 Revised plan

Design for Improving the Harbour of Bristol
  By Wm. Jessop Civil Engineer. W. White Surveyor. Engraved by W. Faden . . . 1803.
Engraved plan   45 × 108 cm   (scale 1 inch to 10 chains).
BM, Bristol R.O.
  Bristol Dock Co. Minute Book, 8 March 1803: Mr Faden to proceed with engraving the plan.

778 Second state: Deposited plan
As No. 777 with street plan of Bristol.
BM, HLRO, Bristol Library
  The HLRO copy is endorsed 8 July 1803. From Bristol Dock Company Letter Book it is known that, in all, 800 copies of the first and second state plans were printed. An accompanying *Explanation of the Design* was issued separately. *See* BRISTOL DOCKS.

779 Third state
Title, date, etc as No. 777 but showing several alterations, as built, and naming the two iron bridges over the cut.
BM, Bristol Library
  Issued *c.* 1806.

780 *Report on the intended Navigation from the Eastern to the Western Sea, by Inverness and Fort William; including Estimate of the Expence of completing the same.*
  By William Jessop, Engineer.
Signed: W. Jessop. Newark, 30 Jan 1804.
                [In]
[First] *Report of the Commissioners for the Caledonian Canal.*
  House of Commons 1804. pp. 17–20.

See JESSOP and TELFORD. Reports on Caledonian Canal, 1804.

781 *Report of Mr Jessop, on the Improvement of the Navigation of the River Wye.*
Hereford: E. G. Wright, Printer.
Signed: W. Jessop. Bristol, 21 Aug 1805.
fol   32 cm   2 pp.
Hereford City Library

See JESSOP and TELFORD. Report (and estimate) on Aberdeen harbour, 1805 (and 1810).

See JESSOP and TELFORD. Annual reports on the Caledonian Canal, 1805–1812.

782 *Report of Mr William Jessop, to the Right honourable the Lords Commissioners of the Admiralty* [on Plymouth Breakwater].
Signed: William Jessop. Plymouth, 23 Aug 1806.
                [In]
Papers relating to Plymouth Sound.
  House of Commons, 1812. pp. 12–13.

783 *A Report on the Present State of the Piers of Sunderland Harbour, and the Means Recommended for their Improvement.*
  By W. Jessop.
Sunderland: Printed by James Graham, 1808.
Signed: W. Jessop. Butterley, 19 Dec 1807.
8vo   20 cm   (i) + 22 + (2) pp.   Folding engraved plan.
Plan of the Piers of Sunderland Haven, with the proposed Extensions.
  W. Jessop, Eng.ʳ Lambert sc.
BM, Sunderland Central Library
  Sunderland Harbour Minute Book, 31 Dec 1807: Mr Jessop having made his report, 300 copies to be printed.

784 *Observations on Roads and Carriages.*
  By William Jessop, Esq. of Butterley, in Derbyshire: in a Letter addressed to Sir John Sinclair.
Signed: W. Jessop, 26 Jan 1809.
                [In]
Third Report from the Committee on Broad Wheels and the Preservation of the Turnpike Roads and Highways.
  House of Commons, 1809. pp. 160–161.

See TELFORD. Report on proposed railway from Glasgow to Berwick; Mr Jessop's opinion thereon, 1810.

JESSOP, William, GOTT, John and WRIGHT, Elias

785 *A Plan of the Intended Navigable Canal from the River Calder near Heath Hall in the Parish of Warm-*

field, to Barnby Bridge, in the Parish of Silkstone, in the West Riding of the County of York.

Messrs Jessop, Gott & Wright, Engineers. J. Teal, Surveyor, 1792. Engr'd by C. Livesey.
Engraved plan   line entered in colour
  56 × 73 cm   (scale 3 inches to 1 mile).
BM, BDL

JESSOP, William and HUDSON, John

786 *Report on the Means and Expence, of making Navigable the Kyme Eau and River Slea, from the Witham to Castle Causeway, above the Town of Sleaford.*
n.p.
Signed: W. Jessop, J. Hudson. Sleaford, 25 Nov 1791.
fol   31 cm   3 pp.
Spalding Gentlemen's Society

JESSOP, William and RENNIE, John

787 *The Report of William Jessop and John Rennie.*
p. (2)
[In]
At a General Meeting of the Company of Proprietors of the Lancaster Canal Navigation, held at the Town Hall in Lancaster, on Tuesday the 7th Day of July 1801.
Lancaster: Jackson, Printer.
fol   37 cm   (2) pp.
ULG
  The report was submitted in May 1801.

JESSOP, William, RENNIE, John, et al.

788 *The Report of Messrs Jessop, Rennie, Maxwell, and Hare, on the Drainage of Deeping Inclosed Fens, and the Commons adjoining thereto.*
Spalding: Printed by J. L. and T. Albin.
Signed: W. Jessop, John Rennie, Geo. Maxwell, Edwd. Hare. Spalding, 11 Aug 1800.
4to   28 cm   7 + (1) pp.   Docket title on p. (8).
LAO, Sutro
  Note in a letter to Lord Brownlow: 700 copies of the report are to be printed, September 1800.

JESSOP, William and STAVELEY, Christopher

789 *A Plan of the intended Navigation from Loughborough to Leicester and the proposed Water Level and Railways from Loughborough to the Coal Mines at Swannington, Cole-Orton and Thringston Commons; and the Lime Works at Cloud Hill, Barrow Hill and Gracedieu. Also of the intended Navigation from the Leicester Navigation to Melton Mowbray.*
Willm Jessop Eng^r Survey'd in 1790 by Chris^r Staveley Junr
Engraved plan, the railways entered in colour
  63 × 120 cm   (scale 2 inches to 1 mile).
BM, Science Museum, Sutro

790 *A Plan (and Profile) of the intended Navigation from Melton Mowbray in the County of Leicester to Oakham in the County of Rutland.*
Wm Jessop Engineer. Survey'd in 1792 by C. Staveley Junr Engraved by W. Faden 1793.
Engraved plan, the lines entered in colour
  49 × 81 cm   (scale 2 inches to 1 mile).
BM, Sutro

JESSOP, William and TELFORD, Thomas

791 *Report and Estimate respecting Locks.*
Signed: W. Jessop, Tho.⁵ Telford. London, 14 June 1804.
[In]
Second Report of the Commissioners for the Caledonian Canal.
House of Commons, 1805. pp. 28–29.

792 *Report of Messrs Jessop and Telford, on the Instructions of the Commissioners,* 1804.
Signed: W. Jessop, Tho.⁵ Telford. 29 Nov 1804.
[In]
Second Report of the Commissioners for the Caledonian Canal.
House of Commons, 1805. pp. 31–34.

793 *Report of Messrs Telford and Jessop on the Harbour of Aberdeen.*
Signed: W. Jessop, Thos. Telford. Edinburgh, 14 Oct 1805.
[and]
Estimate by Messrs Telford and Jessop, of the Expense of executing their plan of Improvements for the Harbour of Aberdeen.
Signed: W. Jessop, Thos. Telford. London, 7 April 1810.
[In]
Reports . . . upon the Harbour of Aberdeen, 1834, pp. 47–50 and p. 56. *See* ABERDEEN HARBOUR.

794 First annual report on Caledonian Canal.
Report of Messrs Jessop and Telford.
Signed: W. Jessop, Tho.⁵ Telford. November 1805.
[In]
Third Report of the Commissioners for the Caledonian Canal.
House of Commons, 1806. pp. 26–29.
Subsequent annual reports, all titled as above and signed by Jessop and Telford, dated 1806–1812, are printed in the Fourth to the Tenth Commissioners' Reports 1807–1813.

JESSOP, William *junior* [*c.* 1783–1852] M.Soc.C.E.

795 *Mr Jessop's Report on the Railways. To the Subscribers to the Midland Counties Railway.*
Signed: William Jessop. Butterley Hall, 25 Nov 1832.
pp. (13)–27.
[Printed with]
Joseph GLYNN's Report on Loco-Motive Engines. 5 Nov 1832 [q.v.].
[In]
Midland Counties Railway. Prospectus of the Projected Railway from Pinxton to Leicester, with Reports of the Estimated Cost of that Undertaking, and on the Application of Loco-Motive Steam Power to Railways Generally.
Alfreton: Printed by George Coates. 1833.
8vo   22 cm   42 pp.   2 folding litho plans (E. Wild).
[1] *A Plan shewing the Line of the intended Midland Counties Railway from Leicester to Pinxton, and to Derby and Nottingham, also shewing other proposed Railways to communicate therewith.*
[2] *Comparative Sections of the Lines of Railway from Swannington, the proposed Line of the Midland Counties Railway from Leicester to Pinxton, and the Line from Manchester to Liverpool.*
CE

JOHNSON, Edward and DICKINSON, J.

796 *A Plan of the Level of Ancholme, in the County of Lincoln.*
As Surveyed in the Years 1767 & 1768 by E. Johnson & J. Dickinson. Reduced to the present Size by E. Johnson & I. Dalton in 1791. Engraved by J. Cary.
Engraved plan, partly hand coloured
35 × 166 cm   (scale 3 inches to 1 mile).
BM

JONES, *General Sir* Harry David [1791–1866] R.E., Assoc.Inst.C.E.

See BURGOYNE *et al.* Reports of the Commissioners for the Improvement of the River Shannon, 1837 and 1839.

JONES, Samuel

See JESSOP. Plan of railway from Croydon to Reigate, surveyed by Samuel Jones & George Wildgoose, 1802.

797   *A Plan of the Intended Harbour and Wet Dock in St. Nicholas Bay in the Isle of Thanet, and the Canal leading from thence to Canterbury.*
   Surveyed by Samuel Jones Aug.! 1810. W.F. 1811.
Engraved plan   11 × 30 cm   (scale 1 inch to 1 mile).
BM

KELK, George

See GRUNDY. Plan of low grounds in Laneham, Rampton, etc. Surveyed by George Kell. 1769. 'Kell' is an error for Kelk.

798   *A Plan of the Intended Navigable Canal from Stainforth Cut to the River Trent at Althorp.*
   By G. Kelk 1772.
Engraved by T. Jefferys.
Engraved plan   23 × 60 cm   (scale 2 inches to 1 mile).
BM

799   *A Map of Low Grounds in Stockwith, Misterton, Gringley, Everton and Scaftworth, in the County of Nottingham.*

By George Kelk. Engraved by Jefferys and Faden, 1773.
Engraved plan   25 × 80 cm   (scale 1¾ inches to 1 mile).
BM
   Incorporating a scheme for draining the above-mentioned low grounds possibly by John Grundy, 1768.

KENNET & AVON CANAL

800   *An Authentic Description of the Kennet & Avon Canal.*
London: Printed for J. M. Richardson . . . 1811.
8vo   23 cm   (iii) + 30 pp.
BM, CE, Elton

KINDERLEY, Nathaniel [1673–1742]

801   *The Present State of the Navigation of the Towns of Lyn, Wisbeech, Spalding, and Boston. The Rivers that pass through those Places, and the Countries that border thereupon, truly, faithfully and impartially Represented. And Humbly Proposed to the Consideration of the Inhabitants of those Places and Countries, and to the Corporation of Adventurers for draining the vast Level of the Fens . . . With a Way laid down how to Remedy all Inconveniences and Defects which they now labour under.*
Bury St. Edmunds: Printed for the Author by T. Baily and W. Thompson.
8vo   21 cm   20 pp.
LCL, ULG, BDL (Gough)
   Known to have been written and published *c.* 1721 by Kinderley, whose christian name is often given erroneously as Charles.

802   *Mr Humphrey Smith's Scheme for the Draining of the South and Middle Levels of the Fens Examined and Compared.*

By Nathaniel Kinderley.
London: Printed in the Year 1730.
8vo   20 cm   16 pp.
CE, BDL (Gough)

See MACKAY. Plan of River Dee estuary, 1732, with 'Mr Kinderley's New Channel'.

KINDERLEY, Nathaniel *junior*

803   *The Ancient and present State of the Navigation of the Towns of Lynn, Wisbeach, Spalding, and Boston; of the Rivers that pass through those Places, and the Countries that border thereupon, truly, faithfully, and impartially represented. And Humbly Proposed to the Consideration of the Inhabitants of those Places and Countries. With a Way laid down how to remedy all the Inconveniences and Defects which they now labour under.*
The Second Edition enlarged.
London: Printed for J. Noon . . . and T. Hollingworth, at Lyn. 1751.
Signed: Nathaniel Kinderley. Cork Street, 19 April 1751.
8vo   21 cm   xviii + (ii) + 108 pp.   Folding engraved map.
A Map of the great Level of the Fens, together with the Rivers that pass thro' the said Level, into the Bay call'd Metaris Aestuarium.
   W. H. Toms sc. 1751.
BM, SML, CE, LCL, ULG, BDL (Gough), CUL
   A much enlarged version of Kinderley's 1721 pamphlet, edited by his son. The list of subscribers implies a sale of at least 400 copies.

KING, James [d. 1744]

804   *One of the five Ribs of the Center on which the Middle Arch of Westminster Bridge, was turned, extending 76 Feet.*

Designed & Executed by James King.   James King junior del.   P. Fourdrinier sc.
Published April 14th 1743.
Price 1s.
Engraved elevation   23 × 41 cm   (scale 1 inch to 8 feet).
RS (Smeaton)

KING'S LYNN

805   *An Answer, Paragraph by Paragraph, to A Report of the present State of the Great Level of the Fens . . . and of the Port of Lynn . . . And also to a Scheme propos'd by Mr Charles Bridgman, for the effectual Draining those Fens, and reinstating that Harbour or Port. Drawn from Authentick Testimonies . . . and from a Survey made in the Years 1723, 1724.*
n.p.   [Printed December 1724].
fol   36 cm   14 + 14 + [15–17] pp.   Hand-coloured engraved plan.
A Mapp of the Fenn-Rivers and of the New Cutts proposed in the following Scheme.
   S. Parker sc.
CE, BDL (Gough), CUL
   Includes (pp. 1–14 recto) a reprint of Bridgeman's report and (pp. 15–17) a reprint of Armstrong's report. The *Answers* (pp. 1–14 verso), submitted by a committee set up by Lynn Corporation, may have been compiled by Badeslade who, in his *Ancient and Present State of the Navigation of King's Lyn*, 1725 (p. 107) gives details of the proceedings and publications involved.

KITCHIN, Thomas and PHILLIPS, John

806   *A New Map of England & Wales. Drawn from several Surveys &c on the New Projection; Corrected from Astronomical Observations . . .*

By Thos Kitchin Geog.^r *The Canals inserted to 1792* by J. Phillips, Surveyor.

London: Published by Laurie & Whittle . . . 12th May 1794.

Engraved map in 4 sheets each 64 × 55 cm (scale 1 inch to 9 miles), county boundaries and canals entered in colours. Inset, a table of lengths and falls of thirty-six canals.

BM, RS

LABELYE, Charles [1705–?1762]

807 *A Description of the Carriages made use of by Ralph Allen, Esq; to carry Stone from his Quarries, situated on the top of a Hill, to the Water-side of the River Avon, near the City of Bath.*

By Charles de Labelye.

pp. 274–276 with 3 engraved plates.

[In]

DESAGULIERS. *A Course of Experimental Philosophy*. Vol. 1. London, 1734 [q.v.].

808 [*An Account of Mr Stubner's pretended Demonstration of the new Opinion relating to the Forces of Bodies in Motion.*]

Signed: Charles de Labelye. 15 April 1735.

pp. 77–90.

[In]

DESAGULIERS. *A Course of Experimental Philosophy*. Vol. 2. London, 1744 [q.v.].

809 *The Result of a Calculation made to estimate the Fall of Water at the intended Bridge at Westminster, and some Conjectures of the Effects it will probably have on the Navigation of the River Thames.*

By Charles de Labelye.

pp. 18–21.

[In]

HAWKSMOOR. *A Short Historical Account of London Bridge; with a Proposition for a Stone-Bridge at Westminster*. London, 1736 [q.v.].

810 *A Mapp of the Downes much more correct than any hitherto published. Shewing the true Shape & Situation of the Coast between the North & South Forelands,* & *of all the adjacent Sands together with the Soundings at Low Water . . .*

By Charles Labelye Engineer, late Teacher of the Mathematicks in the Royal Navy.

Published March 1737.

Engraved plan   66 × 90 cm   (scale 2 inches to 1 mile).

BM

The survey made in November and December 1736.

811 *A Plan of the Intended Harbour between Sandwich Town & Sandown Castle. This Harbour is laid down of such Dimensions as will comodiously admitt 150 of the largest Merchant Ships . . . and is propos'd to be dug, so as to have 12 Feet Water, at a low Ebb of a Spring Tide,* [etc.].

Car.^s Labelye del.   J. Harris sc.

Engraved plan   40 × 64 cm   (scale 1 inch to 350 feet for plan, 1 inch to 85 feet for section).

BM, BDL (Gough)

Corresponds exactly to the 'Intended Harbour' outlined in the *Mapp of the Downes*, 1737.

812 *A Short Account of the Methods made Use of in Laying the Foundations of the Piers of Westminster Bridge. With an Answer to the chief Objections that have been made thereto . . . To which are annex'd, the Plans, Elevations and Sections belonging to a Design of a Stone-Bridge, adapted to the Stone Piers which are to support Westminster-Bridge.*

By Charles Labelye, Engineer.

London: Printed by A. Parker, for the Author. 1739.

Signed: Charles Labelye. Westminster-Bridge Foot, 24 April 1739.

8vo   21 cm   (iii) + vi + 82 pp.

BM, CE, RIBA, ULG, BDL (Gough), CUL

The plate of plans and elevation was published separately: *see* the following entry.

813 *A Design of a Stone Bridge, adapted to the Stone Piers which are to support Westminster Bridge.*

By Charles Labelye Engineer.   P. Fourdrinier sc.   Published 12 May 1739.

Engraved elevation and part plans   30 × 125 cm   (scale 1 inch to 25 feet).

BM, RS (Smeaton)

814 *The Present State of Westminster Bridge. Containing a Description of the said Bridge . . . With a True Account of the Time already Employed in the Building, and of the Works which are now done.*

In a Letter to a Friend.

London: Printed for J. Millman . . . 1743. Price Six Pence.

Dated: Westminster, 8 Dec 1742.

8vo   21 cm   30 pp.

BM, NLS, CE, RIBA, ULG, CUL

Attributed to Labelye.

See LABELYE *et al*. Report [to House of Commons] on the harbour proposed to be made from Sandwich to the Downs, 1745.

815 *The Result of a View of the Great Level of the Fens, taken at the Desire of His Grace the Duke of Bedford . . . and the Gentlemen of the Corporation of the Fens, in July 1745.*

By Charles Labelye, Engineer.

London: Printed by George Woodfall . . . 1745.

Signed: Charles Labelye. Crown Court, Westminster, 8 Aug 1745.

4to   22 cm   vi + (i) + 74 pp.

BM, CE, CRO, BDL (Gough), CUL

816 *The Result of a Particular View of the North Level of the Fens, Taken in August, 1745.*

London: Printed by C. and J. Ackers . . . 1748.

Signed: Charles Labelye. Crown Court, Westminster, 9 Sept 1745.

8vo   21 cm   32 pp.

BDL (Gough), Sutro

817 *The Result of a View and Survey of Yarmouth Haven, Taken in the Year 1747.*

By Charles Labelye, Engineer.

Norwich: Printed by W. Chase. 1775.

Signed: Charles Labelye. Crown Court, King-street, Westminster, 18 Jan 1747/8.

8vo   21 cm   64 pp.

CE, Norfolk R.O., BDL

818 *An Abstract of Mr Charles Labelye's Report, relating to the Improvement of the River Wear, and Port of Sunderland, made in July 1748.*

Newcastle: Printed for Isaac Thompson and Co. by John Gooding. 1748.

4to   24 cm   8 pp.

BM, Sunderland Central Library, CE

Sunderland Harbour Accounts, 23 March 1748/9: Mr Isaac Thompson for coppying of Mr Labelye's plan & abstracting of his report & printing 100 copies.

819 *A Description of Westminster Bridge. To which are added, an Account of the Methods made use of in laying the Foundations of its Piers. And in Answer to the chief Objections, that have been made thereto. With an Appendix, containing Several Particulars relating to the said Bridge, or to the History of the Building thereof. As also its Geometrical Plans, and the Elevation of one of the Fronts, as it is finished.*

By Charles Labelye.

London: Printed by W. Strahan, for the Author, 1751.

Signed: Charles Labelye. Westminster, 30 April 1751.

8vo   21 cm   (iii) + iv + 119 pp.

BM, NLS, CE (Telford), RIBA, ULG, IC, BDL (Gough), CUL

The plate of plans and elevation was published separately: *see* the following entry.

820 *The Geometrical Elevation of the North Front of Westminster Bridge.* [with] *The Plan of the Foundations.* [and] *The Plan of the Superstructure.*

Carolus Labelye Designavit.   P. Fourdrinier sc.

Published 20 May 1751.

Engraved elevation and plans   30 × 126 cm (scale 1 inch to 25 feet).

BM, RS (Smeaton), Soane Museum

LABELYE, Charles *et al.*

821 *Report, and Estimate relating to the Harbour proposed to be made from Sandwich to the Downs.*

[House of Commons, 1745].

Signed: W. Whorwood, John Redman, John Major, Thos. Slade, Ch. Labelye, R. Charles. 24 Jan 1744/5.

fol   33 cm   11 pp.

LANDALE, Charles [d. 1836] M.Inst.C.E.

822 *Report submitted to the Subscribers for Defraying the Expense of a Survey to ascertain whether it is practicable and expedient to construct a Railway between the Valley of Strathmore and Dundee.*

By Charles Landale, Civil Engineer.
pp. 1–10.
Signed: Chas. Landale. Dundee, 30 Sept 1825.
[with]
pp. 1–5 (second part). Observations . . . by Matthias DUNN . . . 1 Oct 1825 [q.v.].
Dundee: Printed by David Hill . . . 1825.
4to 30 cm 10 + 5 + (1) pp. Folding litho plan, hand coloured.
Reduced Plan and Section of the Proposed Railway from Dundee to Newtyle.
Surveyed by C. Landale, Civil Engineer. Wm. Corsar, land Surveyor, 1825.
Robertson & Ballatine's Lithog.
CE (Telford), ULG

LANDMANN, *Col.* George [1779–1854] R.E., M.Soc.C.E., M.Inst.C.E.

823 *Plan of the Proposed Line of the London and Greenwich Rail Road* as Projected by Geo Landmann Esq.ʳ Engineer to the Company. Reduced from the one prepared to be laid before Parliament. [with] *Elevation of the Arches.*
Litho plan 19 × 31 cm (scale 3 inches to 1 mile).
[In]
London and Greenwich Rail-Road Company. [Prospectus].
London: At the Company's Office, Sept. 1832.
fol 33 cm (3) pp.
BM

824 *Plan of the Line of the London and Greenwich Rail Way* as Projected by Geo. Landmann Esq.ʳ Engineer to the Company. Reduced from the one laid before Parliament. [with] *Elevation of the Arches.*
Rowsell & Son litho.
Litho plan 19 × 30 cm (scale 3 inches to 1 mile).
[In]
London and Greenwich Railway Company.
Incorporated 1833.
London: At the Company's Office, 17 May 1833.
fol 35 cm (4) pp. Docket title on p. (4).
BM

825 *Plan of the Railway from Preston to the River Wyre.*
G. Landmann, Esq., Civil Engineer, 1834.
R. Cartwright lithog.
Litho plan 40 × 24 cm (scale 1½ inches to 1 mile).
[In]
Prospectus of the Preston and Wyre Railway, and Improvement of the Harbour of Wyre.
London: Printed by James & Luke G. Hansard.
fol 42 × 26 cm 4 pp.
BM

LANGLEY, Batty [1696–1751]

826 *A Design for the Bridge at New Palace yard, Westminster . . . composed of Nine Arches, independent of each other . . . With Observations on the several Designs published to this Time . . .* [etc.].
By B. Langley.
London: Printed for the Author, and J. Milan . . . 1736.
8vo 20 cm (i) + 30 pp. Folding engraved plate. (I. Carnritham sc.)
BM, CE, RIBA, ULG

827 *A Reply to Mr John James's Review of the several Pamphlets and Schemes, that have been offer'd to the Publick, for the Building of a Bridge at Westminster . . .* [etc.].

By B. Langley.
London: Printed for the Author, and Sold by J.
Millan . . . 1737.
8vo 20 cm (ii) + 54 pp. Folding engraved
plate.
BM, CE, RIBA, ULG, BDL (Gough)

LARCOM, *Sir* Thomas [1801–1879] R.E., F.R.S.

See DRUMMOND *et al.* Map of railways in England and Ireland, 1837.

LAURIE, John

See WHITWORTH. Plan of the Great Canal from Forth to Clyde by Robert Whitworth and John Laurie, 1785.

LAXTON, William

828 *Map of the Hull, Lincoln & Nottingham Railway.*
  W. Laxton, Engineer. 1836.
T.P. Fay, lithog.
Litho map 42 × 28 cm (scale 1 inch to 15 miles).
[In]
Hull, Lincoln, & Nottingham Railway, via Newark.
[Prospectus]
[London]: Printed by C. & W. Reynell.
fol 45 cm 4 pp.
BM

See PRICE and LAXTON. Map of proposed railway between Gloucester and Hereford, 1836.

LEACH, Edmund

829 *A Treatise of Universal Inland Navigations, and the use of all sorts of mines . . . To which is added a Supplement.*
  By Edmund Leach, Surveyor.
London: Printed for Alex. Hamilton . . . 1791.
8vo 23 cm (iii) + vi + (2) + 201 pp. 12 tables. 5 engraved plates.
BM, CE, ULG
  On half-title: Price Five Shillings. Stitched in blue covers.

LEACH, Stephen [*c.* 1777–1842]

830 *Map of the River Thames Navigation as far as the same extends westward Viz, from London Bridge to Lechlade, showing the several Poundlocks thereon, and the Canals, Roads, principal Towns etc, adjacent thereto. 1813.*
  Stephen Leach del. Gale & Butler sc.
Engraved map 19 × 53 cm (scale 1 inch to $3\frac{1}{2}$ miles).
RGS, BDL

LEAFORD, John

831 *Some Observations made of the Frequent Drowned Condition of the South Level of the Fenns, and of the Works made for Draining the same: with a Scheme for Relieving that Level.*
London: Printed in the Year of our Lord 1740.
Signed: John Leaford.
8vo 21 cm 24 pp.
CE, BDL (Gough)

LEATHER, George [1787–1870] M.Inst.C.E.

832 *A Plan and Section of the line of an intended canal from the River Derwent at or near East Cottingwith to Pocklington in the East Riding of the County of York.*

Drawn from an actual survey by Geo. Leather Jun.ʳ Bradford 1814. Butterworth, Livesey sc.
Engraved plan   30 × 58 cm   (scale 2 inches to 1 mile).
BM

833 *Report of Mr. Leather, Civil Engineer, on the Projected Line of Railroad between Newcastle and Carlisle.*

Signed: George Leather. April 1829.
pp. 226–229.
[In]
Copy of the Evidence, taken before a Committee of the House of Commons, on the Newcastle & Carlisle Railway Bill ... To which is added the Report of Mr. Leather.
Newcastle: Printed by Wm. Boag ... 1829.
8vo   22 cm   (i) + 229 pp.
ULG

834 *Level of Hatfield Chase. The First & Second Reports of Mr. Leather on the Better Drainage of the Level, by Means of Steam Power.*

Doncaster: Sold by C. and J. White ... 1830.
Signed: Geo Leather. Doncaster, 14 Oct 1830 and 6 Nov 1830.
8vo   22 cm   20 pp.
Nottingham University Library

835 *Level of Hatfield Chase. The Third Report of Mr. Leather [etc].*

Doncaster: Sold by C. and J. White ... 1831.
Signed: Geo Leather. Leeds, 3 Jan 1831.
8vo   22 cm   12 pp.
Nottingham University Library

836 *[Report] to the Proprietors of Fen Lands in the Parishes of Nocton, Potterhanworth, and Branston, in the County of Lincoln.*

Lincoln: J. W. Drury, Printer.

Signed: George Leather. Leeds, 5 Dec 1831.
fol   40 cm   3 pp.
LCL

LECOUNT, *Lieut.* Peter [1794–1852] R.N.

837 *An Examination of Professor Barlow's Reports on Iron Rails.*

By Lieut. Peter Lecount, R.N., F.R.A.S.
London: Simpkin, Marshall & Co ... [1836].
Dated: Constantine Cottage, Wellington Road, Birmingham, Feb 1836.
8vo   22 cm   192 pp.   2 folding litho plates.
BM, SML, ULG (Rastrick), Elton

838 *Remarks on the Cheapest Distance for Railway Blocks [etc].*

By Lieut. Peter Lecount, R.N., F.R.A.S.
London: Simpkin, Marshall & Co ... [1836].
Dated: Constantine Cottage, Wellington Road, Birmingham, Aug 1836.
8vo   21 cm   93 pp.
ULG (Rastrick)

839 *The History of the Railway connecting London and Birmingham: containing its progress from the commencement. To which are added a Popular Description of the Locomotive Engine: and a Sketch of the Geological Features of the Line.*

By Lieut. Peter Lecount ... one of the Engineers on the Line, who has been engaged on the Railway from its commencement.
London: Published by Simpkin, Marshall & Co. and Charles Tilt ... [1839].
8vo   22 cm   (iv) + 118 + (2) pp.
BM, ULG (Rastrick), Elton
See also ROSCOE, 1839.

840 *A Practical Treatise on Railways, explaining their Construction and Management ... being the article 'Railways' in the Seventh Edition of the Encyclopaedia Britannica, with Additional Details.*

By Lieut. Peter Lecount . . . of the London and
Birmingham Railway.
Edinburgh: Adam & Charles Black . . . 1839.
Dated: 10 June 1839.
8vo   21 cm   (vii) + 422 pp. + errata leaf.   10
folding plates, numerous woodcuts in text.
BM, SML, CE, ULG (Rastrick), Elton
Issued in publisher's blind-stamped cloth.

LESLIE, James [1801–1889] F.R.S.E., M.Inst.C.E.

841   *Report to the Subscribers for Procuring a Survey of
the South Coast of Fife, with a view to fixing on the most
eligible site for a Low-Water Pier.*
By James Leslie, Civil Engineer.
Cupar: Geo. S. Tullis, Printer . . . 1839.
Signed: James Leslie. Dundee, Dec 1838.
fol   34 cm   (i) + 19 pp.   Folding litho plan,
hand coloured.
*Plan and Chart of the Coast of Fife from Burntisland to
Dysart, shewing the proposed sites for a Low Water Pier.*
By James Leslie, Civil Engineer, 1838. Nichol
lithog.
NLS

LOCKE, Joseph [1805–1860] F.R.S., M.Soc.C.E., M.Inst.C.E.

See STEPHENSON and LOCKE. Observations
on comparative merits of locomotive and fixed
engines, 1830.

See STEPHENSON, George. Map of proposed
railway from Liverpool to Chorlton in
Cheshire, [1831].

842   [Report] *To Henry Booth, Esq. Treasurer of the
Liverpool and Manchester Railway.*
Signed: Joseph Locke. Liverpool, 17 Jan 1835.
pp. (21)–32.

[Printed with]
Report by Charles VIGNOLES, 21 Jan 1835
[q.v.].
[In]
*Two Reports, addressed to the Liverpool & Manchester
Railway Company, on the Projected North Line of Rail-
way from Liverpool to the Manchester, Bolton, and Bury
Canal, near Manchester, exhibiting the extent of its cut-
tings and embankings, with estimates of the cost of com-
pleting the said Railway.*
By Charles Vignoles, Esq. and Joseph Locke,
Esq. Civil Engineers.
Liverpool: Printed by Wales and Baines . . . 1835.
8vo   22 cm   32 pp.
SML, CE, ULG (Rastrick)

843   *London and Glasgow Railway, through Lancashire. [Report] To the Directors of the Grand Junction
Railway Company.*
n.p.   [1836]
Signed: Joseph Locke. Liverpool, 27 Jan 1836.
fol   32 cm   3 pp.
BM

844   *Map of Grand Junction Railway and Adjacent
Country.*
Published by Order of the Directors. Joseph
Locke, Engineer.
Published July 1st 1836 . . . by Charles F. Chef-
fins.
Litho   map,   hand-coloured   137 × 95 cm
(scale 1 inch to 2 miles).
Includes: *Section Shewing the Inclinations of the Grand
Junction Railway* and *Section Shewing the Inclina-
tions of the Liverpool and Manchester Railway.*
BM, SML, CE, BDL

845   [Report] *To the Directors of the London and
Brighton Railway Company.*
Signed: Joseph Locke. London, 8 Dec 1837.
pp. (2)–(3).
[Printed with]
Report by John U. RASTRICK, 7 Dec 1837
[q.v.].
[In]
*The London and Brighton Railway . . . Reports of I. U.
Rastrick and Joseph Locke, Esqs., Engineers.*
London: W. Lewis and Son, Printers.

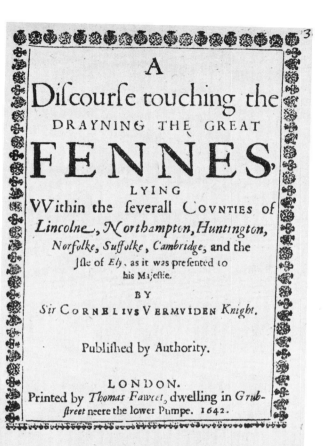

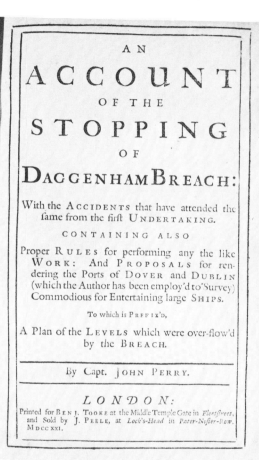

Report by Vermuyden on draining the Fens, 1642.

2. Perry's Account of Stopping Dagenham Breach, 1721.

Colonel *ARMSTRONG's*
REPORT,
WITH
PROPOSALS
FOR
Draining the FENNS
AND
Amending the HARBOUR of *LYNN*, 1724.

SIR,

THE Harbour of *Lynn* being almost choaked up with Sand, and in great Danger of being totally lost, I have according to your Desire viewed the same, in order to find out the true Cause of such visible Decay of so useful and beneficial a Harbour in that part of the Kingdom.

AND being then at the Request of the Right Honourable the Earl of *Lincoln*, &c. viewing the Fens in those Parts, to find out the true Cause of the Difficulty in draining a great Part of those Fens, very much incommoded with Water, I did apply myself very attentively to discover that Cause also.

IN my Perambulation to these Purposes with several Gentlemen of that Country, I have discovered the true Cause of both those great Evils, *viz.* That for a considerable Time past, there hath not been a sufficient Indraught for the Tides to flow up the *Ouse*, as they formerly did, for many Miles together, into that River and the other Rivers that fall into it below *Ely*. By means of which upon their Reflux, it being always half Ebb at *Lynn* before it has done flowing at *Salters Lode*, these Tides constantly returning by a narrow Channel, continually scower'd out the Sands in the said River, and preserved a Channel, I am sufficiently informed, ten Feet deeper than it is at this Time.

A

THEN

3. First page, with caption title, of Armstrong's report on King's Lynn harbour, 1724.

A

SCHEME

FOR THE

Reftoring and making perfect

THE

NAVIGATION

OF THE

River *WITHAM*

FROM

BOSTON *to* LINCOLN, and alfo for
draining the Low-Lands and Fenns
contiguous thereto.

By JOHN GRUNDY, *Sen.* and *Jun.*
of *Spalding*, ENGINEERS.

Printed in the Year MDCC XLIV.

4. Report by John Grundy senior and junior on the
River Witham, 1744.

Thos. Telford

A

DESCRIPTION

OF

Weftminfter Bridge.

To which are added,
An Account of the Methods made ufe of
in laying the Foundations of its Piers.
AND
An Anfwer to the chief Objections,
that have been made thereto.
WITH
An APPENDIX,
CONTAINING
Several Particulars, relating to the faid BRIDGE,
or to the Hiftory of the Building thereof.
AS ALSO
Its Geometrical Plans, and the Elevation of
one of the Fronts, as it is finifhed,
Correctly engraven on two large COPPER-PLATES.

*Quod optanti divum promittere nemo
Auderet, volvenda Dies en attulit ultro.* VIRG.

By *CHARLES LABELYE.*

LONDON:
Printed by W. STRAHAN, for the AUTHOR.
MDCCLI.

5. Labelye's Description of Westminster Bridge, 1751.

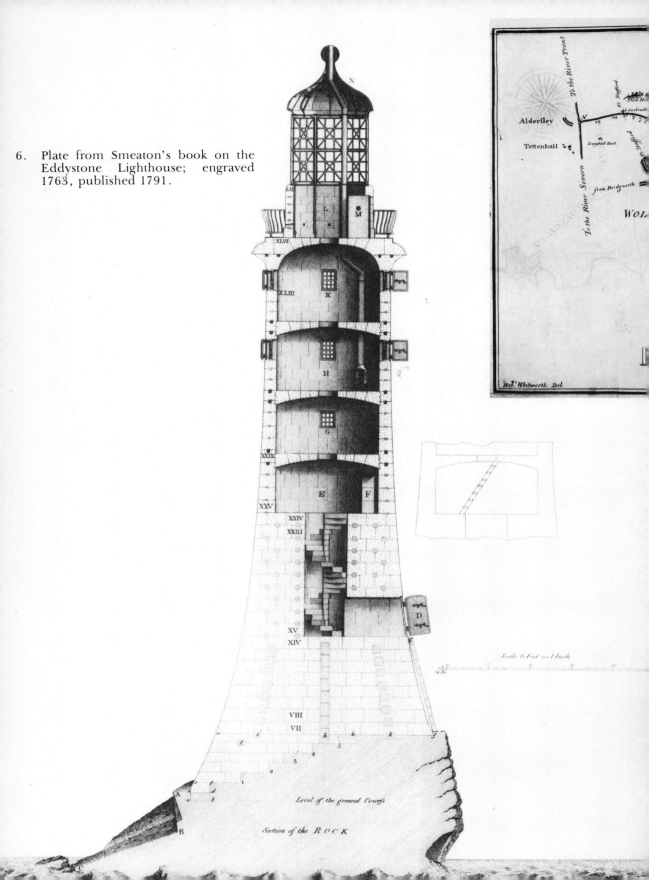

6. Plate from Smeaton's book on the Eddystone Lighthouse; engraved 1763, published 1791.

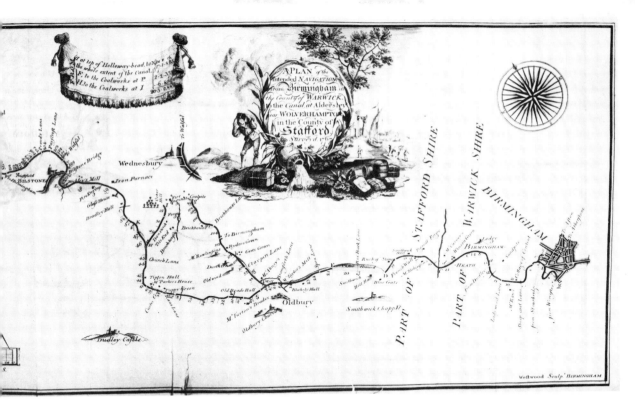

Plan of the Birmingham Canal, 1767. James Brindley engineer, drawn by Robert Whitworth.

A

REVIEW

OF

SEVERAL MATTERS

RELATIVE TO

THE FORTH AND CLYDE NAVIGATION,

as now settled by ACT of PARLIAMENT:

WITH

SOME OBSERVATIONS on the REPORTS

OF

Meſſ. BRINDLEY, YEOMAN, and GOLBURNE.

By JOHN SMEATON, CIVIL ENGINEER, and F. R. S.

[Published by order of a General Meeting of the Company of Proprietors
of the FORTH and CLYDE Navigation (1ſt November 1768.)
for the uſe of the Proprietors.]

Printed by R. FLEMING and A. NEILL.
M,DCC,LXVIII.

8. Smeaton's Review of
Several Matters relative
to the Forth and Clyde
Canal, 1768.

3

REPORTS

of the

ENGINEERS appointed by the
Commissioners of the

NAVIGATION

of the Rivers

THAMES AND ISIS

To Survey the State of the said Navigation,
from Lechlade to Days Lock.

Illustrated by a Plan

To which are added some ORDERS of
the Commissioners on the said Reports

and

AN ACCOUNT of the TOLLS Payable
on the said NAVIGATION from
Cricklade to Stanes

Printed in 1791

An Estimate added at the end

9.   Reports by Jessop and Mylne on the Thames Navigation, 1791.

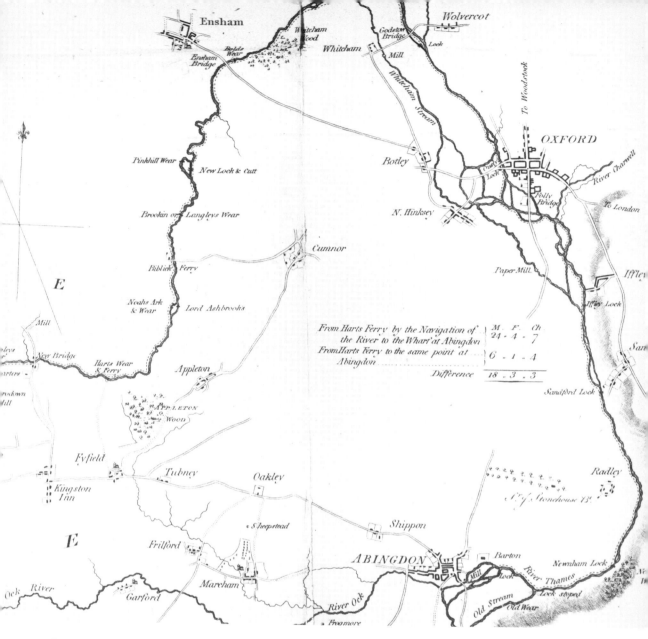

Part of a plan of the Thames from Lechlade to Abingdon, annexed to the 1791 report.

THE REPORT OF

OHN HUDSON, ENGINEER.

ON THE PROBABLE EFFECT

THE PROPOSED CUT FROM

Eau-Brink, to Lynn,

ILL HAVE ON THE BANKS AND DRAINAGE OF

BEDFORD SOUTH LEVEL.

11.   Docket title of Hudson's report on the Eau Brink Cut, 1792.

REPORT

OF

WILLIAM CHAPMAN, ENGINEER,

On the Means of

DRAINING THE LOW GROUNDS

IN THE VALES OF THE

DERWENT AND HERTFORD,

IN THE

NORTH AND EAST RIDINGS

OF THE

COUNTY OF YORK.

NEWCASTLE:

PRINTED BY E. WALKER, PILGRIM-STREET.

1800.

12. Report by Chapman on draining the low grounds adjacent to the River Derwent in Yorkshire, 1800.

fol   32 cm   (4) pp.   Docket title on p. (4).
CE

Issued from the London and Brighton Railway
Office, London . . . 8 Dec 1837.

846   Reprint
[In] *The London and Brighton Railway Company . . .*
Report of the Directors, with copies of Reports of I. U.
Rastrick and Joseph Locke, Esqs., Engineers.
London: W. Lewis and Son, Printers.
fol   32 cm   (4) pp.   Docket title on p. (4).
CE

Issued 19 Dec 1837.

847   [Report] *To J. H. Johnstone, Esq. M.P., and*
the Provisional Committee for promoting the Railway
from Carlisle, through the Vale of Evan, to Edinburgh
and Glasgow.
Signed: Joseph Locke, Liverpool, 21 Nov 1837.
pp. (3)–8.
[Printed with]
Report of the Committee appointed to obtain a
Survey of the Country from Carlisle to Edin-
burgh and Glasgow, by Beattock, with a view
to ascertaining the practicability of a Railway
by that route . . . January 1818.
[and]
Table shewing the Gradients in the proposed
Line of Railway from Johnstone Bridge to
Edinburgh, according to a Survey made under
the direction of Joseph Locke, Esq. C.E. By
Messrs D. M'Callum and J. F. Dundas. 1837.
Signed: D. M'Callum. Dumfries, 7 March 1838.
[In]
Reports on the Proposed Line of Railway from Carlisle to
Glasgow and Edinburgh, by Annandale.
n.p.   [1838].
8vo   24 cm   11 + (1) pp.
BM, NLS, Elton

Issued in yellow printed wrappers.

LOGAN, David [d. 1839] M.Inst.C.E.

848   *Report on the improvements of the River Clyde*
and Harbour of Glasgow, ordered by the Parliamentary
Trustees, with Tables of Soundings and Borings.
By David Logan, Civil Engineer.
Glasgow: Printed by Bell & Bain . . .1835.
Signed: David Logan. Glasgow 5 Oct 1835.
4to   27 cm   22 pp,   2 folding hand-coloured
litho plans (Allan & Ferguson Lith.).
[1] *Longitudinal Section of the River Clyde, extending*
from Glasgow to Port-Glasgow.
[2] *Plan of the River Clyde . . . showing the Proposed*
Improvements 1835.
NLS, CE

849   *Copy of Mr Logans Letters* [on] *a Design for*
finishing the Head of the West Pier.
Signed: David Logan. Glasgow, 11 July [and] 27
Aug 1836.
p. (38) and litho plan.
[In]
Plans . . . for the Improvement of Whitehaven Harbour,
1836. *See* WHITEHAVEN HARBOUR.

LONDON BRIDGE

850   *An Abstract of Proceedings and Evidence, relative*
to London Bridge: taken from the Reports of the Select
Committee of the House of Commons, made in the Years
1799, 1800, and 1801; and from the Journals of the Court
of Common Council, and the Committee for Letting the
Bridge House Estates.
Corporation of London, 1819.
fol   32 cm   iii + 140 pp.
CLRO, CE

Though dated 1819 on the title-page, the date
of issue must have been 1821 as the report of
March 1821 by MOUNTAGUE, RENNIE *et*
al. [q.v.] is included with continuous pagination.

851 *A Professional Survey of the Old and New London Bridges, and their Approaches, including Historical Memorials of both structures; with Remarks on the Probable Effects . . . on the Navigation of the Thames.*
London: M. Salmon . . . 1831.
8vo   22 cm   (iii) + 46 pp.   Portrait frontispiece (John Rennie) + 13 wood-engraved plates and many illustrations in text.
BM, CE, ULG
   The plates are mostly made from drawings by Christopher Davy.

LONDON–EDINBURGH ROAD

852 *Comparison between the Two Proposed New Lines of Road between London and Edinburgh; the one by Jedburgh, and the other by Wooler.*
Edinburgh: Printed for William Blackwood. 1822.
8vo   21 cm   32 pp.
NLS, ULG

LONGBOTHOM, John [d. 1801] M.Soc.C.E.

853 *A Plan of the Intended Navigable Canal from Liverpool to Leeds.*
   J. Longbottom Eng.ʳ del.   Forrest sc. [?1768]
Engraved plan   45 × 98 cm   (scale 1 inch to 2 miles).
BM, BDL

854 *A Plan of the Intended Navigable Canal, between Liverpool and Leeds.*
   J. Longbottom Eng.ʳ del.   J. Butterworth sc. [1770].
Engraved plan   46 × 94 cm   (scale 1 inch to 2 miles). With tables of lengths and falls.
BM, RS (Smeaton)
   Smeaton's copy is endorsed 'Plan of Canal from Leeds to Liverpool 1770'.

855 *A Plan of the Intended Navigable Canal from Leeds to Selby proposed to communicate in the Township of Holbeck, with the Canal now making between Leeds and Liverpool.*
   J. Longbothom Eng.ʳ del.   J. Butterworth sc. [1772].
Engraved plan   36 × 58 cm   (scale 1 inch to 1 mile).
BM

856   Another edition
Title as No. 855 with *a View of the Rivers Aire & Calder*.
   J. Butterworth sc. [1772].
Engraved plan   34 × 72 cm   (scale 1 inch to 1 mile).
BM, RS (Smeaton), MCL
   Smeaton's copy is endorsed 'Selby Canal 1772'.

LONGRIDGE, Michael [1785–1858]

857 *Specification of John Birkinshaw's Patent, for an improvement in the construction of Malleable Iron Rails, to be used in Rail Roads; with Remarks on the Comparative Merits of Cast Metal and Malleable Iron Rail-Ways.*
Newcastle: Printed by Edward Walker . . . 1821.
8vo   22 cm   10 + (1) pp.   Folding engraved plate, showing rail sections and horse-drawn train of waggons on the 'patent rails'.
BM

858   Second edition
As No. 857 but 1822.
8vo   22 cm   14 + (1) pp.   Folding engraved plate.
SML, ULG, Elton

859   Third edition
As No. 857 but 1824.
8vo   23 cm   14 + 8 + (1) pp. Folding engraved plate, as in No. 857 but with a Locomotive Engine and tender in place of the horse.
SML (lacking plate), ULG

860 *Remarks on the Comparative Merits of Cast Metal and Malleable Iron Rail-Ways; and an Account of the Stockton and Darlington Rail-Way, &c.*
Newcastle: Printed by Edward Walker . . . 1827.
8vo   23 cm   22 + 39 + 26 pp.   3 folding litho plates, including views of *Bedlington Iron Works* showing the railway (drawn by W. A. Thompson 1827, printed by C. Hullmandel) and the *Stockton and Darlington Rail Road*.
SML, Elton
The first section is virtually identical to No. 859.

861   Second edition
Remarks on the Comparative Merits . . . and an Account of the Stockton and Darlington Rail-Way, and the Liverpool and Manchester Rail-Way, &c.
Newcastle: Printed by Charles Henry Cook . . . 1832.
8vo   23 cm   26 + 39 + 38 + (1) pp.   2 plates, *Bedlington Ironworks* and *A Locomotive Engine similar to the Planet* (J. Kindar del., W. Miller sc.).
SML, ULG, Elton

862   Third edition
Title as No. 861 but 'By Michael Longridge'.
Newcastle: Printed . . . by J. Blackwell & Co. 1838.
Dated: Bedlington Iron Works, Jan 1836.
8vo   23 cm   32 + 39 + 38 + (1) pp.   2 litho plates.
SML, Elton

McADAM, *Sir* James [1786–1852]

863   *Report on the Surrey and Sussex Turnpike Trust.*
Signed: James McAdam. Office of Roads, Charing Cross, 25 Nov 1838.
pp. 220–228.
[In]
Second Report from the Select Committee of the House of Lords . . . [on] Turnpike Returns, 1833.

McADAM, John Loudon [1756–1836]

864   *Observations on the Highways of the Kingdom.*
By John Loudon Macadam, Esq.
Dated: Bristol, 13 May 1811.
[In]
Report from Committee on the Highways and Turnpike Roads in England and Wales. House of Commons, 1811. pp. 27–32.

865   *Observations on the Highways of the Kingdom, In a Report to the House of Commons.*
By John Loudon M'Adam, Esq.
Worcester: Printed by T. Holl and Son . . . [1816].
8vo   22 cm   23 pp.
NLS, ULG
Reprint of No. 864 with letter from McAdam dated: Office of Roads, Bristol, 20 July 1816.

866   *Memorial on the Subject of Turnpike Roads.*
By John Loudon M'Adam.
n.p.   [1817].
4to   28 cm   12 pp.
ULG

867   Second issue
As No. 866 with some revisions.
n.p.   [1818].
4to   28 cm   9 pp.
CE (Telford)

868   *A Practical Essay on the Scientific Repair and Preservation of Public Roads.*
Presented to the Board of Agriculture, by John Loudon M'Adam.
London: Printed by B. McMillan . . . 1819.
8vo   21 cm   18 pp.
BM, SML

869   *Remarks on the Present System of Road Making; with observations, deduced from practice and experience, with a View to . . . the Introduction of Improvement in the Method of Making, Repairing, and Preserving Roads, and Defending the Road Funds from Misapplication.*
By John Loudon M'Adam, Esq. General Surveyor of the Roads in the Bristol District.

Bristol: Printed and Published by J. M. Gutch . . .
1816.
8vo   22 cm   32 pp.
BM, CE, ULG, BDL

870   Second edition
Title as No. 869 but *Second Edition, carefully revised,
with considerable Additions and an Appendix.*
Bristol: Printed and Published by J. M. Gutch . . .
1819.
8vo   22 cm   47 pp.
BM, ULG

871   Third edition
*Remarks on the present System of Road Making; with
Observations, deduced from practice and experience, with
a view to . . . the introduction of improvement in the
Method of Making, Repairing, and Preserving Roads,
and defending the road funds from misapplication.*
Third Edition, carefully revised, with considerable additions, and an Appendix.
By John Loudon M'Adam, Esq. General Surveyor of the Roads in the Bristol District.
London: Printed for Longman, Hurst [etc] . . .
1820.
8vo   22 cm   196 pp.
BM, CE, ISE, ULG, IC
Includes reprint of Report of Select Committee
on the Highways of the Kingdom, 1819, with
verbatim evidence.

872   Fourth edition
As No. 871 but minor misprints corrected and
Fourth Edition . . . 1821.
BM, NLS, ULG

873   Fifth edition
As No. 872 but *Fifth Edition . . .* 1822, and
author's name now spelt McAdam (as in all
subsequent editions).
BM, SML, BDL

874   Sixth edition
As No. 872 but *Sixth Edition . . .* 1822.
BM, CE, ULG

875   Seventh edition
*Remarks on the Present System of Road Making; with
Observations, deduced from practice and experience . . .*
Seventh Edition, carefully revised, with an
Appendix, and Report from the Select Com-
mittee of the House of Commons, June 1823,
with Extracts from the Evidence.
By John Loudon McAdam, Esq. General
Surveyor of the Roads in the Bristol District.
London: Printed for Longman, Hurst [etc] . . .
1823.
8vo   22 cm   (iii) + vii + viii + (i) + [5]–236 pp.
BM, NLS, BDL

Issued in grey boards, printed paper label on
spine: McADAM ON ROADS.

876   Eighth edition
As No. 875 but *Eighth Edition . . .* 1824 and a dif-
ferent sequence of prelims, without half-title.
8vo   22 cm   (iii) + vii + viii + [5]–236 pp.
BM, NLS, ULG
Price 7s. 6d. in boards.

877   Ninth edition
As No. 875 but *Ninth Edition . . .* 1827.
BM, NLS

878   *Observations on the Management of Trusts for the
Care of Turnpike Roads, as regards the Repair of the
Road, the Expenditure of the Revenue, and the Appoint-
ment and Quality of Executive Officers . . . Illustrated by
Examples from a Practical Experience of Nine Years.*
By John Loudon McAdam.
London: Printed for Longman, Hurst [etc] . . .
1825.
8vo   22 cm   (iii) + iv + 148 pp.
BM, NLS, SML, ULG, BDL
Issued in grey boards, printed label on spine:
McADAM ON ROADS (Second Part). Price
6s.

879   *Report by Mr M'Adam to the Subscribers to the
Expense of the Survey of the Carter Fell.*
Signed: Jno. Loudon M'Adam. Office of Roads,
Bristol, 5 Jan 1829. pp. 3–8. With folding
engraved plan.

Plan of the Present and Proposed Road between Edinburgh and Newcastle by Jedburgh.
    Surveyed under the direction of John L. Macadam Esq. by Grainger & Miller, Edin. & Nic. Weatherly, Belford. 1828. Kirkwood sc.
        [In]
Statement by the Committee appointed upon the Line of Road between Caitha, in the County of Roxburgh, by Jedburgh, and the Carter Fell, to Newcastle.
Jedburgh: Printed by Walter Easton. 1829.
4to   28 cm   19 + (1) pp.   Folding engraved plan (as above).
NLS, ULG (Rastrick)
    Issued in brown printed wrappers.

MACKAY, John

880   *An Abridged Plan of the River Dee & Hyle Lake.*
    Surveyed in the Year 1732 by John Mackay, Math.
Engraved plan   45 × 55 cm   (scale 1 inch to $1\frac{1}{4}$ miles).
BM, BDL (Gough)
    With soundings in the Dee estuary. Shows 'Mr Kinderley's New Channel'.

MACKELL, Robert [d. 1779]

See MACKELL and WATT. Report on proposed canal from the Forth to the Carron, 1767.

881   *Plan of the Great Canal from Forth to Clyde with the new intended line of Alteration.*
    Alex^n Baillie sc. [1770].
Engraved plan, the line of canal entered in colour   35 × 110 cm   (scale $1\frac{1}{4}$ inches to 1 mile).
NLS, RS (Smeaton), MCL
    Mackell's revision of the line west of Kirkintilloch, surveyed by John Laurie and agreed by

Smeaton, who prepared the estimates. Forth & Clyde Canal Minute Book, 5 Nov 1770: Mackell submits approved plan. *Ibid*, 23 Nov 1770: Mr Baillie will engrave the plan.

MACKELL, Robert and WATT, James

882   *An Account of the Navigable Canal, proposed to be cut from the River Clyde to the River Carron, as surveyed by Robert Mackell and James Watt.*
London: Printed in the Year 1767.
4to   27 cm   18 pp.   Folding engraved plan.
A Sketch of the Intended Canal from the River Clyde to the River Carron.
BM, NLS, RS (Smeaton, plan only)

MACKENZIE, Murdoch [1712–1797] F.R.S., M.Soc.C.E.

883   *A Treatise of Maritim Surveying, in Two Parts: with a Prefatory Essay on Draughts and Surveys.*
    By Murdoch Mackenzie, Senior; Late Maritim Surveyor in his Majesty's Service.
London: Printed for Edward and Charles Dilly, 1774.
4to   24 cm   xxiii + (1) + 119 + (1) pp.   4 folding engraved plates (J. Lodge sc.).
BM

884   Second edition
A Treatise on Marine Surveying.
    By Murdoch Mackenzie, Sen . . . Corrected and republished, with a Supplement, by James Horsburgh, F.R.S. Hydrographer to the Hon. East India Company.
London: Printed for the Editor . . . 1819.
8vo   21 cm   xvi + 183 pp.   8 engraved plates.
BM, BDL

885 *An Account of the River Stour, in Kent, with Observations on Messrs Dunthorne and Yeoman's Proposal for Draining the Levels along that River.*
By Murdock M'Kenzie, Sen.
Canterbury: Printed by Simmons and Kirkby.
Signed: Murdock Mackenzie Sen. Hampstead in Middlesex, 27 July 1775.
4to   25 cm   11 pp.   Folding engraved plan.
A Map of the River Stour and Sandwich Haven from Fordwich Bridge to the Sea.
BM

McKERLIE, *Rear-Adm.* John [1774–1848] R.N.

See TELFORD and McKERLIE. Report on communications between the North of England and Ireland, 1808.

MACLAREN, Charles [1782–1866] F.R.S.E.

886 *Railways compared with Canals & Common Roads, and their Uses and Advantages explained; being the substance of a series of papers published in the Scotsman, and now republished with additions and corrections.*
Edinburgh: Published by Archibald Constable & Co . . . 1825.
Signed: C. M.   Edinburgh, February 1825.
8vo   18 cm   66 pp.
BM, NLS, SML, CE, ULG

McLAREN, James

See GRIEVE and McLAREN. Report on railway from Edinburgh to Dalkeith, 1824.

MACNEILL, *Sir* John [1793–1880] F.R.S., M.R.I.A., M.Soc.C.E., M.Inst.C.E.

887 *Report of Mr Macneill* [on measurements of the force of traction on the road from London to Shrewsbury].
Signed: John Macneill. Stowe, May 1830.
[In]
Seventh Report of the Holyhead Road Commissioners.
House of Commons, 1830. pp. 21–45.

See TELFORD. Map and sections of the present and proposed roads between Chirk and Ketley, surveyed by Macneill, 1830.

888 *Road Indicator, or Instrument for ascertaining the Comparative Merits of Roads, and the State of Repair in which they are kept.*
Invented by John Macneill, Mem.Inst.Civ.Eng. . . . and Resident Engineer to the Parliamentary Commissioners of the Holyhead and Liverpool Roads.
London: Roake and Varty . . . 1833.
4to   27 cm   6 + (6) pp.   Docket title on p. (6)
CE (Page)
Reprint of a pamphlet entitled *Pirameter, or Instrument for ascertaining the Power required to draw Carriages over Roads*, which is included as pp. (1)–(5).

889 *Canal Navigation. On the Resistance of Water to the passage of Boats upon Canals, and other Bodies of Water*, being the Results of Experiments, made by John Macneill, M.R.I.A., Mem.Inst.Civ.Eng.
London: Roake and Varty . . . 1833.
Signed: John Macneill. Strand, May 1833.
4to 28 cm (vii) + 55 pp. Litho frontispiece + 3 folding engraved plates (drawn by G. Turnbull, engraved by E. Turrell) + 4 other plates.
NLS, SML, CE (Page), ULG

890 *Tables for Calculating the Cubic Quantity of Earth Works in the Cuttings and Embankments of Canals, Railways, and Turnpike Roads.*
By John Macneill, Civil Engineer.

London: Roake and Varty . . . 1833.

Dated: London, 1 July 1833.

8vo   22 cm   xxxvii + 254 pp.   4 engraved
plates (E. Turrell sc.)

BM

A second edition, considerably enlarged, was
published in 1846.

891 *Report of Mr John Macneill* [on the road from
London to Shrewsbury].

Signed: John Macneill. Highgate, 9 April 1835.
[In]

Twelfth Report of the Holyhead Road Commissioners.
House of Commons, 1835. pp. 5–39.

This is the first of nine annual reports by Macneill printed in the 12th to 20th Commissioners
Reports, 1835–1843.

892 *Report on the Stirling Canal.*

By John Macneill, Civil Engineer.

Signed: John Macneill. Parliament Street, London, 13 Dec 1835.

With 2 folding engraved plans.

[1] *Map or Plan of a Navigable Cut or Canal from
Southfield near the Town of Stirling, to the Forth &
Clyde Canal at or near Wynford; with a . . . Branch
Canal . . . to Castlecary.*

By John Macneill, Civil Engineer, 1835.
Swan sc.

[2] *Section of a Navigable Cut or Canal . . . [etc].*
[In]

Prospectus of the Stirling Canal. 1835.

4to   29 cm   (iii) + 16 pp.   2 folding engraved
plans (as above) + folding litho sketch map.

NLS, CE

893 *Report on the Means of making the Lower Bann
River navigable, and of preventing the Waters of Lough
Neagh from rising above their summer level.*

Signed: John Macneill. St. Martin's Place,
Trafalgar Square, London, 8 Oct 1836.

4to   27 cm   pp. [(13)–25].

CE

Apparently an offprint from some larger publication.

894 *On the Means of comparing the respective advantages of different Lines of Railways; and on the use of
Locomotive Engines.*

Translated from the French of M. Navier . . .
By John Macneill, Civil Engineer.

London: Roake and Varty . . . 1836.

8vo   22 cm   xv + 97 pp.

BM, Elton

895 *Report on the Present State of the River Arun, and
the means proposed for the Drainage of the Adjacent
Country.*

By John Macneill, M.R.I.A., Mem.Inst.
Civ.Eng.

London: Roake and Varty . . . 1837.

Signed: John Macneill. St. Martin's Place.

4to   28 cm   19 pp.

CE

896 *Report on a Proposed Railway between the City of
Cork and the Town of Cove, in the South of Ireland.*

By John Macneill . . . Civil Engineer.

London: Printed by Robson, Levey and Franklin
. . . 1837.

Signed: John Macneill. St. Martin's Place,
Westminster, 28 Feb 1837.

4to   28 cm   10 pp.   Folding litho plan.

CE

897 *Report on the Several Lines of Railway through
the North and North-Western Districts of Ireland, as laid
out under the direction of the Commissioners.*

By John Macneill, Civil Engineer, F.R.S.,
M.R.I.A.

Dated: 8 Jan 1838.
[In]

Second Report of the Commissioners [on] *Railways in
Ireland.*

Dublin, 1838.

Appendix A, pp. 34–49.

The lines run from Dublin to Armagh and
Enniskillen.

898 [Plans of the Lines] *Through the North and
North-Western Districts of Ireland.*

By John Macneill, Civil Engineer.

Title pages + index map + 5 plans, forming 9
  sheets each   45 × 90 cm   (the plans on a
  scale of 1 inch to 1 mile)   Engraved by J. & C.
  Walker.
[In]
*Irish Railway Commission. Plans of the Several Lines
laid out under the direction of the Commissioners.* 1837.

899 [Sections of the Lines] *Through the North and
North Western Districts of Ireland.*
  By John Macneill, Civil Engineer.
Title pages + Diagrams of Clivities + 27 sections
  + Diagram Sections, forming 34 sheets each
  45 × 90 cm (the sections on a scale of 6 inches
  to 1 mile).
Engraved by J. & C. Walker.
[In]
*Irish Railway Commission. Sections of the Several Lines
laid out under the direction of the Commissioners.* 1837.

900 *Drogheda Harbour. Mr Macneill's Report. 1840.*
Drogheda: Printed by Patrick Kelly. 1843.
Signed: John Macneill. Dundalk, 22 Sept 1840.
12mo   12 pp.
NLS

MACQUISTEN, Peter   Assoc. Inst. C.E.

901   *Report by Peter Macquisten, Civil Engineer in
Glasgow, relative to a New Line of Road, from the Con-
fines of the Counties of Ayr and Renfrew, at Drumby and
Floak, to the City of Glasgow.*
[Glasgow]: Hedderwick and Son, Printers.
Signed: Peter Macquisten. Glasgow, 3 Sept 1828.
4to   26 cm   12 pp.
NLS

902   [Report] *To the Trustees and Others interested
in the Proposed New Approach from the Two Great
North Roads to the City of Glasgow.*
[Glasgow]: Hedderwick and Son, Printers.

Signed: Peter Macquisten. Glasgow, 31 March
  1829.
4to   26 cm   6 pp.   Folding litho plan, hand
  coloured.
*Plan of a proposed New Approach from the Two Great
North Roads to the City of Glasgow, 1829.*
NLS, ULG (Rastrick)

MACQUISTON, Bryce

903   *Report of Mr Bryce Macquiston.*
Signed: Bryce Macquiston. Mossfarm, 20 Aug
  1813.
pp. (19)–26.
[Printed with]
Report by Charles ABERCROMBIE, July 1813
  [q.v.].
[In]
*Reports respecting a New Line of Road, from Hamilton
to Elvanfoot.*
Glasgow: Printed by J. Hedderwick and Co . . .
  1813.
4to   28 cm   26 pp.   Folding engraved plan.
*Plan of different Lines of Roads, between Hamilton and
Elvan-Foot . . . from the Respective Surveys of Messrs
Abercrombie and Macquiston.*
Drawn by Peter Fleming, engraved by J. Hal-
  dane.
NLS, CE

MAHAN, *Professor* Dennis Hart [1802–1871] Hon.Mem.A.S.C.E.

904   *An Elementary Course of Civil Engineering.*
  By D. H. Mahan, Professor of Military and
  Civil Engineering in the Military Academy of
  the United States. Edited by Peter Barlow,
  F.R.S., F.R.A.S. . . [etc].
Glasgow: A. Fullarton . . . 1838.

Dated: U.S. Military Academy, 1 June 1837.
Preface by editor dated: Royal Military Academy,
  Woolwich, 27 April 1838.
4to   28 cm   xiii + 211 pp.   15 engraved plates.
BM, ISE

905   Reissue
As No. 904 but published 1839.
P

MARQUAND, Charles

906 *Remarks on the Different Constructions of
Bridges, and Improvements to secure their Foundations on
the different Soils where they are intended to be Built.
Which hitherto seems to have been a Thing not sufficiently
consider'd.*
London: Printed for the Author; and Sold by J.
  Jolliffe . . . 1749.
Signed: Charles Marquand. Horse-Ferry, West-
  minster, 5 May 1749.
4to   22 cm   15 pp.   4 engraved plates, of
  which 3 are folding.
BM, RIBA

MARSHALL, Robert

907 *An Examination into the respective merits of the
Proposed Canal and Iron Railway, from London to Portsmouth.*
London: Printed and Sold by Joseph Robbins . . .
  1803.
Signed: Robert Marshall. Godalming, 24 Oct
  1803.
8vo   19 cm   20 + (1) pp.
CE (Vaughan), ULG, Elton

MATHER, William

908 *Of Repairing and Mending the Highways. In
Five Sections; touching, I. Removing Obstructions in the
Highways, and Scouring the Ditches next adjoining. II.
Draining the Highways, and Repairing them. III, IV, V.
Providing Material, Labourers, Carriages. Published for
the Use and Instruction of Young Surveyors.*
  By William Mather, a Late Surveyor of the
  Highways in Bedford.
London: Printed for Samuel Clark . . . 1696.
8vo   18 cm   32 pp.
BM

MATHEW, Francis

909 *Of the Opening of Rivers for Navigation. The
Benefit exemplified, by the Two Avons of Salisbury and
Bristol. With a Mediterranean Passage by Water for
Billanders of Thirty Tun, between Bristol and London.
With the Results.*
London: Printed by James Cottrel, 1655.
Signed: Francis Mathew, 9 July 1655.
4to   18 cm   (iv) + 11 + (1) pp.   Woodcut
  diagram on p. (12).
BM

910   Second edition
Title as No. 909.
London: Printed by G. Dawson, 1656.
Signed: Francis Mathew. 1 Jan 1656.
4to   18 cm   (viii) + 16 pp.
BM, NLS, ULG

911 *To his Highness, Oliver, Lord Protector of the
Common-wealth . . . Is Humbly presented A Mediter-
ranean Passage by Water Between the Two Sea Towns
Lynn & Yarmouth, upon the Two Rivers the Little Ouse,
and Waveney. With farther Results Producing the Pas-
sage from Yarmouth to York.*
London: Printed by Gartrude Dawson, 1656.
Signed: Francis Mathew.
4to   18 cm   15 pp.
BM, NLS, BDL (Gough)

912 *To the Kings Most Excellent Majesty, and the Honorable Houses of Parliament. A Mediterranean Passage by Water, from London to Bristol, &c. And from Lynne to Yarmouth, and so consequently to the City of York: for the great Advancement of Trade & Traffique.*

By Francis Mathew, Esquire.

London: Printed by Thomas Newcomb, 1670.

4to   19 cm   (v) + 12 pp.

BM, ULG, BDL (Gough)

MATTHEWS, William [1771–1836]

913 *Hydraulia; an Historical and Descriptive Account of the Water Works of London, and the Contrivances for Supplying other Great Cities, in different ages and countries.*

By William Matthews.

London: Simpkin, Marshall, and Co . . . 1835.

Dated: April 1835.

8vo   23 cm   xx + 454 pp.   12 litho plates + 5 folding litho maps.

BM, SML, CE, ULG, IC

MAXWELL, George [*c.* 1744–1816] Hon.M.Soc.C.E.

914 *Observations as to the Present State of the Lands in the Neighbourhood of Spalding, in the County of Lincoln; with a view to an Improvement in the Outfal of the River Welland, and to the Drainage of the Lands on both sides thereof.*

Peterborough: Printed by J. Jacob, 1791.

Signed: M——, 6 June 1791.

4to   25 cm   21 pp.

LAO

Known to be by Maxwell.

See HUDSON and MAXWELL. Report on the drainage of South Holland, 1792.

915 *An Essay on Drainage and Navigation, occasioned by the scheme now in agitation for Improving the Outfal of the River Ouse.*

By George Maxwell.

Peterborough: Printed by J. Jacob, 1792 . . . Price One Shilling.

4to   24 cm   15 pp.

BM, CRO, ULG, BDL (Gough), CUL, Sutro

See HUDSON, GOLBORNE and MAXWELL. Reports on the outfall of the River Welland, 1792 and 1793.

916 *Observations on the Advantages to be derived from an improved Outfall at the Port of Lynn; with Answers to the Objections which it is supposed will be used against that Measure.*

Appendix: pp. 37–47.

[In]

General View of the Agriculture of the County of Huntingdon . . . with an Appendix containing an Account of the Advantages [etc].

By George Maxwell of Fletton.

London: Printed by J. Nichols, 1793.

4to   25 cm   47 pp.

BM, ULG, BDL (Gough), CUL

917 *Reasons attempting to shew the Necessity of the Proposed Cut from Eau Brink to Lynn; with Extracts from the Reports of Engineers and other Writers on the Subject . . . Addressed to all Persons interested in the Drainage or Navigation of the River Ouse.*

By a Member of the Committee.

London: Printed 1793.

8vo   23 cm   (vii) + 32 pp.

BM, CE, CRO, CUL, Sutro

Attributed to Maxwell.

See MAXWELL and HARE. Report on the drainage of Deeping . . . Spalding and Pinchbeck Commons, 1800.

See JESSOP, RENNIE *et al.* Report on Deeping Fen, 1800.

MAXWELL, George and HARE, Edward

918 *The Report of Messrs Maxwell and Hare, on the Drainage of Deeping, Langtoft, Baston, Crowland, Cowbit, Spalding, & Pinchbeck Commons, in the County of Lincoln.*
Spalding: Printed by J. L. and T. Albin.
Signed: Geo Maxwell, Edw.ᵈ Hare. 24 Feb 1800.
4to   27 cm   8 pp.
LCL, Sutro

MAY, George [1805–1867] M.Inst.C.E.

919 *General Report of the State of the Caledonian Canal Works to May 1830.*
By George May, Resident Engineer.
Signed: George May. Inverness, 1 May 1830.
[In]
Twenty-Sixth Report of the Commissioners for the Caledonian Canal.
House of Commons, 1830. pp. 10–13.
George May's annual reports continue until 1867.

920 *Mr May's Report [on the Caledonian Canal] in reply to Mr Borron.*
Signed: George May. Inverness, 1 Nov 1837.
[In]
Report from the Select Committee on the Caledonian and Crinan Canals.
House of Commons, 1839. pp. 120–144.

MEDHURST, George [1789–1827]

921 *Calculations and Remarks, tending to prove the practicability, effects and advantages of a Plan for the*
Rapid Conveyance of Goods and Passengers upon an Iron Road . . . by the Power and Velocity of Air.
By G. Medhurst, Inventor and Patentee, Denmark Street, Soho, London.
London: Printed by D. N. Shury . . . 1812.
8vo   21 cm   18 pp.
BM, SML, CE, ULG, Elton
The first proposals for an atmospheric railway.

922 *A New System of Inland Conveyance, for Goods and Passengers, capable of being applied and extended throughout the country . . . without the aid of Horses or any Animal Power.*
By George Medhurst, Civil Engineer, Denmark Street, Soho, London.
London: Printed by T. Brettell . . . 1827.
8vo   21 cm   (i) + 34 pp.   Frontispiece + 4 litho plates.
BM, SML, CE, ULG, Elton

MEYNELL, Thomas

923 *A Report relative to the opening a communication by a Canal or a Rail or Tram Way, from Stockton, by Darlington, to the Collieries.*
Stockton: Printed by T. Eeles, 1818.
Signed: Thomas Meynell, Chairman.
8vo   20 cm   16 pp.
BM, SML
Includes on pp. 5–9 a summary of a report by the engineer George Overton.

924 *A Further Report of the intended Rail or Tram Road, from Stockton, by Darlington, to the Collieries, with a Branch to Yarum.* February 1821.
Darlington: Printed at the Office of M. Appleton.
Report signed: Thomas Meynell, Chairman.
8vo   23 cm   22 pp.   Hand-coloured folding engraved plan.
Plan of the Darlington Railway.
SML, CE, ULG (Rastrick), Elton
The report is on pp. 13–16, with Observations pp. (7)–11 and an Appendix. The plan shows Overton's proposed line, surveyed by David Davies.

MILBOURN, Isaac

See GRUNDY, John. Plan of the Driffield Canal from a survey and levels by Milbourn, 1766.

See GRUNDY, John. Survey of Cod-Beck by Milbourn, 1767.

925 *A Plan of the Low Grounds between Muston and Malton, which adjoin the Rivers Derwent and Harford, and the Course of those Rivers.*
   Surveyed by Isaac Milbourn.
   Engraved by Faden and Jefferys 1773.
   Engraved plan   18 × 48 cm   (scale 1 inch to 1 mile).
   BM, Hull University Library
      Referred to in Thomas TOFIELD's report of 25 Sept 1773 [q.v.].

MILLAR, Archibald

926 *A Plan of the Great [Forth & Clyde], and Monkland Canals. With the Reservoirs & Feeders.*
   By Arch.ᵈ Millar.
   Engraved by J. Lumsden. [c. 1786]. Engraved plan, hand coloured   47 × 81 cm   (scale 1 inch to 1 mile).
   Scottish R.O., MCL

MILLAR, James and VAZIE, William

927 *Observations on the Advantages and Practicability of Making Tunnels under Navigable Rivers, particularly applicable to the Proposed Tunnel under the Forth. With an Appendix, containing Reports of Surveys, and Opinions of Engineers, relative to this Undertaking.*
   By James Millar, M.D., F.S.A.S. and William Vazie, Esq.
   Edinburgh: Printed for Mundell, Doig, & Stevenson . . . 1807.

Dated: Edinburgh, 2 March 1807.
8vo   ix + (2) + 126 pp.   3 folding engraved plans.
NLS
   The Appendix includes John GRIEVE's report of November 1805 [q.v.].

MILLER, John [1805–1883] F.R.S.E., M.Inst.C.E.

See GRAINGER and MILLER. Reports and plans, 1825–1840.

928 *Report of John Miller, Esq., Engineer, to the Directors of this Company.*
   Signed: J. Miller. Edinburgh, 6 Feb 1839.
   pp. 11–13.
                    [In]
   Report of the Directors to the Shareholders of the Glasgow, Paisley, Kilmarnock, and Ayr Railway Company.
   Glasgow: Printed by George Richardson . . . 1839.
   8vo   22 cm   21 + (2) pp.
   CE

MILLS, James   M.Inst.C.E.

See TELFORD. Map of the roads between Girvan and Stranraer, surveyed by James Mills 1824.

See TELFORD. Map of the road between Edinburgh and Morpeth, surveyed by James Mills, published 1825.

See TELFORD. Map of the mail roads from London to Retford, surveyed by James Mills 1826.

See TELFORD. Report on Liverpool & Manchester Railway, 1829. Field investigations by Mills.

MILNE, John

929 *Plans for the Floating Off, of Stranded Vessels, and for Raising those that have Foundered; with an improved method of carrying vessels over banks in shallow water.*

By John Milne, Teacher of Architectural and Mechanical Drawing, Edinburgh.
Edinburgh: Published by Daniel Lizars . . . 1828. Price Two Shillings.
Dated: St. James Square, 3 Dec 1827.
8vo   22 cm   29 pp.   2 engraved plates (W. H. Lizars sc.).
BM, NLS

930 *Observations on the Tidal Currents at Leith; with Plans for the Improvement of the Port.*

By John Milne, Teacher of Architecture and Engineering. James Square, Edinburgh.
Edinburgh: Sold by the Author, also by Adam and Charles Black . . . 1835. Price One Shilling.
8vo   21 cm   31 pp.   Folding litho plan, hand coloured.
Chart of the Frith of Forth between Hound Point & Portobello shewing the relative Positions of the Proposed Harbours and some of the Soundings at low Water.
Leith & Smith Lithog.
NLS, CE

931 Second edition
Plans for the Improvement of the Port of Leith; with Observations on the Tidal Currents.

By John Milne . . . Second Edition.
Edinburgh: Adam and Charles Black . . . 1835. Price One Shilling.
8vo   21 cm   32 pp.   Folding litho plan (as in 1st edn.).
NLS, CE

932 *Description of Plans for the Extension of Leith Harbour towards the East, to $12\frac{1}{2}$ Feet Water at Ordinary Ebbs; with detailed Estimate of Expense, &c.*

Addressed to the Commissioners . . . by John Milne, Civil Engineer.
Edinburgh: Sold by Messrs M'Lachlan & Stewart . . . Price One Shilling.

Dated: Prince's Street, Edinburgh, Sept 1838.
8vo   22 cm   (i) + 34 pp.   Folding litho plan, hand coloured (Leith & Smith lith.).
NLS

MILTON, Thomas

933 *A Geometrical Plan and North Elevation of His Majesty's Dock Yard at Woolwich, with Part of the Town.*
Tho.ˢ Milton Surv. et del.   P. C. Canot sc.
Published . . . 18 June 1753.
Engraved plan   48 × 65 cm   (scale 1 inch to 100 feet).
BM

934 *A Geometrical Plan & North East Elevation of His Majesty's Dock Yard at Deptford, with Part of the Town.*
Tho.ˢ Milton Surv. et del.   P. C. Canot sc.
Published . . . 30 July 1753.
Engraved plan   48 × 65 cm   (scale 1 inch to 100 feet).
BM

935 *A Geometrical Plan & West Elevation of His Majesty's Dock Yard near Portsmouth, with Part of the Common.*
Tho.ˢ Milton Surv. et del.   P. C. Canot sc.
Published . . . 29 April 1754.
Engraved plan   48 × 65 cm   (scale 1 inch to 160 feet).
BM, CE (Rennie)

936 *A Geometrical Plan & West Elevation of His Majesty's Dock Yard and Garrison at Sheerness, with the Ordnance Wharfe.*
Tho.ˢ Milton Surv. et del.   P. C. Canot sc.
Published . . . 14 April 1755.
Engraved plan   48 × 65 cm   (scale 1 inch to 100 feet).
BM

937 *A Geometrical Plan & North West Elevation of His Majesty's Dock Yard, at Chatham, with yᵉ Village of Brompton adjacent.*

T. Milton Surv. et del.   P. C. Canot sc.   Published . . . 2 Sept 1755.
Engraved plan   48 × 65 cm   (scale 1 inch to 180 feet).
BM, CE (Rennie)

938   *A Geometrical Plan and West Elevation of His Majesty's Dock Yard, near Plymouth; with the Ordnance Wharfe.*
    T. Milton Surv. et del.   P. C. Canot sc.   Published . . . 2 Feb 1756.
Engraved plan   48 × 65 cm   (scale 1 inch to 150 feet).
BM, CE (Rennie)

MITCHELL, Joseph [1803–1883]
F.R.S.E., M.Inst.C.E.

939   *Answer from Mr Joseph Mitchell to Lord Colchester.*
Signed: Jos. Mitchell. Office of Highland Roads and Bridges, Inverness, 26 Jan 1828.
pp. 23–41   with folding engraved plan, the roads entered in colours.
Sketch of the Present Highland Road between Inverness and Perth, shewing also Proposed Alterations.
    By Joseph Mitchell 1828. A. Arrowsmith sc.
                [In]
Fourteenth Report of the Commissioners for Repair of Roads and Bridges in Scotland.
    House of Commons, 1828.
    See also in TELFORD. *Life of Thomas Telford.* London, 1838. pp. 448–469.

940   *Notices of the Improved State of the Highlands of Scotland, since the Commencement of the Public Works executed under the direction of the Parliamentary Commissioners.*
Signed: Jos. Mitchell. Office of Highland Roads and Bridges, Inverness, 6 March 1828.
pp. 57–64   with engraved plate.

Map of Scotland containing Sketch of the Roads, Bridges and Harbours, Made or Improved by the Parliamentary Commissioners 1803–1828.
    A. Arrowsmith sc.
                [In]
Fourteenth Report of the Commissioners for Repair of Roads and Bridges in Scotland.
    House of Commons, 1828.

941   [*Report and Estimates of the several Bridges and other Works destroyed or partially injured by the floods of August last*].
Signed: Jos. Mitchell. Inverness, 1 March 1830.
                [In]
Sixteenth Report of the Commissioners for Repair of Roads and Bridges in Scotland.
    House of Commons, 1830.
    pp. 6–11.

942   *Report and Description of Churches and Manses.*
Signed: Joseph Mitchell. Inverness, 1830.
pp. 9–32   with engraved map.
Map of Scotland shewing the Site of Additional Places of Worship and Manses built in the Highlands and Islands 1827–1830.
    A. Arrowsmith del.
                [In]
Sixth Report of the Commissioners for Building Churches in the Highlands and Islands of Scotland.
    House of Commons, 1831.
    For the designs of these churches and manses *see* TELFORD, 1825.

943   [*Annual Report on Works and Repairs*].
Signed: Jos. Mitchell. Inverness, 14 March 1833.
                [In]
Nineteenth Report of the Commissioners for Repair of Roads and Bridges in Scotland.
    House of Commons, 1833. pp. 3–5.
    Mitchell's annual reports, some with plans, continue for many years.

944   [*Report on Roads between Inverness and Nairn*].
Signed: Jos. Mitchell. Inverness, 26 Jan 1835.
pp. 4–5   with litho plate and folding plan (S. Arrowsmith lithog.).

Timber Bridge over the River Dee at Ballater.
Map shewing the Present Roads between Inverness and Nairn.

[In]

Twenty-First Report of the Commissioners for Repair of Roads and Bridges in Scotland.
    House of Commons, 1835.

945   [*Report on the Great Highland Road from Inverness to Dunkeld*].
Signed: Joseph Mitchell. Inverness, 1 March 1837.
pp. 4–6   with folding litho plan, the lines entered in colours.
Map of the Present Road from Inverness to Dunkeld, with the Contemplated Alterations and Improvements, surveyed . . . 1836.
    By Jos. Mitchell, Civil Engineer.
    S. Arrowsmith lithog.

[In]

Twenty-Third Report of the Commissioners for Repair of Roads and Bridges in Scotland.
    House of Commons, 1837.

MOORE, *Sir* Jonas [1617–1679] F.R.S.

946   *A true Mapp of yᵉ great Levell of the Fenns as itt now lyeth drayned, with yᵉ great works, made at yᵉ cost & charges, of the . . . Earle of Bedford his participants and yᵉ Adventurers, for yᵉ perfect drayning thereof.*
    Described by Jonas Moore [*c.* 1654].
Engraved map   53 × 56 cm   (scale 1 inch to 2 miles).
BM
    Moore was appointed Surveyor of the Great Level in 1650 and given permission to print and publish his map in 1654.

947   *A Mapp of yᵉ Great Levell of yᵉ Fenns extending into yᵉ Countyes of Northampton, Suffolke, Lyncolne, Cambrig & Huntington & the Isle of Ely as it is now drained, described by Sʳ Jonas Moore Survey.ʳ genˡˡ.*
    London. Sold by Moses Pitt . . . 1684.
Engraved map   141 × 196 cm,   in   16   sheets (scale 2 inches to 1 mile).
BDL (Gough)
    The 'Large Map . . . by Sir Jonas Moore, now New Printed and Enlarged by Moses Pitt' as mentioned in FORTREY's History of the Great Level, 1685 [q.v.].

948   Second issue
As No. 947 but
Printed and Sold by Christop.ʳ Browne . . . London. [*c.* 1705].
BM, CRO, CE, RGS
    Bedford Level Corporation ordered the reprinting of a large map in July 1805.

MOORSOM, *Captain* William Scarth [1804–1863] M.Inst.C.E.

949   Deposited plan
Plan of the proposed Birmingham and Gloucester Railway. 1835.
    Captⁿ W. S. Moorsom, Engineer.
Lithographed by J. Gardner.
Title + 12 sheets of plans each 45 × 65 cm (scale 4 inches to 1 mile).

[and]

Section of the proposed Birmingham and Gloucester Railway. 1835.
    Captⁿ W. S. Moorsom, Engineer.
Lithog.ᵈ by T. Underwood.
Title + 16 sheets of sections each 45 × 65 cm (scale 4 inches to 1 mile and 1 inch to 120 feet).
HLRO

MORGAN, James

950 *Regent's Canal. Plan and Sections of an intended Navigable Canal from the Grand Junction Canal at Paddington to the Thames at Limehouse . . . September 1811.*
Engraved by C. Smith.
Engraved map   39 × 51 cm   (scale 1 inch to 1 mile).
BM
> Map drawn by Morgan, who became engineer for the canal after the Act was obtained in July 1812.

951 *Plan of the Regent's Canal in the County of Middlesex. Shewing such parts of the Works as are Finished, In hand [or] Not Begun.*
Engraved map   27 × 58 cm   (scale 2½ inches to 1 mile).
BM
> With MS. endorsement: 2 Dec 1818.

952 Second state.
As No. 951 with minor changes and locks shown. Tyler sc.
BM
> With MS. endorsement: 31 May 1819.

953 *Report to the United Committee for the Stour Navigation and Sandwich Harbour.*
Signed: James Morgan. Sandwich, 11 March 1824.
pp. (3)–5   with hand-coloured litho plan.
A Plan or Map of an intended Navigation from Canterbury through Fordwich and Sandwich to the Sea. James Morgan, Engineer.
[In]
Stour Navigation and Sandwich Harbour. Prospectus; Sketch of the Plan; the Report of the Engineer, with the Estimates; and a Statement of the Revenue.
Canterbury: Wood, Printer . . . 1824.
fol   37 cm   7 pp.   Folding litho plan (as above).
ULG

MORRIS, Thomas

954 *Copy of the Report, of Thomas Morris, Esquire, relative to the making of a New Dock, at Kingston upon Hull.*
n.p.
Signed: Thomas Morris. Liverpool, 30 July 1793.
fol   38 cm   3 pp.   Docket title on p. (4).
Hull City Archives

MOSELEY, *Professor* Henry [1801–1872] F.R.S.

955 *A Treatise on Hydrostatics and Hydrodynamics.*
By Henry Moseley, B.A.
Cambridge: Printed for T. Stevenson . . . 1830.
8vo   23 cm   xi + errata leaf + 290 pp.   4 folding engraved plates.
BM, RS, CE

956 *A Treatise on Mechanics, applied to the Arts; including Statics and Hydrostatics.*
By the Rev H. Moseley, B.A.
London: John W. Parker . . . 1834.
12mo   17 cm   xxxviii + 360 pp.   Woodcuts in text.
BM, NLS, RS, BDL

957 *On the Equilibrium of the Arch.*
By the Rev Henry Moseley, B.A.   Professor of Natural Philosophy at King's College, London.
Cambridge: Printed for the Pitt Press . . . 1835.
4to   29 cm   23 pp.
CE
> Reprint with new title page from *Trans. Cambridge Phil. Soc.* Vol. 5.

MOUNTAGUE, James
[*c*. 1776–1853]

See DANCE, CHAPMAN *et al.*, 1814 and MOUNTAGUE, RENNIE *et al.*, 1821. Reports on London Bridge.

MOUNTAGUE, William
[1773–1843]

See MOUNTAGUE, RENNIE *et al.* Report on London Bridge, 1821.

MOUNTAGUE, William, RENNIE, John *et al.*

958 *Report of the Clerk of the City's Works, Mr Rennie, Mr Chapman, and Mr James Mountague, on the expediency of enlarging the Water-way under London Bridge in the manner recommended by Mr Dance and others, in the Year 1814.*
Signed Wm Mountague, Clerk of the City's Works, John Rennie, Wm Chapman, Civil Engineers, James Mountague, Surveyor to the Committee for improving the Port of London. London, 12 March 1821.
[In]
An Abstract of Proceedings and Evidence, relative to London Bridge.
Corporation of London, 1819. pp. 132–140. *See* LONDON BRIDGE.

MULLINS, Bernard [1772–1851]

959 *Thoughts on Inland Navigation, with a Map, and Observations upon Propositions for Lowering the Waters of the Shannon and of Lough Neagh* . . .
By Bernard Mullins.
Dublin: Printed by C. Hope . . . 1832.
Signed: Bernard Mullins. Fitzwilliam Square, July 1832
8vo   21 cm   54 pp.   Folding hand-coloured litho map.
Sketch of a Map exhibiting the lines of Still water and River Navigation now in operation in Ireland.
J. Allen Lithog.
ULG

MURRAY, James [d. 1807]

See RENNIE. Plan of proposed Polbrook Canal 1796 and Plan of proposed London Canal 1802, surveyed by Murray.

MYLNE, Robert [1733–1811] F.R.S., M.Soc.C.E.

960 *A View of Mr Mylne's elegant Design of a New Bridge, to be built from Black Fryers to the opposite Shore, approved by the Committee in Common Council, 1760.*
Printed for T. Kitchin . . .
Engraved elevation, with pictorial foreground
17 × 52 cm   (scale 1 inch to 55 feet).
BM, SML, RS (Smeaton), CE (Rennie)
This elevation only, to the same scale and with almost identical title, was published as a folding plate in *The London Magazine* for March 1760.

See BALDWIN. Elevation and plan of Blackfriars Bridge, 1766.

961 *Mr Mylne's Report and Opinion of the Design for keeping the Ships afloat at all Times in the Harbour of Bristol. And also, Mr Smeaton's Opinion upon Plans laid before him for that Purpose.*
[Bristol, 1767].
Signed: Robert Mylne. Bristol, 12 Jan 1767.
fol   31 cm   3 pp.   Docket title on p. (4).
Bristol R.O.
    Smeaton's *Opinion*, dated Austhorpe 31 Dec 1766, is a brief statement printed on p. 3.

962 [Report relating to the Petition of the Proprietors of the London-Bridge Water-Works] *To the Right Honourable the Lord Mayor, Aldermen, and Commons, in Common-Council Assembled.*
Signed: Robert Mylne. Arundel-street, 27 Feb 1767.
pp. 5–7.
[In]
Kite, Mayor. A Common-Council holden . . . on Wednesday the 25th Day of February 1767.
Corporation of London, 1767.
fol   33 cm   7 pp.
CLRO

963 *Mr Mylne's Report respecting Tyne Bridge. With his Plan for a Temporary Bridge: also Mess. Rawlings and Wake's Abstract of the Borings into the Bed of the River Tyne.*
Newcastle: Printed by I. Thompson . . . 1772.
Signed: Robert Mylne. Gateshead, 12 March 1772.
8vo   21 cm   24 pp.   Folding engraved plate [plan of piers and temporary bridge] R. Beilby sc.
BM, CE

See BLACKFRIARS BRIDGE. Corporation of London, 1774.

964 *Mr Mylne's Report, on his Survey of the Harbour, &c. of Wells, in Norfolk.*
[London, 1781].
Signed: Robert Mylne. New River Head, 28 April 1781.
fol   33 cm   7 pp.   Docket title on p. (8).
BM, ULG, Sutro

965 *A Letter from Robert Mylne, Esq., to the Right Worshipful John Patteson, Esq. Mayor of Norwich, on*
the State of the Mills, Water-Works, &c. of this city, commonly called the New Mills. [etc].
Norwich: Printed by J. Crouse and W. Stevenson. 1789.
Signed: Robert Mylne. 1 Jan 1789.
4to   25 cm   (i) + 17 pp.
CE

966 *Mr Mylne's Report to the Commissioners, for improving the Navigation of the Rivers Thames and Isis.*
Signed: Robert Mylne. New River Head, 8 May 1791.
pp. 25–56   with folding engraved plan.
A Plan and Survey of the River Thames from the End of the Thames and Severn Canal in the Parish of Kempsford . . . to Abingdon in the County of Berks.
[Printed with]
Mr JESSOP's first (and Second) Report . . . August 1789 [q.v.].
[In]
Reports of the Engineers appointed by the Commissioners of the Navigation of the Rivers Thames and Isis to Survey the State of the said Navigation from Lechlade to Days Lock.
n.p.   Printed in 1791.
8vo   22 cm   60 pp.   Folding engraved plan (as above).
BDL (*see also* the following item)
    Stitched as issued in grey wrappers. Thames Commissioners Minute Book, 10 May 1791: ordered that 500 copies, with map annexed, be forthwith printed.

967 *An Estimate of the Works proposed to be done agreeable to Mr Mylne's Opinion and Report.* [1791].
8vo   22 cm   7 pp.
Separately printed, but bound with foregoing Reports.
BM, CE (Page), ULG
    Thames Commissioners Minute Book, 31 May 1791: Mr Mylne was desired to make an Estimate . . . and to get 500 copies printed.

968 *Report the Second, by Mr Mylne, Surveyor and Engineer on the Navigation of the River Thames, between Lechlade and Whitchurch.*
n.p. 1791.
Signed: Robert Mylne, 10 Aug 1791.

8vo   22 cm   24 pp.

BM, CE (Page), ULG

Mylne's survey of the navigation, described from Lechlade to Abingdon in the first report of 8 May 1791, is here continued from Abingdon to Whitchurch. Ordered to be printed 20 Aug 1791.

969   *The Report of Robert Mylne, Engineer, on the Proposed Improvement of the Drainage and Navigation of the River Ouze, by executing a Straight Cut, from Eau-Brink to King's Lynn.*

London: Printed by Henry Baldwin . . . 1792.

Signed: Robert Mylne, New River Head, 26 Oct 1791.

4to   24 cm   (iii) + 52 [error for 51] pp.

BM, CE, ULG, BDL, CUL

970   Another edition

Title as No. 969 but

Lynn: Printed by W. Whittingham. 1792.

4to   23 cm   (i) + 52 pp.

P

Different setting of type throughout, and correct pagination.

971   *Report on a Survey of the River Thames from Boulter's Lock to the City Stone near Staines and on the best Method of improving the Navigation of the said River, and making it into as compleat a state of perfection as it is capable of.*

By Robert Mylne, F.R.S. Engineer.

n.p.

Signed: Robert Mylne. London, 20 Aug 1793.

8vo   21 cm   53 pp.

BM, CE, ULG

972   [Report] *To the Gentlemen of the Committee of Subscribers to the Proposed Canal from Bristol to Cirencester.*

n.p. Printed in the Year 1793.

Signed: Robert Mylne, 27 Sept 1793.

4to   23 cm   15 pp.   Folding engraved plan.

Plan of the Proposed Canal from Cirencester to Bristol.

P

See WHITWORTH and MYLNE, 1794. On the proposed London Canal from Boulter's Lock to Isleworth.

973   *Report of Mr Mylne, on the State of the River Thames and its Bed; on the Structure of London Bridge, and as to the Navigation of the River above and below it &c. With an Appendix and Drawings.*

Signed: Robert Mylne. New River Head, 30 May 1800.

[Printed in]

Third Report from the Select Committee upon the Improvement of the Port of London.

House of Commons, 28 July 1800. pp. 25–38.

974   A sheet of 2 drawings, soundings and other measurements, published separately as Plate I in *The Several Plans and Drawings referred to in the Third Report from the Select Committee . . .* 28 July 1800.

The plate, engraved by Laurie & Whittle, shows (to a scale of 1 inch to 25 feet) Mylne's observations on the foundations of London Bridge in 1762 and his soundings of the river bed in 1767.

975   Second printing

The report and drawings are reprinted in *Reports from Committees of the House of Commons* Vol. 14 (1803) pp. 550–554 and Plates 29 and 30 (engraved to same scale as in the original large single plate).

976   Third printing

The report only, reprinted in *An Abstract of Proceedings and Evidence relative to London Bridge.* Corporation of London. 1819. pp. 40–52. *See* LONDON BRIDGE.

977   *A Scheme or Outline of a Plan for a new Bridge, adapted to the Situation and Circumstances of London Bridge.* [etc].

Signed: Robert Mylne. 23 June 1800.

[In]

Third Report from the Select Committee upon the Improvement of the Port of London.

House of Commons, 28 July 1800. pp. 51–56.

978 Second printing
In *Reports from Committees of the House of Commons*
Vol. 14 (1803) pp. 562–564.

979 *Mr Mylne's Answers. To the Select Committee of the Bridge-House Lands, appointed to consider a Report of the Select Committee of the House of Commons, for the Improvement of the Port of London, by rebuilding London-Bridge, &c.*
Signed: Robert Mylne, 15 May [and] 30 Oct 1801.
pp. 6–17.
[In]
At a Sub-Committee . . . held at Guildhall, on Wednesday the 19th Day of December 1801.
Corporation of London, 1801.
fol 32 cm 17 pp.
CLRO

See PAGE, Thomas. Correspondence on the Subject of the Eau Brink Cut, between Sir Thomas Hyde Page and Mr Mylne, 1802.

980 *Report by Mr Mylne, Engineer, on Three New Cuts, for the Improvement of the Navigation of the River Thames, above Oxford.*
London: Printed by C. Baldwin . . . 1802.
Dated: 15 May 1802.
8vo 21 cm 18 pp. 2 folding engraved plans.
[1] *A Plan of the Proposed New Cut from Tadpole Bridge to the River Thames near Shifford.*
[2] *A Plan of a Proposed New Cut from the River Thames at the upper end of Cumnor Meadow to the Old River.*
Both plans: Surveyed by W. Rutt and E. Kelsey, Oxford 1802.
BM, CE (Page)
It is possible that the plans were issued separately.

981 *Report, by Mr Mylne [and Second Part of the Report] on the Present State of the Navigation of the River Thames, between Maple-Derham and Lechlade.*
London: Printed by C. Baldwin . . . 1802.
Signed: Robert Mylne. 1 June 1802 [and] 16 June 1802.
8vo 21 cm 71 pp.
BM, CE (Page)

MYLNE, William Chadwell [1781–1863] F.R.S., M.Soc.C.E., M.Inst.C.E.

See WALKER, James and W. C. MYLNE. Report on the Eau Brink Cut, 1825.

See RENNIE, Sir John and W. C. MYLNE. Report on Thames Embankment, 1831.

NICHOLL, Andrew [1804–1866]

982 *Five Views of the Dublin and Kingstown Railway, from Drawings taken on the Spot by Andrew Nichol. With a Description of this Important National Work.*
Dublin: William Frederick Wakeman . . . 1836.
4to 35 cm 8 pp. 5 hand-coloured aquatint plates (J. Harris and S. G. Hughes sc.).
BM, SML, Elton

NICHOLSON, Peter [1765–1844]

983 *The Guide to Railway Masonry; comprising a complete treatise on the Oblique Arch. In three parts. Part First. History of the Oblique Arch . . . Part Second. Practical Construction . . . Part Third. An Appendix, containing various articles connected with railways . . . [etc].*
By Peter Nicholson.

London: John Weale . . . and by . . . Finlay and
  Charlton, Newcastle. 1839.
8vo   23 cm   10 + xiv + lvi + 50 pp. with extra
  leaf.   39 + (1) engraved plates.
ULG (Rastrick), Elton

984   Second edition
*The Guide to Railway Masonry; containing a complete
treatise on the Oblique Arch. In Four Parts. Part First.
Containing as much Practical Geometry as will be requis-
ite . . . Part Second. The First Principles of Descriptive
Geometry . . . Part Third. The Principles of Calculation
. . . Part Fourth. Practical Construction . . . With an
Appendix.*
  By Peter Nicholson. Second Edition.
London: Published by Richard Groombridge . . .
  and by . . . Finlay and Charlton, Newcastle.
  1840.
8vo   23 cm   10 + lii + 50 pp. with 3 extra
  leaves.   Portrait frontispiece + 38 + (2)
  engraved plates.
BM, RIBA, ISE
  A third edition appeared in 1846.

NICHOLSON, Robert [1808–1855]
M.Inst.C.E.

985   *South Durham Railway.*
  Rob! Nicholson, Engineer   Lambert sc.
Engraved map   33 × 46 cm   (scale 1 inch to 1½
  miles).
[In]
Prospectus of the South Durham Railway.
Durham: Printed by F. Humble.
Dated: 18 Jan 1836.
fol   40 cm   (4) pp.   Docket title on p. (4).
BM, CE
  The CE copy is an earlier version of the Pros-
pectus, with the same map.

NICHOLSON, William [1753–1815]

986   *Experimental Enquires concerning the . . . Motion
of Fluids: applied to the explanation of Various Hydraulic Phenomena.*
  By Citizen J. B. Venturi . . . Translated from
  the French [by W.N.].
London: Printed for J. Taylor . . . 1799.
Signed: William Nicholson. London, 5 April
  1799.
8vo   21 cm   viii + 75 pp.   2 folding engraved
  plates.
BM, RS

987   *A Letter to the Incorporated Company of Pro-
prietors of the Portsea-Island Water-Works, on the occa-
sion of a proposal lately made . . . for Conveying Water
from Farlington.*
  By William Nicholson, Engineer to the said
  Company.
London: Printed for the Author, by George Sid-
  ney . . . 1810.
Signed: William Nicholson. Charlotte Street, 4
  Jan 1810.
8vo   21 cm   (iii) + 31 + (1) pp.
BM

NICKALLS, Joseph [?1725–1793]
M.Soc.C.E.

988   *Thames Navigation: with the Improvements as
intended by the New Bill.* [1771]
  Engraved plan, details entered in colours
  30 × 91 cm   (scale 1 inch to 2 miles).
SML, BDL
  SML copy endorsed 'River Thames with a Plan
  for Improvement by the Commissioners'. *House
  of Commons Journal*, 21 Feb 1771: Mr Nickalls
  made the Survey and Plan for amending the
  River Thames. *Ibid.*, 4 March 1771: Commissioners present new Bill to Parliament.

989 *To the Honourable Lord Warden, and Assistants of Dover Harbour. The Report of Joseph Nickalls, Engineer, (made in Obedience to an Order received at Walmer Castle, 2 Sept. 1775) on the said harbour. Together with a general Plan of the same, and of such other Works as appear to me the most proper, for rendering it a complete Artificial Harbour.*
London: Printed by W. Adlard . . . 1777.
Signed: Joseph Nickalls. Southwark, 1777.
fol 36 cm 19 + (2) pp. Folding engraved plan.
Plan of Dover Harbour, with the Additional Works Proposed for the Improvement thereof.
Jos[h] Nickalls 1777.
BM

990 *The Report of Joseph Nickalls, Engineer, on the State of Wells Harbour in Norfolk, in 1782.*
Signed: Jos[h] Nickalls, Southwark, 5 July 1782.
5 pp.

[Printed with]

Report of Joseph HODSKINSON, 5 July 1782 [q.v.].

[In]

The Reports of Mess. Hodskinson, Grundy, Hogard, and Nickalls, on the State and Causes of the Decay of Wells Harbour.
[London]: 1782.
fol 33 cm 16 + 5 pp. Docket title on p. (6) Folding engraved plan (annexed to Hodskinson's report).
BM, ULG, Sutro
The Sutro copy stitched as issued.

991 *Report of Jos. Nickalls, Engineer, on the present state of the shoals in the river Severn.*
Signed: Jos. Nickalls. Gravel-Lane, Southwark, 9 April 1785.
pp. 6–8.

[In]

T. R. Nash. *Supplement to the Collections for the History of Worcestershire.*
London: Printed for John White . . . 1799.
fol 42 cm (i) + 104 pp.
BM, BDL

992 *Report of the Consequences which the New Cut, from Eau-Brink, would be attended with to Drainage, Navigation, the Harbour, and Town of Lynn.*
By Joseph Nickalls, Engineer.
London: Printed by E. Hodson . . . 1793.
Signed: Joseph Nickalls, Gravel-Lane, Southwark, 15 Feb 1793.
4to 25 cm 20 pp.
CE, CRO, CUL, Sutro

993 Reprint
As No. 992 but
Huntingdon: Re-Printed by W. Hatfield . . . 1818.
8vo 23 cm 22 pp.
CUL

NIMMO, Alexander [1783–1832] F.R.S.E., M.R.I.A.

994 *The Report(s) of Mr Alexander Nimmo, on the Bogs in the Barony of Iveragh, in the County of Roscommon [and] in various parts of the Counties of Kerry and Cork [and] in that Part of the County of Galway to the West of Lough Corrib.*
Signed: Alexander Nimmo, December 1811, 1812, [and n.d.].
With 8 hand-coloured folding engraved plans (James Basire sc.).

[In]

Fourth Report of the Commissioners on the practicability of draining and cultivating the Bogs in Ireland.
House of Commons, 1814, pp. 27–57, 59–102, 175–186 and pls. 1–8.

995 *Estimate of the Expense of forming a Packet Harbour at Portcullin Cove, near Dunmore, in the County of Waterford.*
House of Commons, 25 May 1814.
Signed: Alexander Nimmo.
fol 32 cm 2 pp. 2 folding engraved plans, the harbour plan being hand coloured (James Basire sc.).
[1] *Design for the Harbour near Dunmore, in Waterford Haven, from Actual Survey.* By Alexander Nimmo, F.R.S.Edin. Civil Engineer.
[2] *Chart of the Haven of Waterford.*
CE

996 *Report on the Means of Improving the River and Harbour of Cork.*

By Alexander Nimmo, Civil Engineer, F.R.S.Edin.

Cork: Printed by Edwards & Savage . . . 1815.

Signed: Alexander Nimmo. Cork, 11 Sept 1815.

4to   24 cm   viii + 30 pp.

CE

997 *Plan of the Works of the Harbour of Dunmore.*

By Alexander Nimmo, F.R.S.E. M.R.I.A. Civil Engineer. May 1818. Neele & Son sc.

Engraved plan   19 × 27 cm   (scale 1 inch to 400 feet).

BM

998 *The Bay and Harbour of Sligo.*

Surveyed for the Commissioners of that Port by Alexander Nimmo, F.R.S.E., M.R.I.A. Civil Engineer 1821.

Engraved plan   33 × 49 cm   (scale 2 inches to 1 mile).

BM

One of several charts by Nimmo published 1821–22 for the use of the Irish fisheries.

999 *Report of Alexander Nimmo, Civil Engineer, M.R.I.A., F.R.S.E., on the Proposed Railway between Limerick and Waterford.*

Dublin: Printed by Thom & Johnston . . . 1825.

Signed: Alexander Nimmo. Dublin, 15 Nov 1825.

8vo   21 cm   (i) + 24 pp.

CE (Telford)

1000   Revised edition

Title, printer, etc as No. 999 but 1826.

8vo   21 cm   39 pp.

ULG (Rastrick)

1001 *The Report of Alexander Nimmo, Engineer, on the improvement of the Harbour of Drogheda.*

Dublin: Printed by Charles Evans.

Signed: Alexander Nimmo. London, 10 May 1826.

8vo   21 cm   (i) + 13 + (1) pp.

NLS, CE

1002 *Bay and Harbour of Youghal.*

Surveyed under the direction of Alexander Nimmo, C.E. &c. Exhibiting the proposed [Embankment] and Bridge. Lithographed by Allen's . . . Dublin.

Litho plan   45 × 58 cm   (scale 3 inches to 1 mile).

BM

Date given in the BM Catalogue as [1830].

1003 *New Piloting Directions for St. George's Channel and the Coast of Ireland.*

By Alexander Nimmo, F.R.S.E. Civil Engineer and Hydrographer.

Dublin: Printed by A. Thom . . . 1832.

8vo   22 cm   iv + (2) + 208 pp.

BM, CE

OLDHAM, Hugh

See BRINDLEY. Plan of the Rivers Irwell and Mersey, 1761. Surveyed by Hugh Oldham.

OUTRAM, Benjamin [1764–1805]

1004 *A Plan of the intended Derby Canals and Railways with a Sketch of the adjacent Canals, Rivers and Roads shewing their relative situations & connections.*

By B. Outram 1792.

Engraved by W. Faden.

Engraved plan   68 × 60 cm   (scale 2 inches to 1 mile).

BM

1005 *The Report of B. Outram, of Butterley Hall, in the County of Derby, Engineer, on the Proposed Canal, from Sir John Ramsden's Canal, at Huddersfield, in the County of York, to join the Canal at Ashton-under-Lyne, in the County of Lancaster.*

n.p.

Signed: Benj.^m Outram. Huddersfield, 22 Oct 1793.
fol   22 cm   2 pp.   Docket title on p. (4).
P

1006 *Plan of the Canal between Huddersfield in the County of York and Ashton under Lyne in the County of Lancaster.*
   B. Outram, Engineer. N. Brown Surveyor. J. Cary sc. [1793].
Engraved map, the line entered in colour
   48 × 97 cm   (scale 2 inches to 1 mile).
   BM, BDL

1007 *Observations on the Brecknock and Abergavenny Canal and Railways.*
n.p.
Signed: Benjamin Outram, 1 July 1799.
fol   42 cm   4 pp.
National Library of Wales

1008 *Report and Estimate of the Proposed Rail-Ways from the Colleries in the Forest of Dean, to the Rivers Severn and Wye.*
Hereford: Printed by D. Walker . . . 1801.
Signed: Benj. Outram, 5 Sept 1801.
12mo   16 cm   16 pp.
B

OVERTON, George

See MEYNELL. Reports on proposed Stockton & Darlington Railway by Overton, 1818 and 1821.

1009 *A Description of the Faults or Dykes of the Mineral Basin of South Wales. Part I. Introductory Observations on the Mineral Basin, Tramroads, Railways, &c.*
   By George Overton, Esq. Civil Engineer.
London: Printed for Knight and Lacey . . . 1825.
Dated: Llanthetty Hall, 31 Dec 1824.
4to   28 cm   79 pp.
BM, Elton
   The chapter on tramroads and railways covers pp. 25–79.

PADLEY, James Sandby
[*c*. 1792–1881]

1010 *Report on the Practicability of a Junction Canal between the Rivers Witham and Ancholme, in the County of Lincoln.*
Lincoln: Edward B. Drury, Printer.
Signed: J. S. Padley. Lincoln, 18 May 1828.
fol   30 cm   10 pp.   Folding litho plan, hand coloured.
Plan shewing the Line of the proposed Canal, from the River Witham to Bishop's Bridge on the River Ancholme: in the County of Lincoln.
   By J. S. Padley 1828. T. Brettell Lithog.
LCL

1011 *Plan & Section of the Navigable Canal from the Town of Louth, to the River Humber in the County of Lincoln, as surveyed Octr 1828.*
   By J. S. Padley, Lincoln. Lithog by J. & J. Jackson.
Litho plan, hand coloured   43 × 100 cm   (scale 3 inches to 1 mile).
LCL

1012 *Louth Canal. A Description of the Locks, Bridges, &c. on the Canal from Tetney to Louth.*
   Printed to accompany Mr Padley's Plan of the Canal as surveyed by him in 1828.
Louth: Printed and Sold by J. and J. Jackson. 1832.
Signed: J. S. Padley, Lincoln, 17 Nov 1828.
8vo   22 cm   (i) + 21 pp.
LAO, LCL

PAGE, Edward

1013 *The Report of Mr Edward Page upon the Better Drainage of the Lands within the Level of Holderness Drainage.*
Beverley: Printed by M. Turner.
Signed: Edward Page. Beverley, 3 May 1831.

8vo 22 cm 15 pp. Folding engraved plan.
Plan of the Proposed Alterations and Improvements in the Holderness Drainage.
May 1831. Edward Page.
HRO

PAGE, *Colonel* Frederick [1769–1834] Hon.M.Soc.C.E., Hon.M.Inst.C.E.

1014 *Observations on the Present State, and Possible Improvement of the Navigation and Government of the River Thames.*
By Frederick Page, Esq.
Reading: Printed by Smart and Cowslade . . . 1794.
8vo 19 cm 41 pp. Folding engraved plan.
Plan of the Thames to illustrate the Observations &c.
CE (Page)

1015 *Observations on the General Comparative Merits of Inland Communication by Navigation or Rail-Roads.*
Bath: Printed by Richard Cruttwell . . . 1825. Price 2*s*.
Signed: A Proprietor of Shares in the Kennet and Avon Canal. 29 March 1825.
8vo 22 cm (iii) + 62 pp. Folding litho plate.
BM, CE
    Attributed to Page.

1016 *A Letter to a Friend, containing observations on the comparative merits of Canals and Railways, occasioned by the Report of the Committee, of the Liverpool and Manchester Railway.*
[London]: Longman & Co . . . and Robert Fenn . . . 1832.
Signed: F. P. London, 1 February 1832.
8vo 23 cm (i) + 32 pp. Folding plate.
Section of the Cromford High Peak Railway [and] *Section of the intended Railway from Manchester to Sheffield.*
BM, CE, ULG
    The ULG copy endorsed 'Col: Page to Mr Priestley' [of Wakefield].

1017 Second edition
A Letter to a Friend, containing observations on the comparative merits of Canals and Railways . . . Second Edition, with Additions arising from the evidence on the London and Birmingham Railway.
[London]: Longman and Co . . . and Robert Fenn . . . 1832.
Signed: F.P. 10 May 1832.
8vo 23 cm 42 pp. Folding plate (as in first edn.)
BM, CE, ULG (Rastrick)
    The CE copy given by the author to the Society of Civil Engineers.

PAGE, *Sir* Thomas Hyde [1746–1821] R.E., F.R.S., M.Soc.C.E.

1018 *Observations on the Present State of the South Level of the Fens, with a Proposed Method, for the better Drainage of that Country, made by the desire of the . . . Master General of the Ordnance, and the Honourable the Corporation of Bedford Level.*
By Lieut. Page (now Sir Thomas Hyde Page, Knt.) of the Corps of Engineers.
London: Printed in the year 1775. Re-printed in 1793 by H. Reynell. [Dated: London, 31 March 1775].
12mo 18 cm 12 pp.
CRO, BDL, CUL
    Date as given in a contemporary MS. copy (CE).

1019 *The Reports or Observations of Sir Thomas Hyde Page, on the Means of Draining the South and Middle Levels of the Fens.*
Printed in 1775 and 1777. Reprinted 1794.
8vo 22 cm (iii) + 38 pp. Folding engraved plan.
A Sketch of the River Ouse, from Erith to Lynn, with a Section of the Channel shewing the obstructions which prevent the draining of the Bedford Levels.
BM, CE (Vaughan), ULG, BDL, CUL, Sutro
    Comprises (pp. 1–11) Observations on the Present State of the South Level, 1775 [second

reprint] and (pp. 13–38) Observations upon the Draining of the South and Middle Levels of the Fens, 1777. The original editions have not been found.

1020 *Considerations upon the State of Dover Harbour, with its Relative Consequences to the Navy of Great Britain.*
By Sir Thomas Hyde Page, Knt. F.R.S. Of His Majesty's Corps of Engineers.
Canterbury: Printed for the Author, by Simmons and Kirby, 1784.
Signed: Tho.ˢ Hyde Page. Dover, 18 March 1784.
4to   26 cm   vi + 29 pp.
BM, BDL (Gough), CUL

1021 *Opinion of Sir Thomas Hyde Page, of the Royal Engineers, upon the Proposed Cut, from Eau-Brink to Lynn.*
[Addressed to Sir Martin Folkes].
Lynn: Whittingham, Printer . . . 1809.
Signed: Thomas Hyde Page. London, 26 Oct 1793.
fol   32 cm   3 pp.   Docket title on p. (4).
P
Apparently a reprint, but 'Investigator' (*see* EAU BRINK CUT, 1794) had seen only a manuscript copy.

1022 *A Letter to Sir Martin Folkes, upon the Eau-Brink Cut.*
n.p.
Signed: Thomas Hyde Page. Beaumaris, 10 March 1794.
fol   33 cm   3 pp.   Docket title on p. (4).
CUL

1023 *Minutes of Evidence of Sir Thomas Hyde Page, Knight, on the Second Reading of the Eau Brink Drainage Bill.*
London: 1794.
Dated: 27 March 1794.
8vo   21 cm   76 pp.
BM, CE, ULG

1024 *Estimate of the Expence of carrying into execution the Plan of Embankment, recommended . . . for the Improvement of the Drainage of the South and Middle Levels, and other Lands having their Drainage through the River Ouse, and for the Improvement of the Navigation of that River . . . Together with the Survey and Measurement of the Several Rivers upon which the Estimate is founded.*
By Sir Thomas Hyde Page, Royal Engineers.
London: Printed 1794.
Signed: Thomas Hyde Page. London, 12 Aug 1794.
8vo   23 cm   (i) + 29 pp.   Folding engraved plan.
A Map showing the Embankment of the Rivers which have their Outfall at Lynn, according to the Plan proposed by Sir Thomas Hyde Page.
BM, SML, ULG, BDL, Sutro
Includes (pp. 9–28) a report by Thomas Cubitt, land surveyor, on measurement of the rivers between Clayhithe and Lynn, 1794.

1025 *An Account of the Commencement and Progress in Sinking Wells at Sheerness, Harwich and Landguard Fort, for supplying those Dock-Yards and Garrisons with Fresh Water. To which is annexed the Correspondence . . . upon the Subject, in the Years 1778, 1781, and 1783.*
London: Printed for John Stockdale . . . 1797.
Price 1s.
Signed: Thomas Hyde Page. St. Margaret's Street, 12 May 1783 [and] Upper Fitzroy Street, 14 April 1795.
8vo   20 cm   42 pp.
CE (Vaughan), ULG, BDL, CUL

1026 *Observations upon the Embankment of Rivers; and Land inclosed upon the Sea Coast.*
By Sir Thomas Hyde Page, Knt. F.R.S. Written in the year 1796.
Tunbridge Wells: printed for the Author, by J. Sprange, 1801.
8vo   21 cm   (i) + 20 pp.
BM, CE (Vaughan), BDL

1027 *Reports relative to Dublin Harbour and Adjacent Coast, made in Consequence of Orders from the . . . Lord Lieutenant of Ireland, in the Year 1800.*
By Sir Thomas Hyde Page, Knt. F.R.S.
Dublin: Printed in the Year 1801.
8vo   19 cm   55 pp.
CE (Vaughan), BDL

Includes four reports dated Dublin, 7 Sept, 23 Sept, 28 Nov, 29 Nov 1800.

1028 Second printing
The four reports reprinted (pp. 3–22) in:
[*Reports on Dunleary Harbour*].
[Dublin, 1802].
fol   32 cm   84 pp.   5 folding engraved plans.
NLS (Rennie)

1029 *Correspondence upon the Subject of the Eau-Brink Cut, between Sir Thomas Hyde Page, and Mr Mylne, in the Years 1801 and 1802.*
Lynn: Printed by Andrew Pigge . . . 1802.
4to   23 cm   65 pp.
CRO, BDL

PALMER, Henry Robinson [1795–1844] F.R.S., M.Inst.C.E.

See TELFORD. Map of proposed Knaresborough Railway, 1818.

1030 *Description of a Railway on a New Principle; with observations on those hitherto constructed. And a Table shewing the comparative amount of resistance on several now in use . . . and a description of an Improved Dynamometer.*
By Henry R. Palmer, Mem.Inst.Civ.Eng.
London: Printed for J. Taylor . . . 1823.
Dated: Abingdon Street, Westminster, 23 May 1823.
8vo   23 cm   vi + (i) + 60 pp.   2 folding engraved plates.
[1] View of *Patent Railway on a New Principle*.
H. R. Palmer del.   E. Turrell & Miss Letitia Byrne sc.
[2] Details   H. R. Palmer del.   E. Turrell sc.
SML, CE, ULG, Elton

1031 Second edition
As No. 1030 but *Second Edition, Revised* . . . 1824.
Dated: Abingdon Street, Westminster, 1824.

8vo   23 cm   viii + 60 pp.   2 folding engraved plates. [1] as in first edition, [2] retouched.
BM, SML, CE, ULG

1032 *Plan of the Harbour of Swansea shewing the proposed Improvements.*
By Henry R. Palmer, Civil Engineer. [*c.* 1830].
T. Bedford lithog.
Litho map   45 × 23 cm.
SML

1033 *Plan of London Docks as completed in 1831.*
Henry R. Palmer, Engineer. E. Turrell sc.
Engraved map   25 × 52 cm   (Scale 1 inch to 420 feet).
PLA

1034 *Observations on the Motions of Shingle Beaches.*
By Henry R. Palmer, Esq. F.R.S. From the Philosophical Transactions. Part II for 1834.
London: Printed by Richard Taylor . . . 1834.
4to   29 cm   (i) + [567–576] pp.   1 engraved and 2 litho plates.
CE (from the author)
Offprint from *Phil. Trans.* with new title page.

1035 *Map of the Line of the South Eastern Railway, intended to connect the Metropolis with the Counties of Kent and Sussex, and afford the most direct & advantageous means of communication between London & Dover by Tunbridge and Ashford.*
Surveyed under the direction of Henry R. Palmer, Civil Engineer, F.R.S. by P. W. Barlow, F. W. Simms and W. Froude. 1835. J. Basire lith.
Litho map   40 × 52 cm   (scale 1 inch to 4 miles).
[In]
London and Dover (South-Eastern) Railway. Statement of the Promoters.
London: Printed by C. Roworth and Sons . . . [1836].
fol   41 cm   (4) pp.   including map.
BM
A smaller version of the map is in the *Prospectus*.

1036 *Report on the Proposed Improvements in the Port of Ipswich.*

By Henry R. Palmer, Civil Engineer, F.R.S.
London: Vacher and Sons . . .[1836].
Signed: Henry R. Palmer, Great George Street,
  Westminster.
8vo   28 cm   (i) + 9 + (i) pp.   Folding
  engraved plan.
Plan of the Proposed Wet Dock at the Port of Ipswich.
  By Henry R. Palmer, Civil Engineer, F.R.S.
  1836. J.⁵ Basire sc.
BM, CE
  Issued in brown printed wrappers.

1037   *Report on the improvement of the Rivers Mersey
and Irwell, between Liverpool and Manchester, describing
the means of adapting them for the navigation of Sea-
Going Vessels.*
  By Henry R. Palmer, F.R.S. Vice-Pres. Inst.
  C.E.
London: Published by John Weale . . . 1840.
Dated: Great George Street, Westminister, 29
  June 1840.
8vo   21 cm   (iii) + 44 pp.   Frontispiece +
  folding engraved plan.
*Plan of the Rivers Mersey and Irwell between Weston
Point and Manchester shewing the Proposed Improve-
ments . . . to render them capable of passing Sea-Going
Vessels from Liverpool through Warrington to Manches-
ter.*
  By Henry R. Palmer. Surveyed & Drawn by
  M. Singleton. Engraved by Jas. Wyld.
BM, CE, ULG

PALMER, William [d. 1737]

1038   *A Survey of the River Dunn in order to improve
the Navigation from Hull to Doncaster and to continue up
to Sheffield. By Will. Palmer & Partners taken Anno
Dom: 1722.*
  Eman. Bowen sc.
Engraved plan   18 × 57 cm   (scale 1 inch to 1½
  miles).
                    [In]

*The Methods proposed for making the River Dunn
Navigable, and the Objections to it answered . . . To
which is annexed, a Mapp of the River, and the Reasons
lately printed for making it Navigable, with the Advantages of it.*
London: Printed in the Year 1723.
4to   20 cm   15 pp.   Folding engraved plan (as
  above).
BM (plan only), BDL, Sutro
  The partners were Joseph Atkinson and Joshua
  Mitchell.

1039   *A Map of the River Ouse, from its Rise near
Ousbourn, in the West Riding of the County of York: to
its Falling into the Trent, and Humber. With the New
Dutch River, and the Mouths of the Navigable River
Ayres, Darwent, Wharf, and Your; together with the
Mouths of other Rivers that may be made Navigable, also
Drains, Banks, Clows, Ferries, Bridges &c. Survey'd
by Will.ᵐ Palmer & Partners in December 1725.*
Engraved plan   22 × 53 cm   (scale 1 inch to 2
  miles).
BM, BDL (Gough)

See ELLISON and PALMER, 1735. Survey of
  the Rivers Swale & Ouze from Richmond to
  York.

PARKER, Joseph

1040   *A Plan of the intended Navigable Canal from
Basingstoke to the River Wey.*
  Surveyed by Joseph Parker. Engraved by W.ᵐ
  Faden [1777].
Engraved plan   28 × 80 cm   (scale 1 inch to 1
  mile).
BM, SML, RS (Smeaton), CE (Page)
  The Basingstoke Canal Act was obtained in
  May 1778.

PARNELL, *Sir* Henry [1776–1842] M.P., Hon.M.Inst.C.E.

1041 *A Treatise on Roads; wherein the Principles on which roads should be made are explained and illustrated by the Plans, Specifications, and Contracts made use of by Thomas Telford, Esq. on the Holyhead Road.*

By The Right Hon. Sir Henry Parnell, Bart.
London: Printed for Longman, Rees . . .1833.
8vo   23 cm   xii + 438 pp.   7 folding engraved plates (drawn by J. F. Dundas, E. Turrell sc).
BM, NLS, SML, ULG, IC, BDL
  Issued in publisher's cloth, price £1 1s.

1042   Second edition
Title as No. 1041 but *Second Edition.*
London: Printed for Longman, Orme . . . 1838.
8vo   23 cm   xii + 465 pp.   9 folding engraved plates.
BM, CE, ULG, BDL

PASLEY, *Sir* Charles William [1780–1861] R.E., F.R.S., M.Soc.C.E., Hon.M.Inst.C.E.

1043   *Course of Instruction originally composed for the use of the Royal Engineer Department. Vol. 1 Containing Practical Geometry and the Principles of Plan Drawing.*

By C. W. Pasley, Capt. R.E.
London: Printed for John Murray . . . 1814.
8vo   21 cm   xvi + 269 pp.   Woodcuts in text.
BM, SML, RS

1044   *Course of Military Instruction, originally composed for the use of the Royal Engineer Department. Vol. II (and Vol. III) Containing Elementary Fortification.*

By C. W. Pasley, Lieut-Colonel Royal Engineers, F.R.S.
London: Printed for John Murray . . . 1817.

8vo   21 cm   **2**, (iii) + 6 + xxxix + 335 pp.   **3**, (iii) + xxix + 335–702 pp.
Woodcuts in text throughout.
BM, RS

1045   Second edition
A Course of Elementary Fortification, including rules, deduced from experiment, for determining the Strength of Revetments . . . Originally published as part of a course of military instruction.

By C. W. Pasley, Lieutenant-Colonel Royal Engineers, F.R.S. Second Edition, Vol. I (and Vol. II).
London: Published by John Murray . . . 1817.
Dated: Chatham, 15 April 1822.
8vo   21 cm   **1**, (iii) + 6 + xxxix + (2) + 335 pp.   Woodcuts in text. 5 folding engraved plates (E. Turrell sc.)   **2**, (iii) + xxix + 335–702 pp.   Woodcuts in text.
BM, CE
  Except for new title-pages and 'Advertisement to Second Edition' these volumes are a reissue of Nos. 1043 and 1044 with addition of 5 plates.

1046   *Observations, deduced from experiment, upon the natural Water Cements of England, and on the Artificial Cements, that may be used as substitutes for them.*

By C. W. Pasley, Lieutenant-Colonel in the Corps of Royal Engineers, F.R.S.
Chatham: Printed by Authority, at the Establishment for Field Instruction, 1830.
Dated: Chatham, 7 July 1830.
8vo   22 cm   (iv) + 12 pp.   1 engraved plate.
BM, RS, CE

1047   *Observations on Limes, Calcareous Cements, Mortars, Stuccos and Concrete, and on Puzzolanas Natural and Artificial. Rules deduced from numerous experiments, for making an Artificial Water Cement* [etc]

By C. W. Pasley, C.B.   Colonel in the Corps of Royal Engineers, F.R.S.
London: John Weale . . . 1838.
Signed: C. W. Pasley. Royal Engineer Establishment Chatham, 17 Sept 1838.
8vo   22 cm   (lxxiii) + 288 + 124 pp.   Woodcuts in text.
BM, RS, CE

PATERSON, James

1048 *A Practical Treatise on the making and uphold-ing of Public Roads . . . and a Dissertation on the utility of Broad Wheels, and other improvements.*
By James Paterson, Road Surveyor, Montrose.
Montrose: Printed by James Watt, for Thomas
  Donaldson, Dundee; Longman . . . 1819.
12mo   18 cm   93 + (1) pp.
BM, NLS, CE, ULG

1049 *A Series of Letters and Communications, ad-dressed to the Select Committee of the House of Commons, on the Highways of the Kingdom* [etc].
By James Paterson, Road Surveyor, Montrose.
Montrose: Printed by James Watt; for Longman
  & Co., London; Constable & Co., Edinburgh
  . . . 1822.
12mo   17 cm   87 pp.
BM, NLS, ULG, BDL

1050 *M'Adam and Roads.*
[Letter] To Sir Thomas Baring, Baronet, M.P.
  and late Chairman of the Select Committee . . .
  to inquire into the claims of Mr M'Adam.
Montrose: Printed by [James] Watt. Sold by
  Longman & Co . . . [1824].
Signed: James Paterson. Montrose, 30 Nov 1824.
8vo   21 cm   20 pp.
CE

PATERSON, John [d. 1823]

1051 *[Report to the Lord Provost of Edinburgh and the Joint Committee for Superintending the Construction of the Wet Docks at Leith].*
[Edinburgh. 1810].
Signed: C. Cuningham (Clerk), John Paterson
  (Resident Engineer).
Edinburgh, 21 Aug 1810.
4to   28 cm   7 pp.
CE
  Ordered to be printed 22 Aug 1810.

1052 *Observations occasioned by the Mineral Survey and Report of Mr Robert Bald, on the proposed Level Line of Canal between Leith, Edinburgh, and Glasgow, Surveyed by Mr Rennie in 1798, in which a Comparison is attempted to be drawn between the Utility of that Canal and of the proposed Union Navigation.*
By John Paterson.
Dated: 12 Sept 1814.
pp. (1)–9
           [Printed with]
Report of a Mineral Survey . . . by Robert BALD,
  28 May 1814 [q.v.].
           [In]
Report of a Mineral Survey along the Track of the Pro-posed North or Level Line of Canal betwixt Edinburgh and Glasgow.
Leith: W. Reid & Co. Printers. [1814]
4to   25 cm   (iii) + 25 + 9 + 2 pp.
NLS, CE, ULG (Rastrick)

PEAR, Thomas

1053 *Preliminary Report and Estimate of the means of Improving the discharge of the North Level Waters from Clow's Cross to the Sea.*
n.p.
Signed: Thomas Pear. Spalding, 5 Sept 1828.
fol   32 cm   3 pp.   Folding litho plan.
Sketch of the North Level, and Great Portsand, and of Proposed New Line of Drainage from Clow's Cross through Wisbech Hundred to Kinderley's Cut.
  Sept 1828. R. Cartwright Lithogr
CRO

PERRY, *Captain* John [c. 1670–1733]

1054 *The State of Russia, under the Present Czar. In Relation to the several great and remarkable Things he has done, as to his Naval Preparations, the Regulating his Army, the Reforming his People, and Improvement of his*

Countrey. Particularly those Works on which the Author was employ'd, with the Reasons of his quitting the Czar's Service, after having been Fourteen Years in that Countrey [etc].

By Captain John Perry.

London: Printed for Benjamin Tooke . . . 1716.

8vo   21 cm   (vii) + 280 pp.   Folding engraved map.

A New Map of the Empire of the Czar of Russia, with the Improvements and Corrections of Capt. John Perry.

By Herman Moll Geographer.

BM, NLS, ULG, BDL, CUL

Perry was engaged on canal and dock works in Russia from 1698 to 1712.

1055   *An Account of the Stopping of Daggenham Breach: with the Accidents that have attended the same from the first Undertaking. Containing also proper Rules for performing any the like Work: and Proposals for rendering the Ports of Dover and Dublin (which the Author has been employ'd to Survey) Commodius for Entertaining large Ships.*

By Capt. John Perry.

London: Printed for Benj. Tooke . . . and Sold by J. Peele . . . 1721.

8vo   20 cm   131 pp.   Folding engraved plan.

A Plan of the Late Breach in the Levells of Havering and Dagenham.

H. Moll sc.

BM, CE, CRO, ULG, IC, BDL (Gough), CUL

The *Proposals for . . . the Ports of Dover and Dublin* (signed: J. Perry, November 1718) are printed on pp. 112–127.

1056   *The Description of a Method humbly proposed for the making of a better Depth coming over the Barr of Dublin: As also for the making of a Bason within the Harbour.*

n.p.

Signed: J. Perry. Dublin, 7 Oct 1721.

4to   23 cm   12 pp.

CUL

1057   *An Answer to Objections against the making of a Bason, with Reasons for the bettering of the Harbour of Dublin.*

By Capt. John Perry.

Dublin: Printed by S. Powell . . . 1721.

Signed: John Perry. Dublin, 6 Dec 1721.

8vo   17 cm   28 pp.

BM, BDL (Gough)

1058   *Remarks on the Reasonings and different Opinions laid down in some late Printed Proposals for Restoring the Navigation of the Ports of King's-Lynn and Wisbich; and Draining of the great Levels, which have the Outfall of their Waters by the River Ouse and the Nean, thro Mortons-Leam. With a Method humbly offered to Consideration, for obtaining a proper Depth to both the said Rivers, and for the future Effectual Draining of the Levels, thro which they have their course.*

n.p.

Signed: J. Perry. 22 March 1724/5.

fol   34 cm   15 pp.

BM

1059   *A Method propos'd for making a safe and convenient Entrance to the Port of Dublin for Ships of Burthen to come in and go out by Day or Night at Low-water; without those Hazards, Loss of Opportunity of Winds and Difficulties they are now expos'd to in their Voyages. Together with an Estimate of the Charge that will be required for effecting the same.*

n.p.

Signed: John Perry. Dublin, 29 Nov 1725.

fol   35 cm   18 pp.

ULG, CUL

1060   [Report] *To the Gentlemen Land-Owners in the Parts of South Holland, in the County of Lincoln.*

[London, 1727].

Signed: J. Perry. London, 25 Feb 1726/7.

fol   33 cm   14 pp.   Folding engraved plan.

A Map of the North Level &c and of the Marshes. A Drain is propos'd to be made through to convey the Waters that now overflow the Level &c to Sea. To which are added Some Particulars explaining Capt. Perry's Report, on his view lately taken of the Drains, Sluices, & Outfalls.

BM, CE, Sutro

PHILLIPS, John

1061 *A Treatise on Inland Navigation: illustrated with a whole-sheet plan, delineating the course of an intended Navigable Canal from London to Norwich and Lynn . . . and a Plan for extending the Navigation from Bishop-Stortford to Cambridge,* [etc].
London: Printed for S. Hooper . . . 1785.
Signed: John Phillips.
4to   28 cm   xii + (viii) + 50 pp.   Engraved
  plate + folding hand-coloured engraved plan.
A Plan for a Navigable Canal from London to Norwich & Lynn . . . and . . . from Bishop Stortford to Cambridge.
  Published 21 Aug 1785.
BM, NLS, CE (lacks plan), ULG, BDL (Gough)

See KITCHIN and PHILLIPS. A New Map of England & Wales with Canals to 1792.

1062 *A General History of Inland Navigation, Foreign and Domestic: containing a complete account of the canals already executed in England . . . To which are added, Practical Observations. The whole illustrated with a map of all the canals in England, and other useful plates.*
  By J. Phillips.
London: Printed for I. and J. Taylor . . . 1792.
4to   27 cm   (i) + xx + 369 + (4) pp.   4 en-
  graved plates + large folding engraved map,
  the rivers and canals entered in colours.
A Map of England, shewing the Lines of all the Navigable Canals; with those which have been proposed.
  Engraved for The General History of Navigation. 1792. T. Conder sc.
NLS, BDL (the map bound separately), Elton

1063   First edition, with Addenda
As No. 1062 but bound with *Addenda to the History of Inland Navigation.* Printed in 1793.
4to   27 cm   (i) + xx + 369+ (4), (i) + 33 pp.
  4 engraved plates + folding engraved map (as
  before).
CE

1064   First edition, with two Addenda
As No. 1062 but bound with the [First] and *Second Addenda to the History of Inland Navigation.* Printed in 1794.

4to   27 cm   (i) + xx + 369 + (4),   (i) + 33,
  [35–185] + 1 pp.   4 engraved plates + folding engraved map (as before).
BM, ULG
  Copies of the two *Addenda*, which were issued as separate publications, are in CE (Telford).

1065   Second edition
Title as No. 1062 but *A New Edition Corrected. With an Addenda, which completes the History to 1792.*
London: Printed for I. and J. Taylor . . . 1793.
4to   27 cm   xx + 366 + (5) + (i) + 33 pp.   4
  engraved plates + folding engraved map (as in
  first edition).
LCL, ULG
  The second edition incorporates some textual changes and rearrangement of material, and the *Addenda* of 1793.

1066   Third edition
A General History of Inland Navigation, Foreign and Domestic . . . To which are added, Practical Observations. The whole illustrated with a Large Map Coloured, Shewing the Lines of all Canals executed, those proposed, and the Navigable Rivers; with other Useful Plates.
  By J. Phillips. A New Edition Corrected, with Two Addenda, which complete the History to 1795.
London: Printed for I. and J. Taylor . . . 1795.
4to 27 cm xx + 366 + (5) + (i) + 185 + (i) pp.
  4 engraved plates + large folding engraved map,
  the rivers and canals entered in colours (titled
  as in first edition but dated 1795).
BM, CE, ULG
  Issued in boards. Price £1 8s.

1067   Fourth edition
A General History of Inland Navigation, Foreign and Domestic: containing a complete account of the canals already executed in England; with considerations on those projected. Abridged from the quarto edition, and continued to the present time.
  By J. Phillips. The Fourth Edition.
London: Printed for J. Taylor . . . and C. and R. Baldwin . . . 1803.
Signed: J. Phillips. London, July 1803.
8vo   23 cm   xix + 598 + (2) pp.   Folding
  engraved plate.

BM, ULG

Issued without the map which could be purchased separately for 5s. coloured, as stated on p. 598. Material of the 3rd edition is here rearranged, in shorter form, together with a chronological list of canal events from 1794 to 1802.

1068   Fourth edition variant

As No. 1067 except for note on p. 598 referring to the map, stated to cost 7/6^d, and a new title page.

The General History of Inland Navigation; containing a complete account of all the canals of the United Kingdom, with their variations and extensions . . . And a brief history of the Canals of Foreign Countries.

By John Phillips. Senr Sometime Surveyor to the Canals in Russia under Mr Cameron, Architect to the late Empress Catherine II. Fourth edition. London: Printed by and for C. and R. Baldwin . . . 1803.

P

A French translation was published in Paris 1819.

1069   Fifth edition

A General History of Inland Navigation, Foreign and Domestic: containing a complete account of the Canals already executed in England; with considerations on those projected.

By J. Phillips, author of the New Builders Price Book. The Fifth Edition.

London: Printed for B. Crosby and Co . . . [1809].

8vo   23 cm   (ii) + xix + 598 + (2) pp.   Folding engraved plate.

SML, CE

As No. 1067 except for title-page and addition of half-title with advertisement on verso. The date 1805 usually ascribed to this edition is not correct. It was issued in October 1809, price 10s. 6d. in boards.

1070   Fifth edition variant

As No. 1069 but title-page in new setting of type and with addition of the date 1089 [i.e. 1809].

P

PHILLIPS, Robert

1071   *A Dissertation concerning the Present State of the High Roads of England, especially of those near London. Wherein is propos'd a New Method of Repairing and Maintaining them.*

Read before the Royal Society 27th January and 3rd February 1736–7.

By Robert Phillips.

London: Printed and Sold by L. Gilliver and J. Clark . . . 1737. Price One Shilling and Six Pence.

8vo   20 cm   xvi + 62 pp.   8 engraved plates.

BM, CE, ULG, BDL (Gough)

PICKERNELL, Jonathan [*c. 1738–1812*]

1072   *A Plan for altering the Course of the River Tees between Stockton and Portrack.*

Stockton: Printed by R. Christopher, 1791.

Dated: 14 March 1791.

Original not found, but the report is reprinted in John Brewster. *History and Antiquities of Stockton-upon-Tees* (Stockton, 1829) pp. 162–166. *See also*: Admiralty Report on Tees Conservatory Bill, 1851.

1073   *Plan of the Town of Whitby* [showing the harbour piers].

Inset (33 × 20 cm; scale 12½ ins to 1 mile) in *Draught of part of the Coast of Yorkshire . . . by Jon^a Pickernell 1791.*

Engraved by J. Walker . . . Published by J. Pickernell 28 Jan 1792.

BM, RGS, Sutro

See PILLEY and PICKERNELL. Grimsby navigation, 1795.

PICKERNELL, Jonathan *junior* [d. 1814]

1074 *A Report on the Present State of the Harbour of Sunderland, in the County of Durham, with Remarks on the Measures to be adopted for its Improvement.* Addressed to the Commissioners of the River Wear, and by them ordered to be printed.
By Jon. Pickernell, Jun.
Sunderland: Printed by James Graham [1798].
Signed: J. Pickernell jnr., 16 April 1798.
8vo   18 cm   16 pp.   Folding engraved plate [plan of the harbour].
CE (Vaughan)

PILLEY, Michael and PICKERNELL, Jonathan

1075   Deposited plan
A Plan of the Navigation, from the Humber to Great Grimsby, in the County of Lincoln; with the proposed Alterations for the Improvement thereof.
By Rob.ᵗ Stickney, 1795.
Engraved plan   41 × 38 cm   (scale 10 inches to 1 mile).
HLRO, RGS
The accompanying estimate (HLRO) is signed: Mich. Pilley, Jon.ⁿ Pickernell.

PINNELL, Thomas

1076   *Plan of the intended Navigable Canal from Berkeley Pill to the City of Gloucester.*
Surveyed by T. Pinnell 1792.
Engraved by W. Faden.
Engraved plan   28 × 47 cm   (scale 1 inch to 1 mile).
BM, CE (Page)

PITT, William

1077   *A Plan of the Wyrley and Essington Canal and Extension, from Wolverhampton to the Coventry Canal at Huddlesford, with Collateral Cuts to Wyrley and Essington Coal Mines, and to Rushall and Hay Head Lime Works;* from Surveys taken in 1791, 1792 & 1793.
By William Pitt.
Engraved plan   36 × 51 cm   (scale 1 inch to 1 mile).
BM, BDL (Gough)

PLYMOUTH BREAKWATER

1078   *Interesting Particulars, relative to that Great National Undertaking, the Breakwater now constructing in Plymouth Sound.*
Plymouth-Dock: Printed by J. Johns. 1821.
8vo   21 cm   (iii) + 34 pp.   3 folding engraved plans (John Cooke sc.).
[1] *Borough of Plymouth.* Publ. April 1820.
[2] *Plymouth Breakwater [Plan] and Transverse Section.* Publ. Aug 1820.
[3] *Cooke's Guide to Plymouth Sound and Breakwater.* Publ. Aug 1819.
CE (Telford), ULG

POCKLINGTON, William

1079   *Report and Estimate of Mr William Pocklington, respecting the Drainage of the East, West, and Wildmore Fens, and the East Holland Towns.*
Boston: Printed by B. B. Kelsey.
Signed: William Pocklington. Sibsey, 23 April 1800.
fol   34 cm   4 pp.   Docket title on p. (6).
NLS (Rennie), LCL
Stitched as issued.

POWNALL, Thomas [1722–1805] F.R.S.

1080 *A Memoir, entitled Drainage and Navigation but One United Work; and an Outfall to Deep Water the first and necessary Step to it. Addressed to the Corporations of Lynn-Regis and Bedford Level.*
By T. Pownall, Esq. M.P.
London: Printed for J. Almon . . . 1775.
Signed: T. Pownall. Richmond, 27 July 1775.
8vo   22 cm   iv + 51 pp.
BM, CE (Page), BDL (Gough)

PRICE, Francis [c. 1704–1753]

1081 *A Series of particular and useful Observations, made with great Diligence and Care, upon that Admirable Structure, the Cathedral-Church of Salisbury.*
By Francis Price.
London: Printed by C. and J. Ackers . . . 1753.
4to   (xiv) + v + 78 pp.   Engraved frontispiece + 13 engraved plates (P. Fourdrinier sc.).
BM, RIBA, BDL
   Includes accurate measurements of foundation settlement and structural deformation.

PRICE, Henry Habberley [1794–1839] M.R.I.A., M.Inst.C.E.

1082 *Reports of a Survey of the River Tees, made by Order of the Tees Navigation Company, in the Year 1824.*
By H. H. Price, Civil Engineer.
Stockton: Printed by T. Jennett.   1825.
Signed: Henry Habberley Price. Stockton-on-Tees. 1 May 1834 [and] 1 Nov 1834.
8vo   21 cm   21 pp.
NLS

1083 *Report on the Improvement of the Harbour of Swansea.*
Signed: H. H. Price. The Rhyddings, near Swansea, 1 July 1831.
pp. (43)–61.
[In]
Reports on the Formation of a Floating Harbour at Swansea, with reference to plans submitted to the Trustees.
Swansea: Printed . . . by W. C. Murray and D. Rees. 1831.
8vo   20 cm   78 pp.
Nat. Library of Wales, SML

1084 *Report on the Harbour of Falmouth.*
By Henry H. Price, Civil Engineer. M.R.I.A., F.R.G.S.
Falmouth: Printed by Jane Trahan . . . [1835].
Signed: Henry Habberley Price. Falmouth, 9 Jan 1835.
fol   32 cm   15 pp.   3 folding litho plans, hand-coloured.
[1] *Chart of Falmouth Harbour.*
[2] *Survey of the Inner Harbour of Falmouth with the Improvement proposed.*
[3] *Sections of the Inner Harbour of Falmouth.*
SML

1085 *Report of Henry Habberley Price, Esq. M.R.I.A., F.R.G.S., &c. Civil Engineer, respecting the completion of the Grand Surrey Canal, to the Thames at Vauxhall, and laying a Railway on the banks thereof . . . with branches to the Elephant and Castle, and to the New Pier, at Deptford.*
[London]: Published by Smith, Elder, and Co . . . 1835.
Signed: Henry Habberley Price. Parliament Street, Westminister, 27 Oct 1835.
fol   31 cm   13 pp.   Folding hand-coloured litho plan + folding litho plate.
Grand Surrey Canal Extension and Railway.
   By Henry Habberley Price., C.E.   J. R. Jobbins Lithog.
Transverse Section and View of the Grand Surrey Canal and Intended Railway.
Drawn on Stone & Printed by J. R. Jobbins.
BM

See PRICE and LAXTON. Map of proposed railway between Gloucester and Hereford, 1836.

1086 *Plan of the Grand Collier Docks, Rotherhithe and Deptford.*
Henry Habberley Price, Engineer. John Newman, Architect & Surveyor.
Litho plan, hand-coloured   24 × 36 cm   (scale 2½ inches to 1 mile).
[In]
Grand Collier Docks, Rotherhithe. 1837. [Prospectus]
[London]: E. Colyer, Printer.
fol   38 cm   (4) pp.   Docket title on p. (4).
CE

PRICE, Henry Habberley and LAXTON, William

1087   *Map of the Gloucester & Hereford Railway.*
W. Habberley Price, William Laxton, Engineers. April 1836.
Litho map   28 × 43 cm   (scale 1 inch to 10 miles)
BM
The initial 'W' is presumably a misprint.

PRICE, John [d. 1736]

1088   *Some Considerations humbly offer'd to the Honourable Commissioners, appointed by Act of Parliament for Building a Bridge over the River Thames, from Fulham to Putney. Together with Proposals relating to a Design drawn for that Purpose.*
[London]: Printed in the Year 1726.
Signed: John Price, 22 July 1726.
8vo   20 cm   15 pp.
ULG, BDL (Gough)

1089   *Some Considerations humbly offered to the Honourable Members of the House of Commons, for Building a Stone-Bridge over the River Thames, from Westminster to Lambeth. Together with some Proposals relating to a Design drawn for that Purpose.*
London: Printed in the Year 1735.
Signed: John Price, 25 Feb 1734–5
8vo   20 cm   16 pp.   Folding engraved plate (P. Fourdrinier sc.).
Design of a Bridge from Westminster to Lambeth.
   By I. Price.
BM, CE (text only), RIBA, ULG, BDL (Gough)

PRIESTLEY, Joseph *of Bradford* [1741–1817] M.Soc.C.E.

1090   *A General Map of the Grand Canal, from Liverpool to Leeds with its different Branches.*
   Carefully laid down from an Accurate Survey, by Joseph Priestley. This Map is Humbly Inscribed to the Proprietors of these Canals by their Obedient Servant Sylvester Forrest.
Engraved map   60 × 170 cm   (scale 1 inch to 1 mile).
BM, MCL, BDL (Gough)
   Published before 1780.

See HUSTLER, BIRKBECK and PRIESTLEY.
   Report on the Leeds & Liverpool Canal, 1789.

PRIESTLEY, Joseph *of Wakefield* [c. 1774–1852]

1091   *Historical Account of the Navigable Rivers, Canals, and Railways, throughout Great Britain, as a Reference to Nichols, Priestley & Walker's New Map of Inland Navigation, derived from Original and Parliamentary Documents in the possession of Joseph Priestley, Esq.*

London: Longman, Rees . . . G. & J. Cary . . . and Richard Nichols, Wakefield. 1831.
Preface signed: Jo. Priestley. Aire and Calder Navigation Office, 1 Oct 1830.
4to   28 cm   (iv) + xii + 776 + (1) + x pp.
Engraved frontispiece map (drawn by J. Walker) + folding hand-coloured engraved plate (Franks & Johnson, Wakefield sc.).
BM, SML, CE (Telford), ULG

1092   Second (octavo) edition
Historical Account of the Navigable Rivers, Canals, and Railways, of Great Britain, as a Reference to Nichols, Priestley & Walker's New Map of Inland Navigation, derived from Original and Parliamentary Documents in the possession of Joseph Priestley, Esq.
London: Longman, Rees . . . G. and J. Cary . . . and Richard Nichols, Wakefield. 1831.
Preface signed: Jo. Priestley. Aire & Calder Navigation Office, April 1831.
8vo   23 cm   xiv + (i) + 702 + (i) + viii pp.
Folding engraved frontispiece map + folding hand-coloured engraved plate (as in quarto edition).
NLS, ULG, BDL
Entirely reset and Addenda material of first edition incorporated in main text. Issued in publisher's blue cloth, printed paper label on spine. Price £1 1s.

Nichols, Priestley and Walker's New Map of Inland Navigation, Canals and Railroads in Great Britain. For details, *see* WALKER, John, 1830.

PROVIS, John   M.Inst.C.E.

1093   *Report. To the Commissioners for improving and maintaining the Road from Holyhead to Shrewsbury* [with] *An Account of the Receipt and Expenditure . . . for the Year ended 1st February 1828.*
House of Commons, 24 March 1828.
Signed: John Provis. Bangor, 26 Feb 1828.
fol   34 cm   10 pp.   Docket title on p. (12).

1094   *Report of Mr John Provis* [on the roads from Holyhead to Shrewsbury and Chester].
Signed: John Provis. Holyhead, 25 May 1836.
[In]
Thirteenth Report of the Holyhead Road Commissioners.
House of Commons, 1836. pp. 9–13.
This is the first of fifteen annual reports by John Provis printed in the 13th to 27th Commissioners Reports, 1836–1850.

PROVIS, William Alexander [1792–1870] M.Inst.C.E.

See TELFORD. Drawings by W. A. Provis of various works, 1810–18.

See TELFORD. Map of mail road between Glasgow and Carlisle, surveyed and drawn by W. A. Provis, 1815.

1095   *Report on the present State of the Road from London by Coventry to Shrewsbury; with the Means of Improvement.* [and] *Mr Provis's Second Report on the Road from Stone Bridge to Shrewsbury.*
Signed: W. A. Provis. London, 7 June 1817 [and] Cernioge Mawr, North Wales, 3 July 1817.
[In]
Fifth Report from the Select Committee on the Roads from Holyhead to London.
House of Commons, 1817. pp. 84–91, 91–95.

1096   [*First and Second*] *Report of Mr W. A. Provis, to the Commissioners, of his proceedings on the Road from Bangor to Shrewsbury.*
Signed: W. A. Provis. Bangor, 1 March 1820. [and] 1 March 1821.
[In]
Annual Report of the Commissioners for the Shrewsbury and Bangor Ferry Road.
House of Commons, 1820. pp. 4–16 and *Ibid.*, 1821. pp. 4–12.

1097 *An Historical and Descriptive Account of the Suspension Bridge constructed over the Menai Strait, in North Wales: with a Brief Notice of Conway Bridge.*
From designs by, and under the direction of Thomas Telford, F.R.S.L. & E. . . .
By William Alexander Provis, the Resident Engineer, and Mem.Inst.Civ.Eng.
London: Printed for the Author, by Ibotson and Palmer . . . 1828.
fol   65 cm   (vii) + 105 pp.   Engraved leaf of signatures + 14 engraved plates of which 4 are large and folding (Edmund Turrell from drawings by G. Merritt, W. T. Hall, E. W. Morris and Thomas Rhodes) + 3 aquatint engraved views (R. G. Reeve after G. Arnald A.R.A.).
BM, NLS, SML, CE
Issued December 1828, price £7 7s. in boards.

See PROVIS and RHODES. Report on Menai and Conway Bridges, 1836.

1098 *Suggestions for improving the canal communication between Birmingham, Wolverhampton, Shropshire, Cheshire, North Wales, and Manchester, by means of a new canal from Middlewich to Altringham.*
London: J. Weale, 1837.
8vo   22 cm   12 pp.   Folding plan.
Not seen.

PROVIS, William Alexander and RHODES, Thomas

1099 *The Report of William Alexander Provis and Thomas Rhodes relative to the Menai and Conway Bridges.*
Signed: W. A. Provis, Thomas Rhodes. Bangor, 5 March 1836.
[In]
Thirteenth Report of the Holyhead Road Commissioners.
House of Commons 1836. pp. 13–15.

RAMSGATE HARBOUR

1100 *A Brief Account of the Proceedings of the Trustees appointed by Act of Parliament for building a Harbour at Ramsgate.*
n.p.   [1754]
8vo   18 cm   27 pp.   2 folding engraved plans.
P

1101 *A True State of Facts, relating to Ramsgate Harbour.*
n.p.   [1755].
4to   21 cm   16 pp.
ULG

RAND, Cater [1749–1825]

1102 *On a Viewed Survey of the part of the River Rother and Levels through which it runs, from and between the Lands situate above Newenden-Bridge . . . and the place of Scott's Float-Sluice.*
London: G. Cooke, Printer.
Signed: C. Rand. Lewes, 11 Nov 1812.
fol   35 cm   3 pp.   Docket title on p. (4).
East Sussex R.O.

RASTRICK, John Urpeth [1780–1856] F.R.S., M.Inst.C.E.

See WALKER and RASTRICK. Report on comparative merits of locomotive and fixed engines, 1829.

1103 *Liverpool and Manchester Railway. Report to the Directors on the Comparative Merits of Loco-Motive and Fixed Engines as a Moving Power.*
By John U. Rastrick, Civil Engineer. Second Edition, Corrected.

Birmingham: Printed and published by R.
Wrightson . . . 1829.
Signed: John U. Rastrick. Stourbridge, 7 March
1829.
8vo   23 cm   vii + 46 pp.   Folding plate.
SML, CE, ULG (Rastrick)

1104   Deposited plan
*Plan and Section of the Intended Grand Junction Railway
commencing at the Bank Quay Branch of the Warrington
and Newton Railway at Warrington in the County of
Lancaster and Terminating at Birmingham in the County
of Warwick. In Two Parts. Part Second from Birming-
ham to the Termination of Part First in the Township of
Chorlton in the County of Chester.*
   Surveyed (and the Levels taken) under the
   Direction of J. U. Rastrick Esq. Civil Engineer.
   R. Martin, Lithog. [1832].
Litho maps and sections (scale 4 inches to 1
   mile) cut and mounted to form two large sheets
   75 and 22 cm × *c.* 450 cm.
HLRO
   The accompanying plans and sections of the
   (northern) Part First, from Warrington to
   Chorlton, surveyed under the direction of
   George Stephenson by Joseph Locke, are hand
   drawn.

1105   *Parliamentary Line of the Grand Junction
Railway, with the Variation Line Proposed by the Com-
pany.*
Signed: John U. Rastrick, 17 Feb 1833.
Litho map, the lines entered in colours
   23 × 100 cm   (scale 1 inch to $1\frac{1}{3}$ miles).
ULG (Rastrick)

1106   *Plan and Section of the Intended Manchester &
Cheshire Junction Railway commencing at Manchester in
the County of Lancaster and terminating by a Junction
with the Grand Junction Railway in the Township of
Crewe in the County of Chester together with Four
Branches from the said Railway.*
   Surveyed under the direction of I. U. Rastrick,
   Esq. Engineer. Printed by Ja:ˢ Wyld & Son.
   [1836].
Litho map, hand coloured   30 × 78 cm   (scale
   1 inch to 1 mile).

BM
   An accompanying note by the Secretary is
   dated Manchester 15 Jan 1836.

1107   [Report on the] *Manchester and Cheshire
Junction Railway.*
Signed: J. U. Rastrick. Manchester, 10 Oct 1836.
Single sheet   42 × 27 cm
BM

1108   [Report] *To the Directors of the London and
Brighton Railway Company.*
Signed: John U. Rastrick. London, 7 Dec 1837.
pp. (1)–(2).
                    [Printed with]
Report by Joseph LOCKE, 8 Dec 1837 [q.v.].
                    [In]
*The London and Brighton Railway . . . Reports of I. U.
Rastrick and Joseph Locke, Esqs., Engineers.*
London: W. Lewis and Son, Printers.
fol   32 cm   (4) pp.
CE
   Issued from the London and Brighton Railway
   Office, London . . . 8 Dec 1837.

1109   Reprint
[In] *The London and Brighton Railway Company . . .
Report of the Directors, with copies of Reports of I. U.
Rastrick and Joseph Locke, Esqs., Engineers.*
London: W. Lewis and Son, Printers.
fol   32 cm   (4) pp.   Docket title on p. (4).
CE
   Issued 19 Dec 1837.

1110   *West Cumberland Railway. Mr Rastrick's
Report to the Provisional Committee.*
Signed: John U. Rastrick. Charing Cross East, 24
   Dec 1838.
pp. (11)–23   with folding litho map.
West Cumberland, Furness & Morecambe Bay Railway.
   1839.
                    [Printed with]
Report on Morecambe Bay embankment, by
   John HAGUE [q.v.].
                    [In]
*West Cumberland, Furness, and Morcambe Bay Rail-
way. Reports of J. U. Rastrick, Esq. and John Hague,
Esq.*

London: J. T. Norris . . . [1839].
8vo   18 cm   23 pp.   Folding litho map (as above).
BM, NLS, ULG (Rastrick)

RENDEL, James Meadows [1799–1856] F.R.S., M.Soc.C.E., M.Inst.C.E.

1111   *Particulars descriptive of the accompanying design for an intended Bridge over the Avon at Clifton.*
   By James M. Rendel.
Plymouth: Rowe . . . 1830.
8vo   23 cm   23 pp.   With folding litho plan.
Elevation and Plan of the intended Suspension Bridge over the Avon at Clifton
   as proposed by J. M. Rendel, Civil Engineer. 17 Nov 1829. Printed by C. Hullmandel.
Bristol City Library

1112   *A Report to the Subscribers to a Survey of proposed Turnpike Road from Plymouth & Devonport to St. Austell, through Looe and Fowey.*
   By James M. Rendel.
Plymouth: Printed by J. B. Rowe.
Signed: James M. Rendel. Plymouth, 30 Jan 1835.
fol   36 cm   8 pp.   Large folding litho plan, the line of road entered in colour.
Plan of the Intended Coast Road from near St. Austell to Torpoint through Fowey and Looe shewing the present circuitous and hilly Road through Liskeard and Lostwithiel.
   Surveyed under the directions of Ja.ˢ M. Rendel Civil Engineer 1834. Hackett, lith.
SML, CE

1113   *Report and Plans for a Breakwater at Brixham, in Torbay.*
   By J. M. Rendel.
London: 1836.
fol   with plans.
   Not seen

1114   *Report on the Practicability of Forming a Harbour, at the Mouth of the Loe Pool, in Mount's Bay, near Helston, in the County of Cornwall.*
   By J. M. Rendel, Civil Engineer.
Plymouth: Printed by J. B. Rowe . . . 1837.
Signed: J. M. Rendel. Plymouth, 30 May 1837.
4to   29 cm   16 pp.   2 folding hand-coloured litho plans (C. Hullmandel lithog.).
Plan of the Loe Pool near Helston.
   Ja.ˢ M. Rendel, Civil Engineer. Surveyed by C. Greaves 1837.
Design for a Harbour, to be formed at the Mouth of the Loe Pool in Mount's Bay.
   By Ja.ˢ M. Rendel, Civil Engineer. 1837.
CE

1115   *Report on the Proposed Steam-Packet Harbour and Docks in the Port of Portsmouth.*
   By James M. Rendel. Civil Engineer.
London: Printed by Mills and Son . . . 1840.
Signed: Jas. M. Rendel. Great George Street, Westminster, 18 July 1840.
8vo   23 cm   15 pp.   Folding engraved plan.
Plan of the Port of Portsmouth and Shore of the Isle of Wight . . . shewing the Proposed Steam Ship Harbour and Docks and Terminus of the South Western Railway at Gosport.
   James M. Rendel, C.E. 1840. Davies sc.
CE

RENNIE, George [1791–1866] F.R.S., M.Soc.C.E., M.Inst.C.E.

1116   *Plan of a Proposed Dock, for Colliers & other Vessels, to be situate in the Isle of Dogs, Middlesex.*
   Designed by George Rennie, Civil Engineer. F.R.S. 1824.
Engraved plan, hand coloured   65 × 49 cm (scale 1 inch to 420 feet).
PLA (Vaughan)

See RENNIE, George and John. Plans of the Liverpool & Manchester Railway, 1826.

1117 *Eau Brink Drainage. Mr George Rennie's Report on the state of Denver Sluice . . . with Resolutions of Quarterly Meeting at Lynn, October 8th, 1829.*
Cambridge: W. Hatfield, Printer.
Signed: George Rennie. London, 6 Oct 1829.
fol   33 cm   3 pp.   Docket title on p. (4).
CRO

See EASTON, John, Plan of proposed line of navigation and drainage from Dunball, near Bridgwater, to Yeovil; approved by George Rennie, 1829.

1118 *Report to the Commissioners of the River Wear on the Formation of Wet Docks at Sunderland.*
By George Rennie, Esq.
London: Printed for John and Arthur Arch . . . 1832.
Signed: George Rennie. London, 24 Nov 1832.
8vo   21 cm   9 pp.
[Printed with]
Report by James WALKER, 24 Nov 1832 [q.v.].
[In]
Sunderland. Report on the Formation of Docks; by George Rennie Esq. Also, Report on the same by James Walker . . . with Plans.
London: Printed for John and Arthur Arch . . . 1832.
8vo   21 cm   (i) + 9 + 17 pp.   2 folding hand-coloured litho plans.
BM, NLS, CE

See COOKE and RENNIE. Old and new London Bridges, 1833.

1119 *Report of George Rennie, Esq. Civil Engineer, on the Midland Counties Railway.*
To the Provisional Committee.
n.p.
Signed: George Rennie. Leeds, 27 Nov 1833.
Single sheet   28 × 20 cm
CE (Page)

1120 *Report to the Chairman and Directing Committee of the Midland Counties Railway.*
Signed: George Rennie. London, 26 Nov 1836.
pp. 3–4.
[In]
Midland Counties Railway, Prospectus.

n.p.
fol   44 cm   4 pp.
BM

1121 *Report on the Progress and Present State of our Knowledge of Hydraulics as a Branch of Engineering.*
By George Rennie, Esq. F.R.S. Part I [and] Part II.
London: Printed by Richard Taylor . . . 1835.
8vo   22 cm   (i) + 153–184, 415–512 pp.
Folding engraved plate.
Section of the River Thames, from the River Kennet to the Nore Light.
J. W. Lowry sc.
RS, IC
Offprints, with new title page, from British Association 3rd and 4th Reports, for 1833 and 1834.

See RENNIE, George and John. Reports on Central Kent Railway, 1836 and 1837.

1122 [Report] *To the Chairman and Directors of the Central Kent Railway and Harbour at Sandwich.*
n.p.
Signed: George Rennie. Holland Street, 23 Feb 1837.
fol   33 cm   2 pp.
CE

RENNIE, George and *Sir* John

1123 *Map of the Country between Liverpool and Manchester, exhibiting the proposed New Line of Railway.*
Surveyed under the directions of Geo. & In.º Rennie Esq.ˢ F.F.R.S. Civil Engineers, 1826.
J. Wyld sc.
Engraved map, the line entered in colour
24 × 40 cm   (scale 1 inch to 2½ miles).
[In]
Liverpool and Manchester Rail-Way Company. New Line. Prospectus.
n.p.

Chairman's statement dated 26 Dec 1825.
fol   42 cm   (2) pp.   Map on p. (3).
BM, BDL

1124   *A Plan and Section of an intended Railway or Tram-Road from Liverpool to Manchester in the County Palatine of Lancaster.*
George & John Rennie, Engineers. Surveyed under the directions of George and John Rennie Esqrs. F.F.R.S. Civil Engineers. By Charles Vignoles. Engraved by J. A. Walker [1826].
Engraved   plan,   partly   hand   coloured   66 × 99 cm   (scale 1 inch to 1 mile).
Liverpool R.O.

1125   *[Report by Messrs George and John Rennie on five proposed lines for a canal from London to Portsmouth].*
Signed: George and John Rennie, London, 12 Oct 1827.
pp. 16–18.
[In]
Imperial Ship Canal from London to Portsmouth. Mr Cundy's Reply.
London: 1828. *See* CUNDY.

1126   *Central Kent Railway. Report of the Messrs Rennie to the Committee of the Central Kentish Railway and Sandwich Harbour Company.*
[London]: Marchant, Printer.
Signed: George & John Rennie. London, 9 Dec 1836.
fol   33 cm   2 pp.   Docket title on p. (4).
BM
For plan and prospectus, *see* CUNDY, 1836.

1127   *Central Kent Railway. [Report] To . . . Directors of the Central Kent Railway, and Sandwich Harbour Company.*
n.p.
Signed: George and John Rennie. London, 5 April 1837.
fol   33 cm   6 pp.
CE

RENNIE, John [1761–1821]
F.R.S.L. & E., M.Soc.C.E.

1128   *A Plan shewing the Line of the Proposed Navigation from Bishops Stortford . . . to the Brandon River, on the border of Norfolk.*
Approved by the General Meeting . . . at Great Chesterford in the Year 1789, & Engraved by order of . . . the City of London. Surveyed by John Rennie Engineer & F.R.S.E. Engraved by Wm Faden . . . March 1790.
Engraved plan, the line entered in colour   52 × 124 cm   (scale 1 inch to 1 mile).
BM, SML, RS (Smeaton), BDL (Gough)

1129   *A Plan shewing the Line of the proposed Rochdale Canal between the Calder Navigation near Sowerby in the County of York and Manchester . . . with its several Branches.*
Accurately survey'd under the direction of John Rennie Engineer and F.R.S.E. by William Crosley, 1791. Engraved by B. Baker.
Engraved plan, the line entered in colour   65 × 88 cm   (scale 1½ inches to 1 mile).
BM, SML, BDL

1130   Revised scheme
A Plan Shewing the line of the proposed Rochdale Canal between the Calder Navigation near Sowerby Bridge wharf . . . and His Grace the Duke of Bridgewater's Canal near Manchester . . . with its Different Branches.
Accurately survey'd under the direction of John Rennie Engineer & F.R.S.E. by William Crosley 1793. Engraved by W. Faden 1794.
Engraved plan, the line entered in colour   42 × 91 cm   (scale 1½ inches to 1 mile).
BM

1131   *Plan of the Proposed Lancaster Canal from Kirby Kendal in the County of Westmorland to West Houghton in the County Palatine of Lancaster.*
Surveyed in the Years 1791 & 1792. By John Rennie, Engineer & F.R.S.E. Engraved by W. Faden 1792.
Engraved plan   50 × 41 cm   (scale 1 inch to 1 mile).
BM

1132 *Plan of the Proposed Crinan Canal, between the Lochs of Crinan and Gilp, in the County of Argyll.*
Surveyed . . . in the Year 1792. By John Rennie, Engineer. Engraved by W. Faden.
Engraved plan, the line entered in colour
51 × 65 cm   (scale 3 inches to 1 mile).
BM

1133 [*Report*] *To the Committee of the proposed Chelmer Navigation.*
n.p.
Signed: John Rennie. New Surry-Street, London, 6 Dec 1792.
fol   32 cm   3 pp.
BM

1134 *Plan of the Proposed Navigation from Chelmsford to Colliers Reach with a Branch to Maldon in the County of Essex.*
Survey'd under the Direction of John Rennie, Engineer & F.R.S.E. by Charles Wedge 1792. Engraved by W. Faden 1793.
Engraved plan, the cuts entered in colour
45 × 89 cm   (scale 3 inches to 1 mile).
BM, ULG

1135 [*Report*] *To the Committee of Land Owners, and others interested in the Improvement of the Outfall of the River Ouze.*
n.p.
Signed: John Rennie. New Surrey Street, 16 Feb 1793.
fol   33 cm   3 pp.   Docket title on p. (4).
BM
Refers to the proposed Eau Brink Cut.

1136 *Plan of the proposed Navigable Canal, between the River Kennet at Newbury . . . and the River Avon at Bath . . . Likewise of a Branch from the said Canal near a place called Marsh Barn to the Towns of Calne & Chippenham, to which is added a Plan of the proposed Somersetshire Coal Canal.*
Surveyed 1793 by John Rennie, Civil Engineer. Engraved by W. Faden 1794.
Engraved plan, the line entered in colour
49 × 197 cm   (scale 1½ inches to 1 mile).

BM, HLRO, BDL (Gough)
The HLRO deposited plan is accompanied by estimates signed; John Rennie, London 3 Jan 1794.

1137   Revised scheme
Plan of the proposed Navigable Canal, between the River Kennet at Newbury . . . and the River Avon at Bath . . . Likewise a part of the Wilts & Berks Canal, from Semington to the Towns of Calne & Chippenham; to which is added a Plan of the proposed Somersetshire Coal Canal.
Surveyed 1793, by John Rennie, Civil Engineer. Engraved by W. Faden 1794.
Engraved plan   49 × 197 cm   (scale 1½ inches to 1 mile).
BM
In this second state the plate gives more topographic detail.

1138   Third state
Title as No. 1136 with addition *Completed in 1810.*
Engraved by W. Faden.
BM
Some minor revisions, showing the canal as built.

1139   *Report of a Survey of the River Thames, between Reading and Isleworth; and of Several Lines of Canals projected to be made between those places: with Observations on their Comparative Eligibility.*
By John Rennie, Civil Engineer.
[London]: Printed July 30, 1794.
Signed: John Rennie. Stamford-Street, 30 July 1794.
8vo   22 cm   (iii) + 55 pp.
BM, NLS, SML, CE (Page), ULG, BDL
The plan accompanying this report appears not to have been published.

1140   *Plan of the proposed Grand Western Canal, from Topsham in the County of Devon, to Taunton in the County of Somerset; with the Branches to Tiverton and Cullompton.*
Surveyed   Anno   1794.   John   Rennie Engineer.   J. Cary sc.
Engraved plan   48 × 126 cm   (scale 1 inch to 1½ miles).
B

1141 *Plan of the Proposed Polbrook Canal, from the Tide-Way of the River Camel, near Wadebridge, to Dunmeer. With a Collateral Branch towards Ruthran Bridge, in the County of Cornwall.*

Surveyed Anno 1796, under the direction of John Rennie, Civil Engineer, by James Murray. Engraved by W. Faden.

Engraved plan, the line entered in colour
32 × 63 cm   (scale 4 inches to 1 mile)

BM

1142 *Estimate of the proposed London Docks in Wapping and of the proposed Canal from . . . Wapping to Blackwall.*
Signed: John Rennie [March 1796].
pp. 225–226.
[Printed with]
Plan of the London Docks, and abstract of the value of the Lands, by Daniel ALEXANDER, 22 April 1796 [q.v.].
[In]
Report from the Committee . . . on the Port of London.
House of Commons, 1796.

1143 Second Printing
[In] *Reports from Committees of the House of Commons* Vol. 14 (1803) p. 437.

1144 *First (and Second) Report by John Rennie on the Harbour of Aberdeen.*
Signed: John Rennie. London, 27 Jan 1797 (and) 30 Dec 1797.
pp. 21–40   with folding hand-coloured litho plan.
Plan of the Harbour of Aberdeen in the year 1797 with Improvements by Mr Rennie.
[In]
Reports upon the Harbour of Aberdeen . . . 1834. See ABERDEEN HARBOUR.

1145 *Report concerning the Different Lines surveyed by Messrs John Ainslie & Robert Whitworth, Jun. for a Canal proposed to be made between the Cities of Edinburgh and Glasgow . . . with an Account of a Running Level, taken for a New Line by Linlithgow and Falkirk.*
By John Rennie, Civil Engineer, and F.R.S.Edin.
n.p.   1797.

Signed: John Rennie. Edinburgh, 14 Sept 1797.
4to   27 cm   (i) + 12 pp.
BM, NLS, CE, ULG (Rastrick)

1146 *Report concerning the Practicability and Expence of the Lines suggested by Messrs John Ainslie & Robert Whitworth, Jun. for a Canal proposed to be made between the Cities of Edinburgh and Glasgow . . . with the Improvements that have been made on these Lines; and also concerning . . . the New Line by Linlithgow and Falkirk.*
By John Rennie, Civil Engineer, and F.R.S.L. & E.
n.p.   1798.
Signed: John Rennie. Edinburgh, 22 Oct 1798.
4to   27 cm   (i) + 27 pp.   Folding engraved plan.
Reduced Plan of the different Lines proposed for a Canal between Edinburgh and Glasgow.
Surveyed under the direction of John Rennie Civil Engineer by John Ainslie.
BM, NLS, CE, ULG (Rastrick)

See ALEXANDER. Revised plan of the London Docks, 1799.

1147 *Report concerning the Drainage of Wildmore Fen, and of the East and West Fens.*
By John Rennie, Civil Engineer.
London: Printed by H. Baldwin and Son . . . 1800.
Signed: John Rennie. Stamford-Street, 7 April 1800.
fol   32 cm   (i) + 22 pp.   Folding engraved plan.
A Sketch of Wildmore Fen, West Fen & East Fen with the Marshes and Highlands adjacent in the County of Lincoln.
Publ. Feb 1800 by A. Arrowsmith.
BM, NLS, LCL, CRO, BDL (Gough)
Issued in marbled wrappers.

1148 *Second Report concerning the Drainage of Wildmore Fen, and of the East and West Fens.*
By John Rennie, Civil Engineer.
London: Printed by H. Baldwin and Son . . . 1800.

Signed: John Rennie. Stamford-Street, 1 Sept 1800.
fol   32 cm   (i) + 8 pp.
LCL

See JESSOP, RENNIE *et al.*, Report on drainage of Deeping Fens, 11 Aug 1800.

1149   *Report concerning the Improvement of Boston Haven, by John Rennie, Esq.*
Signed: John Rennie. Stamford Street, 6 Oct 1800.
pp. 5–10.
[In]
Captain Huddart's Report upon the Improvement of the Port and Harbour of Boston . . . accompanied by Mr Rennie's Report thereon.
London: Luke Hansard & Sons . . . 1800.
fol   30 cm   10 pp.   Folding engraved plan.
BM, CE

See JESSOP and RENNIE. Report on Lancaster Canal, 1801.

1150   *Answers [to the Questions respecting the Construction of a Cast Iron Bridge, of a Single Arch, 600 Feet in the Span, and 65 Feet Rise]*
By Mr J. Rennie.
Signed: John Rennie. Stamford-Street [May 1801].
pp. 58–62.
[In]
Report from the Select Committee upon Improvement of the Port of London.
1801.

1151   Second printing
[In] *Reports from Committees of the House of Commons*
Vol. 14 (1803) pp. 625–627.

1152   *Mr Rennie's Reports on the State of the Harbour of Port Patrick.*
Dated: Port Patrick, 12 Aug 1801 [and] London, 4 June 1802.
[In]
Report from the Committee on the Communication between England and Ireland.
House of Commons, 1809. pp. 58–61.

1153   *The Report of John Rennie, Esq. on the Ancholme Drainage and Navigation.*
Hull: Printed by Robert Peck.
Signed: John Rennie, London, 9 Nov 1801.
fol   32 cm   8 pp.   Docket title on p. (10).
Sutro

1154   *Further Report of John Rennie, Esq. on the Ancholme Drainage and Navigation.*
London: Printed by Luke Hansard.
Signed: John Rennie, Stamford Street, 12 Feb 1802.
fol   32 cm   3 pp.   Docket title on p. (4).
Sutro

1155   *The Report of John Rennie, Esq. on building Bridges over the Straits of Menai, and over the Conway River.*
Signed: John Rennie. London, 16 Feb 1802.
pp. 41–48 with 4 engraved plates (J. Basire sc.).
[1]   *Plan of part of the Straits of Menai shewing the two different Situations for the proposed Bridges and Roads leading thereto.*
[2] and [3]   *Designs for a Bridge proposed to be Built over the Straits of Menai, at Ynys y Moch [and] at the Swelly Rocks. 1802. John Rennie*
[4]   *Design for a Bridge proposed to be built over the Conway River, opposite the Castle. 1802. John Rennie.*
[In]
Second Report from the Committee on Holyhead Roads and Harbour.
House of Commons, 1810.

See ALEXANDER. Plan of the London Docks, 1802 [as built].

1156   *Plan of the Proposed London Canal from the Grand Junction Canal at Paddington to the London Docks in the County of Middlesex.*
Surveyed under the direction of John Rennie Civil Engineer by Ja.ˢ Murray 1802. W.F. 1802.
Engraved plan   42 × 117 cm   (scale $1\frac{1}{2}$ inches to 1,000 feet).
BM

1157 [Reports] *To the Directors General of Inland Navigation in Ireland.*
Signed: John Rennie. London, 24 March (and) 20 July 1802.
pp. 58–80   with folding engraved plan.
[In]
Reports [on the Improvement of the Harbour of Dublin].
[Dublin *c.* 1802].
fol   32 cm   84 pp.   6 folding engraved plans.
NLS

1158   *The Report for Improving the Harbour, and Making Wet Docks, at Greenock.*
By John Rennie, Esq. Civil Engineer.
Greenock: Printed by John Davies . . . 1802.
Signed: John Rennie. London, 12 July 1802.
8vo   19 cm   (i) + 12 pp.
CE, BDL

1159   [Report] *To the Commissioners for Navigation on the River Witham.*
Sleaford: Thornhill, Printer [1806].
Signed: John Rennie, London, 1 Dec 1802.
fol   32 cm   5 + (1) pp.
Sutro
   Printed with record of a meeting held at Boston, 24 Oct 1806. p. (6).

1160   *A Plan of the River Witham between Boston and Lincoln, with the Intended Improvement of the Navigation.*
   Surveyed under the Direction of John Rennie Civil Engineer. By A. Bower 1803. [with] *Profile of the River Witham.*
Anthᵞ Bower 1804. W. Faden 1804.
Engraved plan, line of the river entered in colour   53 × 178 cm   (scale 2 inches to 1 mile and 1 inch to 12 feet vertical).
BM, Sutro
   An exceptionally fine river survey.

1161   *Report and Estimate of a Canal proposed to be made between Croydon and Portsmouth. By means of which, and the Croydon Canal, an Inland Navigation will be opened between London and Portsmouth.*
   By John Rennie, Civil Engineer.
London: Printed by C. and R. Baldwin . . . 1803.

Signed: John Rennie. London, 16 July 1803.
fol   33 cm   (i) + 22 pp.
CE, ULG

1162   *Plan of a Canal proposed to be made between Croydon and Portsmouth; and which, by means of the Croydon Canal, will form a Communication with the River Thames at Rotherhithe.*
Engraved by W. Faden, 1803.
Engraved plan, the line entered in colour   57 × 174 cm   (scale 1 inch to 1 mile).
BM

1163   *Report of John Rennie, Esq. Civil Engineer . . . upon a Survey & Plan of a Canal from the River Clyde at the City of Glasgow, to the West Coast . . . at or near the Harbour of Saltcoats, by John Ainslie, Surveyor.*
Air: Printed by J. & P. Wilson, 1804.
Signed: John Rennie. London, 1 Aug 1803.
4to   28 cm   14 pp.
NLS
   Issued in grey wrappers.

1164   *Plan of a Canal, Proposed to be made between the River Clyde at the City of Glasgow; and the Harbour of Saltcoats, with a Branch to Paisley.*
   Surveyed under the direction of John Rennie Civil Engineer, by John Ainslie 1803.
Engraved plan, the line entered in colour   36 × 62 cm   (scale 1 inch to 1 mile).
BM

See RENNIE and WALKER, Ralph. Plans of the East India Docks. 1803 and 1804.

1165   *Mr Rennie's Report to the Commissioners for the Navigation on the River Witham, and to the Proprietors of the Foss Navigation, and the Landowners draining through the same.*
Boston: Kelsey, Printer.
Signed: John Rennie. London, 26 Oct 1803.
fol   33 cm   7 pp.   Docket title on p. (8).
NLS (Rennie), Sutro

See SPECIFICATION of the locks for the Royal Canal in Ireland, 1803.

1166   [*Plan of Rennie's proposed improvements to Dublin Harbour, 1804*].

Heneey & Fitzpatrick sc.
Engraved plan   36 × 33 cm   (scale 1½ inches to
   1 mile).
[In]
*Representation for the Improvement of Dublin Harbour,
submitted by the Directors General of Inland Navigation
to . . . the Lords of His Majesty's Treasury.*
Dublin: Printed by A. B. King . . . 1805.
fol   32 cm   40 + (1) pp.   2 folding engraved
   plans.
NLS

1167   *River Lee. Mr Rennie's Report.*
Hertford: Printed by S. Austin. 1804.
Signed: John Rennie. London, 17 Sept 1804.
8vo   22 cm   26 pp.
CE

1168   *Report of John Rennie, Esq. on the State of the
Works of Drainage of the East, West and Wildmore
Fens.*
Boston: Hellaby, Printer.
Signed: John Rennie. London, 1 Jan 1805.
fol   32 cm   3 pp.   Docket title on p. (4).
LCL

1169   *Report and Estimate of the Expence of Enlarg-
ing and Improving the Harbour of Peterhead, agreeably to
a Plan and Sections thereof.*
   Designed by John Rennie, Esquire, of London,
   Civil Engineers.
Aberdeen: Chalmers & Co. Printers.
Signed: John Rennie. London, 14 Feb 1806.
4to   25 cm   8 pp.
NLS

See RENNIE and WHIDBEY. Report on
   Plymouth Breakwater, 1806.

1170   *Further Report of Mr Rennie, to the Lords
Commissioners of the Admiralty.*
Signed: John Rennie. London, 24 Sept 1806.
[In]
Papers relating to Plymouth Sound.
   House of Commons, 1812. pp. 13–15.

1171   Reprint
[In] *Interesting Particulars relative to . . . the Break-*

water now constructing in Plymouth Sound.
   Plymouth, 1821. pp. 21–26. *See* PLYMOUTH
   BREAKWATER.

1172   [Report] *To the Trustees of the Holderness
Drainage.*
Hull: Robert Peck, Printer.
Signed: John Rennie. Hall, 3 Jan 1807.
fol   32 cm   3 pp.
HRO
   Holderness Drainage Minute Book, 10 Jan
   1807: 500 copies of the report to be printed.

1173   *Letter from John Rennie, Esq. To be added to
certain Resolutions of the General Commissioners for
Drainage, and the Commissioners for Navigation, by the
River Witham.*
Sleaford: Thornhill, Printer.
Signed: John Rennie. London, 28 April 1807.
fol   31 cm   2 pp.   Docket title on p. (4).
NLS

1174   *Plan of the River Clyde at the City of Glasgow;
with the proposed Dock at the Broomielaw.*
   Surveyed under the direction of John Rennie,
   Esq. C.E. W.F. 1807.
Engraved plan, hand coloured   30 × 66 cm
   (scale 1 inch to 200 feet).
BM

1175   *Reports by Mr Rennie, Civil Engineer, respect-
ing the proposed Improvements at Pettycur, and intended
Ferry-Boat Harbour at Newhaven* [and] *relative to the
Improvements proposed to be made upon the Harbour of
Burntisland.*
Signed: John Rennie. London, 4 Feb 1808 (and)
   13 March 1809.
pp. 8–13 (and) 15–19   with 3 folding engraved
   plans. Surveyed & Engraved by J. Ainslie.
[In]
*Resolutions of a General Meeting of the Trustees for
carrying into execution the Act for improving the com-
munication between the Counties of Fife and Mid-
Lothian, by the Ferries of Kinghorn and Burntisland and
Leith and Newhaven.*
n.p. 1809.
fol   30 cm   19 pp.   3 folding engraved plans.
NLS

See RENNIE and HUDDART. Reports on Howth Harbour, 1808 and 1809.

1176 *Mr Rennie's Reports, presented to the Lords of the Admiralty.*
Signed: John Rennie. London, 17 Jan 1809 and 17 March 1809.
pp. 12 and 5–6 with folding engraved plan.
Sketch of the Bay and Harbour of Ardglass, with the Proposed Pier.
  John Rennie, Esq.ʳ Engineer. J. Basire sc.
    [In]
Report from the Committee on the Petition relating to Ardglass Harbour.
  House of Commons 1809.

1177 *The Report of John Rennie, Civil Engineer, F.R.S. &c. concerning the Practicability and Expence of making a Navigable Canal through the Weald of Kent, to join the Rivers Medway, Stour and Rother.*
London: printed by E. Blackader.
Signed: John Rennie. London, 19 July 1809.
4to   26 cm   (i) + 9 pp.   Folding engraved plan.
Plan of a Navigable Canal, proposed to be made through the Weald of Kent, to join the Rivers Medway, Stour & Rother.
  Surveyed in the Years 1802 & 3. Engraved by J. Barlow.
CE

See SPECIFICATION for a new quay wall on the River Clyde, 1809.

1178 *Report, by Mr John Rennie, Engineer, respecting the proposed Rail-Way from Kelso to Berwick.*
Kelso: Printed by A. Ballantyne . . . 1810.
Signed: John Rennie. London, 14 Nov 1809.
4to   26 cm   16 pp.   Folding engraved plan.
Plan of a Proposed Line of Iron Railway from Kelso to Berwick.
Engraved by J. Barlow.
NLS

1179 Another issue
As No. 1178 but
London: Printed by Strahan and Preston . . . 1810.
B

1180 Reprint
As No. 1178 but
Edinburgh: Printed by M. Anderson, 1824.
4to   27 cm   16 pp.   Folding litho plan (title as in No. 1178), Robertson & Ballantine, lithog.
ULG

1181 *Estimate of the probable Expense of building a Pier at Holyhead, extending 650 Feet beyond the Perch, and forming a Road, and excavating Part of the Interior Harbour.*
Signed: John Rennie. London, 19 March 1810.
p. 2   with folding engraved plan.
Plan of the Harbour of Holyhead, with the proposed New Road & Pier.
J. Basire sc.
    [In]
Report from Committee on Holyhead Roads and Harbour.
  House of Commons, 1810.

1182 *Report and Estimate of the Grand Southern Canal, proposed to be made between Tunbridge and Portsmouth: by means of which and the River Medway, an Inland Navigation will be opened between the River Thames and Portsmouth.*
  By John Rennie, Civil Engineer, and F.R.S. &c.
London: Printed by E. Blackader . . . 1810.
Signed: John Rennie. London, 10 May 1810.
4to   29 cm   (i) + 10 pp.   Folding engraved plan.
Plan of the Grand Southern Canal proposed to be made between Tunbridge and Portsmouth. . .
  Surveyed in 1803. Cooper sc.
BM, CE, ULG, BDL (plan only), CUL

1183 *New Shoreham Harbour. The Report of John Rennie, Esq.*
London: Printed by W. Smith . . . 1810.
Signed: John Rennie, 21 July 1810.
8vo   19 cm   22 pp.
ULG

1184 *Report and Estimate on the Improvement of the Drainage and Navigation of the South & Middle Levels of the Great Level of the Fens.*
  By John Rennie, Civil Engineer.
London: Printed by E. Blackader . . . 1810.

Signed: John Rennie. London, 7 Aug 1810.
4to   27 cm   (i) + 18 pp.   9 folding hand-coloured engraved plans (Neele sc.).

[1] *Plan of the Bedford Level with the intended lines of Drainage, 1810.*

[2]–[4] *Transverse Sections of the River Ouze, between Eaubrink and Denver Sluice.*

[5] *A Section on the East Side (and West Side) of the Ouze from Eaubrink to Denver Sluice. [and] Plan of the River Ouze from Eaubrink to Denver Sluice.*

[6] *A Section of the Ouze and Hundred Foot Rivers, from Lynn Deeps to the Hermitage Sluice at Earith.*

[7] *A Section of Vermudens or Forty Foot Drain, from its Junction with the Old Bedford River to Ramsey Bridge.*

[8] *A Section of the Rivers Ouze and Cam, from Denver Sluice to Clayhithe.*

[9] *A Section of the Old Bedford River from the Old Bedford Sluice to the River Ouze at Earith.*

Nos. [2]–[9] all signed: Anth.ᵞ Bower, 3 Oct 1810.

BM, LCL, CRO, CE, ULG, BDL, CUL.
   Issued in grey boards.

1185 *A Plan of the Proposed Canal from Bristol to Taunton. Together with Collateral Branches to Chedder and Nailsea in the County of Somerset.*

   Surveyed under the Direction of John Rennie, Civil Engineer, F.R.S. &c. 1810. Cooper sc.
Engraved plan, the line entered in colour
   50 × 189 cm   (scale 2 inches to 1 mile).
BM, BDL

1186 *Line of the proposed Ship Canal from Bridgewater to Seaton.*

   Surveyed under the Direction of J. Rennie Esq.ʳ by Messrs W. Bond & J. Dean in 1810. Engraved by C. Smith.
[with 2 inset plans] *Plan of Seaton & Beer Bay [and] Plan of Beer Cove.*
   Surveyed . . . by Netlam & Francis Giles 1810.
Engraved plan   56 × 102 cm   (scale 1 inch to 1 mile).
BM
   A smaller version exists, scale 1 inch to 4 miles, undated.

1187 *A Plan of the Proposed Extension of the Kennet and Avon Canal from the City of Bath to the City of Bristol.*

   Surveyed under the Direction of John Rennie, Civil Engineer & F.R.S. 1810. W.F. 1811.
Engraved plan   30 × 146 cm   (scale 5 inches to 1 mile).
BM

1188 *Plan of the Docks at Liverpool, with the Proposed Alterations & Additions 1810.* W.F. 1810.
Engraved plan, hand coloured   20 × 52 cm (scale 1 inch to 180 feet).
HLRO, Mersey Docks & Harbour Office, BDL
   The HLRO deposited plan is accompanied by engineering estimates signed: John Rennie, 11 May 1811.

1189 *Mr Rennie's Survey of the proposed Line of Road from Dumfries to Port Patrick.*
Signed: John Rennie. London, 8 April 1811.
pp. 5–7   with folding engraved plan, the roads entered in colours.
Plan of Improvements in the Road between Carlisle and Portpatrick as proposed by Mr Telford and Mr Rennie.   W. A. Provis del.   J. Basire sc.
[In]
Report from the Committee, upon the Roads between Carlisle and Port Patrick.
   House of Commons, 1811.

1190 *Report of John Rennie, Esq. Civil Engineer, F.R.S. &c. on the Proposed Ship Canal between the English and Bristol Channels.*
London: Printed by M. and S. Brooke . . . 1811.
Signed: John Rennie. London, 19 July 1811.
4to   25 cm   14 pp.
CE

1191 *Plan of the Frith of Forth with the Improvements at the Queensferry Piers and Landing Places forming the Great Line of Communication between the North & South of Scotland.*
   By John Rennie, Civil Engineer. W.F. 1811.
Engraved plan, the roads entered in colour
   25 × 30 cm   (scale 1 inch to 1,200 feet).
BM, NLS

1192 *Plan of the Proposed London & Cambridge Junction Canal from the Head of the Stort Navigation at Bishops Stortford to the River Cam near Clay Hithe Sluice . . . and of the proposed Branch from Sawston to the North Road near Whaddon in the County of Cambridge.*

Surveyed under the Direction of John Rennie Civil Engineer F.R.S. F.A.S. &c. by Netlam & Francis Giles 1811.

Engraved by G. Mills.

Engraved plan, partly hand coloured 65 × 140 cm (scale 2 inches to 1 mile).

BM

1193 *Plan of the Proposed Weald of Kent Canal, to Unite the Rivers Medway and Rother, and of the Proposed Branches to Lamberhurst, Cranbrook and the Chalk Hills near Wye.*

Surveyed under the Direction of John Rennie, Civil Engineer, F.R.S.L. & E. & F.R.A.S. &c. By Netlam and Francis Giles. W.F. 1811.

Engraved plan 111 × 157 cm (scale 2½ inches to 1 mile).

BM, BDL

1194 *A Plan of the River Witham between Lincoln and Boston, with the Intended Improvements of the Navigation.*

Surveyed under the Direction of John Rennie, Civil Engineer. By A. Bower, 1812.

Engraved plan 18 × 58 cm (scale 1 inch to 1½ miles).

BM

1195 *[Report on the Hundred Foot Wash].*
London: Printed by Richard Taylor.
Signed: John Rennie. London, 7 Dec 1812.
fol 24 cm 2 pp.
CE

1196 *[Report on St. John's Eau].*
London: Printed by Richard Taylor.
Signed: John Rennie. London, 7 Dec 1812.
fol 24 cm 3 pp.
CRO, CE

1197 *Copy of Report made by John Rennie, Esq. respecting Rye Harbour, and the Upper Levels, on the Banks of the River Rother.*

London: G. Cooke, Printer.
Signed: John Rennie. London, 26 Dec 1812.
fol 28 cm 3 pp. Docket title on p. (4).
East Sussex R.O.

1198 *The Report of John Rennie, Esq. on the Drainage of Hatfield Chace.*
Doncaster: Sheardown, Printer.
Signed: John Rennie. London, 16 Jan 1813.
fol 33 cm 6 pp. Docket title on p. (8).
Nottingham University Library
Stitched as issued.

1199 *A Plan and Elevation of the New Bridge now building over the River Thames near Somerset Place in the Strand.*

John Rennie, Civil Engineer. F.R.S. &c. 1813.

Engraved plan and elevation 58 × 125 cm (scale 1 inch to 30 feet).

CE (Rennie)

Waterloo Bridge, built 1811–1817. Known at first as Strand Bridge, it was renamed in 1816.

1200 *Mr Rennie's Report and Plan for the Improvement of the Outfall to Sea through Moreton's Leam Wash and Wisbeach; and for the Drainage of the North Level, South Holland, and the low Lands adjacent.*

Signed: John Rennie, London, 26 Jan 1814.
pp. 1–16 with folding hand-coloured engraved plan.

A Plan of the River Nene and North Level and Part of South Holland, with the Proposed Lines of Drainage into Wisbeach Upper Eye at Crab Hole.

By John Rennie, Civil Engineer. Anth.ʸ Bower 14 Dec 1813. Neele sc.

[Printed with]

Mr Rennie's Plan for completely Draining South Holland.

Signed: John Rennie. London, 28 Jan 1813.
pp. 17–22.

[and]

Mr BOWER's Statement on the Fens north of Boston, 23 Feb 1814 [q.v.].

[In]

Reports as to Wisbech Outfall, and the Drainage of the North Level and South Holland.

By John Rennie, Civil Engineer.

With a Statement &c. as to the Drainage of the Fens North of Boston.

By Anthony Bower.

London: Printed by Richard and Arthur Taylor . . . 1814.

4to   28 cm   (i) + 33 pp.   2 folding hand-coloured engraved plans.

LCL, CRO, CE (lacking Rennie's plan), ULG, CUL

Issued in blue wrappers.

1201   *Report by Mr John Rennie, Engineer.*
Signed: John Rennie. London, 15 Jan 1814.
pp. 12–15.
[Printed with]
Report by Hugh BAIRD. 24 Jan 1814 [q.v.].
[In]
Subsidiary Report to the General Committee of Subscribers to the intended Edinburgh and Glasgow Union Canal.
Edinburgh: Printed by Oliver & Boyd . . . 1814.
4to   27 cm   15 pp   Folding engraved plan.
NLS, CE

1202   *The Report of John Rennie, Esq., Engineer, for the Enlargement and Improvement of the Harbour of Whitehaven.*
Signed: John Rennie. London, 8 Feb 1814.
p. 12   with 2 litho plans.
Designs for the Extension of Whitehaven Harbour.
By the late John Rennie, Esq. Civil Engineer 1814.
[In]
Plans suggested at different periods for the Improvement of Whitehaven Harbour. WHITEHAVEN HARBOUR, 1836 [q.v.].

See SPECIFICATIONS for a new entrance to the Basin at Deptford and a river wall, 1815.

1203   *Report to the Commissioners for the Construction of a Pier and other Works at Holyhead.*
Signed: John Rennie. London, 21 May 1815.
pp. 6–7   with folding engraved plan.
Plan of the Harbour of Holyhead, on which is represented the Proposed Extension of the Pier, and of a Wet Dock in the Interior Harbour.
J. Basire sc.
[In]

Report from Committee on Holyhead Harbour.
House of Commons, 1816.

1204   [*Report on the Black Sluice Drainage*].
Sleaford: Thornhill and Tindale, Printers.
Signed: John Rennie. Lincoln, 9 Sept 1815.
fol   33 cm   3 pp.
LCL

1205   *Report of the Subscribers to a Canal from Arundel to Portsmouth.*
By John Rennie.
Portsmouth: Printed by Mottley, Harrison and Miller, 1816.
4to   15 pp.
[not seen]

1206   *Plan of the Intended Arundel and Portsmouth Canal, Aqueducts, &c.*
Surveyed under the direction of John Rennie, Esq. C.E. F.R.S. By Netlam & Francis Giles 1815. W.F. 1817.
Engraved plan   25 × 72 cm   (scale 1 inch to 1 mile).
BM, RGS

1207   *Mr Rennie's Report*
Signed: John Rennie. London, 17 June 1816.
pp. 1–12.
[In]
Papers relating to the River Tyne. (Newcastle, 1836).
See TYNE, RIVER.

1208   *Copy of a Letter from John Rennie, Esq. to Henry Yeo, Esq. Secretary to the Commissioners for the Harbour of Howth.*
Signed: John Rennie. London, 30 Dec 1816.
pp. 55–59.
[In]
Fourth Report from the Select Committee on the Roads from Holyhead to London.
House of Commons, 1817.

See SPECIFICATIONS for Dunleary Harbour, 1817.

See RENNIE and AIRD. Plan of proposed new harbour at Dunleary, 1817.

1209 *Dimensions of the Southwark Cast Iron Bridge, erected over the Thames under the direction of John Rennie, Esq.*
[London]: Smith, Printer . . . [*c.* 1817]
12mo   17 cm   12 pp.   with woodcuts in text.
SML

See SPECIFICATIONS of earthwork of the Eau Brink Cut, 1818.

1210 *Communication from Mr Rennie, to the Commissioners of the Navy, on the best Method of preventing the general Shoaling of the River Medway.*
Signed: John Rennie. London, 20 May 1818.
pp. 30–34.
[In]
Report from the Select Committee on the Present State of Rochester Bridge.
House of Commons, 1820.

1211 *Report . . . to the Commissioners of the Haven and Piers at Great Yarmouth, Norfolk, on the State of the Bar and Haven, and the Measures advisable to be adopted for improving the same.*
By John Rennie, Esq. Engineer.
Yarmouth: Printed by C. Sloman . . . 1819.
Signed: John Rennie. London, 29 May 1818.
4to   23 cm   11 pp.
Norfolk R.O.
Yarmouth Haven Minute Book, 8 June 1819: 100 copies of Mr Rennie's report to be printed.

1212 *Letter (and Paper) from Mr Rennie on the subject of a new Bridge at Rochester.*
Signed: John Rennie. London, 27 March 1819 [and] 21 June 1820.
pp. 34–37.
[In]
Report from the Select Committee on the Present State of Rochester Bridge.
House of Commons, 1820.

1213 *The Reports of John Rennie, Esq. Civil Engineer, on the Proposed Bridge over the River Nene, to effect a direct communication from Norfolk and Suffolk, with Lincolnshire . . . and a Letter thereon from the Right Hon. Lord William Cavendish Bentinck, M.P. to His Grace the Duke of Bedford.*
Wisbech: Printed by J. White . . . 1819.
Signed: John Rennie. London, 10 April [and] 23 Aug 1819.
4to   28 cm   19 pp.
CE
Lord Bentinck's letter, dated 27 Sept 1819, is on pp. 15–19.

1214 *Copy of a Letter from John Rennie, Esq. to W. G. Adam, Esq.*
Signed: John Rennie. London, 30 Jan 1821.
pp. 17–21.
[Printed with]
Mr TELFORD's report, 31 May 1821 [q.v.].
[In]
Report of the Proceedings of the Committee for taking into Consideration Mr Rennie's Reports on the Improvement of the Outfall of the River Nene.
Wisbech: Printed by White and Leach. 1821.
8vo   22 cm   (i) + 24 + (1) pp.
BM, Cambs R.O., CE
The reports considered are by RENNIE, 26 Jan 1814 [q.v.] and his letter to Mr Adam. The letter was separately published earlier in 1821 (Wisbech: White and Leach.   fol   3 pp. Cambs R.O.).

See MOUNTAGUE, RENNIE *et al.* Report on London Bridge, 1821.

1215 *A Map and Survey of the River Wear, as far up as Biddick Ford, and of the Port and Haven of Sunderland.*
By John Rennie, Esquire. 1819, 1820, 1821 and 1822. Reduced from the Original Maps made by Francis Giles . . . and ordered to be Engraved 17 May 1826.
Engraved   plan   50 × 108 cm   (scale 10 inches to 1 mile).
Tyne & Wear R.O.

RENNIE, John and AIRD, John

1216 *Design for the New proposed Asylum Harbour at Dunleary, 1817.*

Jn.º Rennie & Jn.º Aird, Engineers.
Engraved plan (with sections) 23 × 29 cm
   (scales 1 inch to 500 feet and 1 inch to 100 feet).
NLS

RENNIE, John and HUDDART, *Captain* Joseph

1217 *Messrs Rennie and Huddart's first Report on the best means to be adopted for improving the Harbour of Howth.* [and] *second Report on their plan for the completion of the Harbour.*
Signed: John Rennie, J. Huddart. Dublin, 27 May 1808 [and] London, 18 May 1809.
pp. 19–22   with 2 folding engraved plans.
1. *Chart of the Shore at Howth.*
2. [*Plan of Howth Harbour*].
                [In]
Report from the Committee on Howth Harbour.
   House of Commons, 1810.

RENNIE, John and WALKER, Ralph

1218 *A Plan of the East India Docks at Blackwall.*
   John Rennie, Ralph Walker, Engineers.
Engraved by B. Baker. London. Published 8th March 1803.
Engraved plan, hand coloured   34 × 49 cm
   with inset plan   (scale 1 inch to 300 feet).
BDL (Gough)

1219 Revised scheme
Plan of the East India Docks.
   John Rennie & Ralph Walker, Engineers. London: Published 1.ˢᵗ December 1804 by Rob.ᵗ Wilkinson.
Engraved plan, hand coloured   37 × 54 cm
   (scale 1 inch to 240 feet).
PLA (Vaughan), BDL (Gough)

RENNIE, John and WHIDBEY, Joseph

1220 *The Report of Mr John Rennie and Mr Jos. Whidbey, to the Right Honourable the Lords Commissioners of the Admiralty.*
Signed: John Rennie, Jo.ˢ Whidbey. London, 21 April 1806.
pp. 3–11 with folding engraved plan.
Chart of Plymouth Sound, with the Situation of the Proposed Breakwater, and of the Anchorage . . . within the Same.
   J. Basire sc.
                [In]
Papers relating to Plymouth Sound.
   House of Commons, 1812.

1221 Reprint
[In] *Interesting Particulars relative to . . . the Breakwater now constructing in Plymouth Sound.*
   PLYMOUTH BREAKWATER, 1821 [q.v.].
   pp. 1–20.

RENNIE, *Sir* John [1794–1874] F.R.S., M.Soc.C.E., M.Inst.C.E.

See TELFORD and RENNIE. Reports on Eau Brink Cut, 1822.

1222 *Report concerning the Improvement of Boston Haven.*
   By John Rennie, Esq.
London: Printed by S. & R. Bentley.
Signed: John Rennie. London, 29 June 1822.
fol   30 cm   12 pp.   Folding litho plan.
Map and Chart of the River Witham & Boston Harbour from the Grand Sluice at Boston to Hob Hole Sluice & Clay Hole.
   Surveyed under the direction of John Rennie, Esq.ʳ . . . by Francis Giles . . .
Also a Plan of the Cut proposed for the improvement of the said Town and Harbour.
   By John Rennie Esq.ʳ Civil Engineer.

B. R. Baker. Lithog . . . July 1822.
CE

See WHIDBEY and RENNIE. Report on Whitehaven Harbour, 1823.

1223 *Report of John Rennie, Esquire, Civil Engineer, on the Ancholme Drainage and Navigation.*
Brigg: Bull, Printer.
Signed: John Rennie. London, 27 July 1825.
fol   33 cm   2 pp.   Docket title on p. (4).
LCL, BDL

See RENNIE, George and John. Plans of the Liverpool & Manchester Railway, 1826.

See SPECIFICATION for sea wall and West Pier, Leith Harbour, 1826.

1224 *Reports by Mr Rennie on the Works in Progress at Whitehaven Harbour.*
Signed: John Rennie. Whitehaven, 11 Sept 1826 (to Sept. 1830).
pp. 19–22.
[In]
Plans suggested at different periods for the Improvement of Whitehaven Harbour. WHITEHAVEN HARBOUR 1836 [q.v.].
CE, ULG

1225 *Report of John Rennie, Esq. Engineer, relative to the Approaches to the new London Bridge.*
Corporation of London, 1827.
Signed: John Rennie. London, 23 July 1827.
fol   32 cm   8 pp.
CLRO, CE
    For reprints *see* Nos. 1226 and 1227.

1226 *Mr Rennie's Second (and First) Report relative to the Approaches to the new London Bridge.*
Signed: John Rennie. London, 14 April 1828 (and 23 July 1827).
pp. 19–21 (and pp. 13–18).
[In]
Report to the Court of Common Council, from the Committee for Rebuilding London Bridge.
    Corporation of London, 1828.
fol   32 cm   24 pp.
CLRO, CE

1227   Reprint of both reports
[In] *Minutes of Evidence taken before the Lords Committee . . . on the Bill . . . for improving the Approaches to London Bridge.* 1829. pp. 41–46 and 46–48.

1228   [Report] *To Sir C. Sharp, Chairman, and to the Commissioners of Sunderland Harbour.*
Sunderland: Reed & Sons, Printers.
Signed: John Rennie. London, 2 April 1831.
fol   33 cm   6 pp.
CE

1229   *Report of Sir John Rennie on the Removal of the Old London Bridge.*
Signed: J. Rennie. London, 24 Oct 1831.
pp. 13–14.
[In]
Copy of the Reports presented to the Corporation of London . . . relative to the Stability of the New London Bridge.
    House of Commons, 1832.

See RENNIE, Sir John and MYLNE, W. C. Report on Thames Embankment, 28 Oct 1831.

1230   *Report of Sir John Rennie, on the subject of Messrs Telford and Walker's Report.*
Signed: John Rennie. London 17 Nov 1831.
pp. 11–13.
[In]
Copy of the Reports presented to the Corporation of London . . . relative to the Stability of the New London Bridge.
    House of Commons, 1832.

1231   *Whitehaven Harbour. Recommendations of a North Pier [and] Proposed Alterations in the Direction of the New North Pier.*
Signed: John Rennie. Whitehaven, 23 Feb 1833 [and] London, 27 March 1834.
pp. 24 and 28   with 2 litho plans.
Design for the New North Pier with a Proposed Termination of the West Pier.
    By Sir John Rennie, 1833.
Sir John Rennie's altered design for the New North Pier, 1834.
[In]

Plans suggested at different periods for the Improvement of Whitehaven Harbour. WHITEHAVEN HARBOUR, 1836 [q.v.].
CE, ULG

1232 [Report] *To His Excellency the Lieutenant-Governor and the Commissioners . . . for Making, Maintaining, and Improving the Harbours of the Isle of Man.*
n.p.
Signed: John Rennie. London, 2 April 1835.
4to   23 cm   17 pp.   Folding litho plan.
Plan and Chart of the Bay and Harbour of Douglas with the Low Water Asylum Harbour.
As proposed by Messrs Rennie, 1835. R. Martin, lithog.
CE

1233 *Reports and Estimates on the improvement of the Navigation of the River Nene, from its outfall to Peterborough; and for the more efficient drainage of Moreton's Leam Wash and Whittlesea Mere, and a portion of the adjoining Lands in the Middle District of the Great Bedford Level.*
By Sir John Rennie.
London: Printed by William Clowes and Sons . . . 1837.
Signed: John Rennie. London, 7 Dec 1836.
4to   24 pp.   2 folding engraved plans.
[1] *Section of the River Nene from Goldiford Staunch above Peterborough to the Sea shewing the improvements* as proposed by Sir John Rennie 1836.
[2] *Map of the . . . Great Level . . . by Samuel Wells.*
To accompany the Report of Sir John Rennie . . . 1836.
CE, CRO (lacking map)

See RENNIE, George and John. Reports on Central Kent Railway, 1836 and 1837.

1234 *The Report of Sir John Rennie, on the Improvement of the River Dee and Port and Harbour of Chester.*
Chester: Printed by F. P. Evans.
Signed: John Rennie. London, 27 July 1837.
8vo   22 cm   (i) + 17 pp.
BM, CE

1235 *River Ouze Outfall Improvement.*
London: Printed by James, Luke G. & Luke J. Hansard.

Signed: John Rennie. London, 1 July 1839.
4to   32 cm   13 pp.   2 folding engraved plans.
[1] *A Plan of the Great Wash . . . Shewing the Outfalls of . . . the Ouse, the Nene, the Welland, and the Witham . . . also the Plan for improving them . . .* by Sir John Rennie.
[2] *Map of the Great Level* (as in No. 1233)
ULG

1236 *Report on the Estuary of the Nene, between the Sutton Bridge Embankment, and the Limits prescribed by the Nene Outfall Act, as to the most advisable Line and Plan for Embanking out the Whole or any Portion of it from the Sea.*
By Sir John Rennie.
London: Printed by William Clowes and Sons . . . 1839.
Signed: John Rennie. London, 10 Nov 1839.
4to   30 cm   5 pp.   Folding engraved plan.
Plan and Sections of the Deserted Channel and the River Nene to accompany Sir John Rennie's Report.
B

RENNIE, *Sir* John and MYLNE, William Chadwell

1237 *Report of Sir John Rennie and W. C. Mylne, Esq. relative to Embanking the River.*
To the Worshipful the Navigation Committee of the Corporation of the City of London.
Signed: John Rennie, William Chadwell Mylne.
London, 28 Oct 1831.
pp. 83–88.

[In]

Report from the Select Committee on Thames Embankment.
House of Commons, 1840.

REYNOLDS, John

See BADESLADE. *The New Cut Canal*, 1736, with *Remarks* by Reynolds on the dam and sluice on the River Dee.

RHODES, Thomas [1789–1868]
M.R.I.A., M.Inst.C.E.

1238 *Report upon Canal and River Navigation from Limerick to Killaloe.*
Signed: Thomas Rhodes. Limerick, 7 March 1832.
With 5 folding hand-coloured litho plans including
Map of the River Shannon and Canal Navigation from Limerick to Killaloe shewing the Proposed Improvements.
> Tho: Rhodes, Civil Engineer. S. Arrowsmith lithog.
> [In]
Papers relating to the River Shannon Navigation.
> House of Commons, 1832. pp. 41–63 and pls. 1 and 6–9.

1239 *Second Report upon the Means of Improving the Shannon Navigation, and of Reducing the Waters of Lough Derg from the Winter or Flooded State, to the Ordinary Summer Level.*
Signed: Thomas Rhodes. Portumna, 14 April 1832.
With folding litho plan.
Plan of the River and Falls of Killaloe shewing the Proposed Improvements in the Navigation.
> S. Arrowsmith lithog.
> [In]
Papers relating to the River Shannon Navigation.
> House of Commons, 1832. pp. 64–66 and pl. 10.

1240 *Report upon the Lower Brusna River, for the purpose of making it Navigable from the Shannon to Croghan Bridge.*
Signed: Thomas Rhodes. Limerick, 26 May 1832.
With 3 folding litho plans including

Brusna River. Plan and Sections of Lock and Weir for making the River Navigable. (S. Arrowsmith).
> [In]
Papers relating to the River Shannon Navigation.
> House of Commons, 1832. pp. 76–82 and pls. 19–21.

1241 *Third Report upon the Present State of the River Shannon, and its Navigation, showing the Means of Improvement; also of reducing the Flood Waters.*
Signed: Tho: Rhodes, Civil Engineer. Limerick, 2 Feb 1833.
With 30 folding hand-coloured litho plates including
1. *Map of the River Shannon from Limerick to its source in Lough Allen.* 1833. (S. Arrowsmith).
16. *Shannon Navigation. Sections showing the number and Rise of each lock from Limerick to Lough Allen with their relative levels above Datum.* 1832.
14. *River Shannon. Elevations of all the Bridges from its source in Lough Allen to Limerick.* (J. Basire).
> [In]
Papers relating to the River Shannon Navigation. [Second Set].
> House of Commons, 1833. pp. 7–82 and pls. 1–9, 11–12, 14–19, 30–42.

1242 *Report upon the Ouse Navigation.* 1834.
York: Printed by Thomas Wilson and Sons ... and sold by H. Sotheran ... Price Two Shillings.
Signed: Tho. Rhodes. 18 Jan 1834.
4to 28 cm 24 pp.
York Central Library
Stitched as issued.

1243 *Longitudinal Section of the River Ouse from Linton Lock to Selby.*
> Drawn by H. Renton. Baird's lithog. [Signed]: Tho. Rhodes, 18 Jan 1834.
Litho plan 23 × 244 cm (scale 3 inches to 1 mile and 1 inch to 20 feet).
HRO

See PROVIS and RHODES. Report on Menai and Conway Bridges, 1836.

See BURGOYNE *et al.* Reports of the Commissioners for the Improvement of the River Shannon, 1837 and 1839.

RICHARDSON, Joshua [1799–1886] M.Inst.C.E.

1244 *Observations on the Proposed Railway from Newcastle upon Tyne to North Shields and Tynemouth.*
By Joshua Richardson, Civil Engineer.
Newcastle: Published by Charles Empson . . . 1831.
Dated: Railway Office, Leicester, 5 Jan [1831].
8vo   23 cm   iv + 48 pp.
CE

1245 *Observations on the Proposed Railway from Newcastle upon Tyne to North Shields.*
Second Edition.
By Joshua Richardson, Civil Engineer, M.Inst.-C.E.
London: Published by Longman and Co., and M. A. Richardson . . . Newcastle upon Tyne. 1834.
Dated: Royal Arcade, Newcastle, May 1834.
8vo   23 cm   vi + 72 pp.   Engraved plan.
Plan & Section of the Proposed Railway from Newcastle upon Tyne to North Shields.
By Joshua Richardson. Engraved by W. Collard.
CE

1246 *A Report on the Best Practical Means of Improving the Navigation of the River Tyne, with an Appendix on the River Clyde.*
By Joshua Richardson, Civil Engineer.
London: Longman . . . [and] Newcastle: M. A. Richardson . . . 1836.
Dated: Arcade, Newcastle, 13 July 1834.
8vo   23 cm   28 pp.
CE

1247 Second printing
The report with minor revisions reprinted in *Papers relating to the River Tyne.* (Newcastle, 1836). pp. 13–19. *See* TYNE, River.

1248 *The First Report on the Tyne, Edinburgh, and Glasgow Railway.*
By Joshua Richardson, Civil Engineer, M. Inst.-C.E.

Newcastle: Printed by John Hernaman.
Dated: Newcastle, 10 June 1836.
fol   34 cm   (4) pp.   Engraved plan, hand-coloured.
General Plan of the Tyne, Edinburgh & Glasgow Railway.
J. Richardson, Engineer. Engraved by W. Collard.
BM

1249 *The Second General Report on the Newcastle-upon-Tyne, Edinburgh and Glasgow Railway.*
By Joshua Richardson, Civil Engineer.
Newcastle: Printed by John Hernaman . . . 1837.
Dated: Bird-Hill, 22 June 1836 [i.e. 1837].
8vo   24 cm   (i) + 32 + 13 pp.   Large hand-coloured folding engraved map (W. Collard) and folding litho section; both by Joshua Richardson and dated 1837.
[1] *General Plan of the Newcastle upon Tyne Edinburgh and Glasgow Railway.*
[2] *Section of the Newcastle upon Tyne, Edinburgh and Glasgow Railway, from Horsley, Reedwater to the River Tweed.*
BM, CE
Issued with *Newcastle, Edinburgh, and Glasgow Railway . . . An Account of the Public Meetings* [etc]. Newcastle . . . 1837. This includes a reprint of Richardson's 1836 report, incorrectly dated.

RICKMAN, John [1777–1840] F.R.S., Hon.M.Inst.C.E.

1250 *Instructions to be observed by Messrs Jessop and Telford, respecting their Proceedings in determining upon the different Works to be executed upon the Line of the Caledonian Canal.*
Signed: Jnᵒ Rickman, 18 June 1804.
pp. 23–24.

[In]

Second Report of the Commissioners for the Caledonian Canal.

House of Commons, 1805.
This and the two following items are among many papers by Rickman written as Secretary to the Commissioners for the Caledonian Canal and for Highland Roads and Bridges.

1251 *Alphabetical List of Roads and Bridges in the Highlands of Scotland* [under the Commissioners].
Signed: Jn.º Rickman.
[In]
Third Report of the Commissioners for Highland Roads and Bridges.
House of Commons, 1807. pp. 62–65.
Includes dimensions, dates, costs and names of the surveyors.

1252 *Statement of the Origin and Extent of the several Roads in Scotland, made wholly, or in part, at the Public Expense . . . Together with Papers relating to the Military Roads.*
House of Commons, 1814.
Signed: Jn.º Rickman, 6 April 1814.
fol   33 cm   66 pp.   Engraved map.
Map of Scotland containing a Sketch of the Military Roads, also of those since made by means of the Highland Road and Bridge Act of 1803.
A. Arrowsmith sc.

See TELFORD. *Life of Thomas Telford,* edited with additional material by John Rickman, 1838.

RIOU, Stephen [1720–1780]

1253 *Short Principles for the Architecture of Stone-Bridges. With Practical Observations, and a New Geometrical Diagram to Determine the Thickness of the Piers to the Height and Base of any given Arch* [etc].
By Stephen Riou, Esq; Architect.
London: Printed for C. Hitch and L. Hawes . . . 1760.
8vo   22 cm   xx + 100 pp.   4 folding engraved plates (J. Mynde sc.).
BM, CE (Telford), RIBA, ULG, BDL

ROADS

1254 *An Essay for the Construction of Roads on Mechanical and Physical Principles.*
London: Printed for T. Davies . . . 1774.
Signed: AE.   London, 1 March 1774.
8vo   20 cm   48 pp.   1 folding engraved plate.
BM, CE, ULG

1255 *General Rules for Repairing Roads, recommended, by the Parliamentary Commissioners for the Improvement of the Mail Coach Road from London, by Coventry, to Holyhead, to the Turnpike Trustees between London and Shrewsbury.*
London: Published by J. Taylor . . . 1820.
8vo   21 cm   8 pp.   Engraved plate.
Tools for making & repairing Roads.   Jn. Easton del.   E. Turrell sc.
BM
Issued in boards with printed label. Price 2s. A reprint (the plate having been re-engraved) from TELFORD's report of 5 June 1820 on the English part of the Holyhead Road [q.v.].

1256   Second issue
As no. 1255 but *Second Edition.*
London: Published by J. Taylor . . . 1820.
P
Includes publisher's catalogue. At least two further issues appeared: the 4th (1823) is in ULG.

1257   Another edition
General Rules for Repairing Roads, published, by order of the Parliamentary Commissioners, for the Improvement of the Mail Coach Roads from London to Holyhead, and from London to Liverpool, for the Use of the Surveyors on these Roads.
A New Edition, Enlarged.
London: Published by J. Taylor . . . 1827.
8vo   21 cm   11 pp.   2 engraved plates.
I. (as in first edition).
II. [Sections of the old and improved roads]. Drawn by I. Macneill.
CE, ULG

ROBERTS, Charles

1258 *Plan of the intended Newcastle Junction Canal, and Railways.*
By Charles Roberts 1797.
Engraved map, line entered in colour
15 × 39 cm   (scale 4 inches to 1 mile).
BM
    i.e. Newcastle under Lyme.

ROBERTS, *Sir* Walter

1259 *An Answer to Mr Ford's Booke, entitled A Designe for bringing a Navigable River, from Rickmansworth in Hartfordshire to St. Giles in the Fields.* [and] *A Proposition for the serving and supplying of London and Westminster, and other places adjoyning; with a sufficient quantity of good and cleare spring water, to be brought from Hoddesdon in Hartfordshire in a close Aqueduct of Bricks, Stone, Lead, or Timber.*
    Printed at London, 1641.
4to   18 cm   (32) pp.   With small woodcut map showing the proposed course of the aqueduct and the existing New River.
BM, ULG
    An Answer is printed on pp. (3)–(13) and *A Proposition* on pp. (15)–(32).

1260 Second edition
An Answer to Mr Forde's Book, entitled, A Design for Bringing a Navigable River from Rickmansworth in Hartfordshire, to St. Giles's in the Fields.
Printed at London, 1641. And Reprinted Anno 1720.
4to   21 cm   pp. (13)–24.
BM, CE, ULG, BDL (Gough)
    Constitutes the second part in FORD's *Design*, London, 1720 [q.v.] but does not include *A Proposition*.

ROBINSON, James

1261 *An Exact Survey of the River Ouse, from Brandon Creek to Denver Sluice; with Observations thereon, in order to discover the true Cause of the present drowned Condition of the Fens in the South Level.*
By J. Robinson, of Ely.
Cambridge: Printed by J. Bentham . . . 1753.
Dated: Ely, 14 Sept 1752.
8vo   20 cm   15 pp.
CUL

ROBINSON, Richard

1262 *An exact and accurate Plan, of the East part of the South Level of the Fens, Situate between the River Ouse and the Hundred Foot River.*
Survey'd A.D. 1758. By Richard Robinson.
Engraved plan   54 × 152 cm   (scale 4 inches to 1 mile).
BM, BDL (Gough)

ROBISON, *Professor* John [1739–1805] F.R.S.E.

1263 *Answers [to the Questions respecting the Construction of a Cast Iron Bridge, of a Single Arch, 600 Feet in the Span, and 65 Feet Rise]*
By Mr J. Robeson, Professor of Natural Philosophy, Edinburgh.
Signed: John Robeson. Edinburgh [April 1801].
[In]
Report from the Select Committee upon Improvement of the Port of London, 1801. pp. 16–35.

1264 Second printing
In *Reports from Committees of the House of Commons*
Vol. 14 (1803) pp. 612–616.

The mistaken spelling of Robison's name remains in this reprint.

1265 *A System of Mechanical Philosophy.*
By John Robison, LL.D. Late Professor of Natural Philosophy in the University of Edinburgh. With Notes, by David Brewster, LL.D., F.R.S.L. & E.
Edinburgh: Printed for John Murray, London. 1822.
8vo 22 cm 4 vols with folding engraved plates (W. H. Lizars sc.).
1, (i) + x + 713 pp. 12 pls. **2**, (iii) + x + 708 pp. 12 pls. **3**, (ii) + 803 pp. 14 pls. **4**, (iii) + 684 pp. 10 pls.
BM, NLS, RS, CE, BDL
> Chiefly comprising articles contributed by Robison to the *Encyclopaedia Britannica*, 3rd edn. (1797) and its *Supplement* (1804). Those of civil engineering interest include: Pumps, Resistance of Fluids, Rivers, Roofs, Strength of Materials, Waterworks, and (in the *Supplement*) Arch, Carpentry, Centres for Bridges, Machinery.

ROBSON, Joseph [c. 1704–1780]

1266 *The British Mars. Containing several Schemes and Inventions, to be practised by Land and Sea against the Enemies of Great Britain* [etc].
By Joseph Robson, Engineer.
London: Printed for the author, and sold by W. Flexney . . . 1763.
8vo 20 cm xii + 210 pp. 11 engraved plates.
BM
> A section of the text dealing with rivers and harbours was reprinted as a 10 pp. 8vo pamphlet at South Shields by J. Bell in 1819 (BM).

1267 *To the Commissioners of the River Wear, and port and haven of Sunderland . . . the following pages, containing reasons against the filling up or lessening the receptacles for the flux of water in navigable rivers . . . are most respectfully dedicated.*

By Joseph Robson, Engineer.
Newcastle: Printed by J. White and T. Saint . . . 1766.
8vo 19 cm 36 pp.
Not seen.

ROCHDALE CANAL

1268 *Papers published in favour of the intended Rochdale Canal, in the Applications made to Parliament in the Sessions 1791, 1792, and 1793.*
Chester: Printed by J. Fletcher.
8vo 20 cm (i) + 79 pp.
CE, BDL

ROGET, Peter Mark [1779–1869] F.R.S., M.Soc.C.E., M.Inst.C.E.

See ROGET, BRANDE and TELFORD. Report on supply of water to the Metropolis, 1828.

ROGET, P. M., BRANDE, W. T. and TELFORD, Thomas

1269 *Report of the Commissioners appointed by His Majesty to inquire into the State of the Supply of Water in the Metropolis.*
Signed: P. M. Roget, William Thomas Brande, Thos. Telford. New Palace Yard, Westminster, 21 April 1828.
[In]
Report of the Commissioners [as above].
House of Commons, 1828. pp. 3–12.
See also in TELFORD. *Life of Thomas Telford.*
London, 1838. pp. 622–632.

ROSCOE, Thomas [1791–1871]

1270 *The Book of the Grand Junction Railway; being a History and Description of the Line from Birmingham to Liverpool and Manchester.*
By Thomas Roscoe, Esq., assisted by the Resident Engineers of the Line.
London: Orr and Co . . . 1839.
8vo   22 cm   Engraved vignette title page + (iii) + 154 + (3) pp.
Folding hand-coloured map with gradient profiles + 3 plans   15 steel-engraved plates (drawn by D. Cox, C. Radclyffe, G. Dodgson; engraved by W. Radclyffe, E. Radclyffe).
BM, SML, BDL, Elton
Price 14s. in blind and gilt stamped cloth.

1271 *The London and Birmingham Railway; with the Home and Country Scenes on each side of the Line.*
By Thomas Roscoe, Esq . . . assisted in the historical details by Peter Lecount, Esq.
London: Charles Tilt . . . [1839].
8vo   22 cm   Engraved vignette title page + (vii) + 192 + (4) pp.
Folding engraved map with gradient profile + plan   24 wood-engraved vignette views in text (S. Williams sc. and others)   16 steel-engraved plates (mostly drawn by Thomas Creswick, George Dodgson; engraved by E. Radclyffe, W. Radclyffe, Le Keux, D. Wilson).
BM, SML, CE, ULG, BDL, Elton
Price 16s. in cloth.

RUSSELL, John Scott [1808–1882] F.R.S.L. & E., M.Inst.C.E.

1272 *Report of the Committee on Waves, appointed by the British Association at Bristol in 1836.*
By John Scott Russell, Esq., M.A., F.R.S.Edin.
London: Reprinted by Richard Taylor, 1838.
8vo   23 cm   (i) + 417–496 pp.   8 plates.
CE (from the author)
Offprint, with new title page, from British Association 7th Report, for 1838.

SAMUDA, Joseph D'Aguilar [1813–1885] M.Inst.C.E.

See CLEGG and SAMUDA, 1840.

1273 *A Treatise on the adaptation of Atmospheric Pressure to the purposes of Locomotion on Railways.*
By J. D'A. Samuda.
London: John Weale . . . 1841.
8vo   22 cm   (iii) + 50 pp.   2 folding engraved plates (J. W. Lowry and S. Porter sc.).
BM
Second issue 1844 (BM, SML, CE).

SANDARS, Joseph

1274 *A Letter on the Subject of the Projected Rail Road, between Liverpool and Manchester, pointing out the Necessity for its Adoption and the Manifest Advantages it offers to the Public, with an Exposure of the exorbitant & unjust Charges of the Water Carriers.*
Liverpool: Printed by W. Wales and Co . . . [1824].
Signed: Joseph Sandars, 6 Oct 1824.
8vo   23 cm   32 pp.
CE, ULG, Elton

1275 Second edition
As No. 1274 with *By Joseph Sandars. Second Edition.*
Liverpool: Printed by W. Wales and Co . . . [1824].
Dated: 29 Oct 1824.
8vo   23 cm   32 pp.
CE (Page)

1276 Third edition
As No. 1274 with *By Joseph Sandars. Third Edition.*
Liverpool: Printed by W. Wales & Co . . . [1825].
Dated: 25 Jan 1825.
8vo   23 cm   46 pp.
BM, SML, ULG (Rastrick)
Two further editions were published in 1825.

SCOTT, John [1730–1783]

1277 *Digests of the General Highway and Turnpike Laws; with the Schedule of Forms, as Directed by Act of Parliament; and Remarks. Also, an Appendix on the Construction and Reparation of Roads.*

By John Scott, Esq.

London: Printed for Edward and Charles Dilley. 1778.

Dated: Amwell, 1778.

8vo 21 cm (xxix) + 352 + (16) pp. Woodcuts in text.

BM, ULG, BDL

Poet and friend of Dr Johnson, Scott was an active trustee of turnpike roads in Hertfordshire.

SCOTTEN, Edmund

1278 *A Desperate and Dangerous Designe Discovered concerning the Fen-Countries.*

London: Printed by G. B. and R. W. and . . . sold by Robert Constable . . . 1642.

Signed: Edmund Scotten.

4to 19 cm (iii) + 26 pp.

ULG

SEAWARD, John [1786–1858] M.Inst.C.E.

1279 *Observations on Suspension Chain Bridges; with an improved Method of forming the supporting Chains or Rods.*

London: Printed by Richard Taylor.

Signed: John Seaward. Lambeth, 11 Dec 1823.

8vo 22 cm 8 pp. Engraved plate (S. Porter sc.).

CE

Offprint from *Phil. Mag.* Vol. 62 (1823).

1280 *Observations on the Re-Building of London Bridge: demonstrating the practicability of executing that work in three flat elliptical arches of stone, each two hundred and thirty feet span: with an examination of the arch of equilibrium proposed by the late Dr Hutton.*

By John Seaward, Civil Engineer.

London: Printed for J. Taylor . . . 1824.

8vo 22 cm xiv + (2) + 143 pp. 7 folding engraved plates.

BM, NLS, CE (Telford), RIBA, BDL

Price 12s. in boards.

SEMPLE, George

1281 *A Treatise on Building in Water. In Two Parts. Part I. Particularly relative to the Repair and Re-building of Essex Bridge, Dublin . . . Part II. Concerning an Attempt to contrive and introduce quick and cheap Methods, for erecting substantial Stone-buildings and other Works, in fresh and salt Water* [etc.].

By George Semple.

Dublin: Printed for the Author, by J. A. Husband . . . 1776.

Signed: George Semple. Dublin, 2 Oct 1775.

4to 27 cm (viii) + 157 pp. 63 engraved plates.

BM, CE (Telford), RIBA, ISE, IC, BDL

1282 Second edition

A Treatise on Building in Water . . . The Second Edition. To which is added Part III. Hibernia's Free Trade; or, A Plan for the general Improvement of Ireland.

By George Semple.

London: Printed for the Author: and Sold by T. Longman . . . and I. Taylor . . . 1780.

4to 28 cm x + (v) + 190 pp. 64 engraved plates.

NLS, RIBA

SHARP, James

See WHITWORTH. Report on proposed canal from Waltham Abbey to Moorfields, with an Address to the Lord Mayor by James Sharp, 1773.

1283 *Extracts from Mr Young's Six Months Tour through the North of England* [in 1769]. *And from the Letter of an unknown Author, published in the London Magazine, for October 1772, on the Subject of Canal Navigation.*
[London], 1774
Signed: James Sharp. Leadenhall-Street, 16 Feb 1774.
8vo   20 cm   (iv) + 29 pp.   6 engraved plates and folding engraved plan.
A Plan of the Duke of Bridgewater's Navigable Canal.
BM, ULG, BDL (Gough)
Arthur Young's *Tour* was published in London 1770. 4 vols.

SHAW, Isaac

1284   *Views of the most interesting Scenery on the Line of the Liverpool and Manchester Railway . . . From Drawings taken on the Spot . . . by I. Shaw.*
Liverpool: Published by I. Shaw . . . 1831
4to   30 cm   16 pp.   8 plates drawn and engraved by I. Shaw jun. (Pls. 1–4) and by I. Shaw (Pls. 5–8).
BM, Elton

SHERRIFF, James

1285   *A Plan of the present and propos'd New Road lying between Wood Brook and the Pigeon House in the Road from Birmingham to Bromsgrove.*
Survey'd by Jas Sherriff in Aug. 1786.

Engraved by J. Cary.
Engraved plan, the lines entered in colour 20 × 47 cm   (scale 10 inches to 1 mile).
[Printed with]
Case, on the Part of the Trustees for the Turnpike Road leading from . . . Birmingham . . . to Bromsgrove.
Birmingham. 1789.
Single sheet.
BM

1286   *A Plan of the Proposed Line of Navigable Canal from Warwick to Birmingham.*
Surveyed by Jas. Sherriff 1792. Engraved by B. Baker.
Engraved plan   30 × 94 cm   (scale 2 inches to 1 mile).
BM, BDL, BDL (Gough)

1287   *A Plan of the Proposed Line of Navigable Canal from Warwick to Braunston.*
Survey'd by Jas. Sherriff 1793.
Engraved by B. Baker.
Engraved plan   27 × 91 cm   (scale 2 inches to 1 mile).
BM, BDL

SIMMS, Frederick Walter [1803–1864] M.Inst.C.E.

1288   *A Treatise on the Principal Mathematical Instruments employed in Surveying, Levelling, and Astronomy: explaining their Construction, Adjustments, and Use.*
By Frederick W. Simms.
London: Printed by T. Bensley . . . and sold, for the author, by Messrs Troughton and Simms . . . 1834.
8vo   22 cm   viii + 100 + (8) pp.   Woodcuts in text.
BM, NLS, RS, BDL

1289   Second edition
Title as No. 1288 but *Second Edition, improved and enlarged.*

London: Sold for the Author by Messrs Troughton and Simms . . . 1836.
Dated: Greenwich, 17 Feb 1836.
8vo   22 cm   xi + 118 + (10) pp.   Woodcuts in
text.
BM
   There are several later editions.

See PALMER, H.R. Map of the line of the South
Eastern Railway surveyed by P. W. Barlow,
F. W. Simms and W. Froude, 1835.

1290 *A Treatise on the Principles and Practice of
Levelling, showing its applications to purposes of civil
engineering, particularly in the construction of roads, with
Mr Telford's Rules for the Same. With an Appendix
containing a description of Mr Macneill's Dynamometer.*
   By Frederick W. Simms, Surveyor and Civil
   Engineer.
[London]: John Weale . . . 1837.
Dated: Greenwich, 3 Dec 1836.
8vo   23 cm   vii + 122 pp.   3 engraved plates.
BM, RS, BDL
   Issued in publisher's cloth, gilt-stamped: price
   6s. Subsequent editions appeared in 1843,
   1846, etc.

1291 *Rules for Making and Repairing Roads, as laid
down by the late Thomas Telford, Esq., Civil Engineer.
Extracted, with Additions, from a Treatise on the Prin-
ciples and Practice of Levelling.*
   By F. W. Simms, Surveyor and Civil Engineer.
[London]: John Weale . . . Price Two Shillings.
8vo   22 cm   29 pp.   2 engraved plates.
BM
   Probably published soon after the *Treatise.*

1292 *Public Works of Great Britain, consisting
of Railways . . . Cuttings, Embankments, Tunnels,
Oblique Arches . . . Cast Iron Bridges . . . Canals . . .
the London and Liverpool Docks . . . and other important
Engineering Works, with Descriptions and Specifications.*
   Edited by F. W. Simms, C.E.
London: John Weale . . . 1838.
Signed: F. W. Simms. Greenwich, 20 Jan 1838.
fol   55 cm   xii;   *Division I,*   Railways.
   72 pp.   83 engraved plates including 2 frontis.

(aquatint C. Rosenberg sc.), engraved title and
5 vignettes; *Division II,* Canals, Bridges,
River Walls, Port of Liverpool   32 pp.   29
engraved plates (Pls. 84–112); *Division III,*
Turnpike Roads &c.   24 pp.   19 engraved
plates (Pls. 113–131); *Division IV,* Survey of
the Port of London by James Elmes   70 pp.
22 engraved plates including frontis. (map by
J. & C. Walker), engraved title and vignette.
BM, NLS, SML, CE, ULG, IC, BDL
   From publisher's advertisement: bound in half
   morocco price £4 4s. or with twenty of the
   plates carefully coloured an additional £1 1s. is
   charged.

SIMPSON, James [1799–1869]
M.Soc.C.E., M.Inst.C.E.

1293 *Filtration of Thames Water at the Chelsea
Waterworks.*
   Communicated by Mr James Simpson, Resident Engineer.
pp. 645–647   with woodcut in text.
                    [In]
TELFORD. *Life of Thomas Telford.* London, 1838
[q.v.].

SMEATON, John [1724–1792] F.R.S.,
M.Soc.C.E.

1294 *Report on the Drainage of Lochar Moss, near
Dumfries; Drawn up for Charles Duke of Queensberry
and others.*
   By J. Smeaton.
Signed: J. Smeaton, Dumfries, 21 Sept 1754.
pp. (491)–502.
                    [In]
*General View of the Agriculture, State of Property, and
Improvements, in the County of Dumfries . . .*

By Dr [William] Singer.

Edinburgh: Printed by James Ballantyne and Co
... 1812.

8vo   xxvi + 696 pp.   Folding map + 9 engraved plates.

BM, NLS, SML, ULG

See WESTON, R. H. Letters [from Smeaton and
others 1755–1759] and documents relative to
the Edystone Lighthouse, 1811.

1295   [Begins] *From a Survey of the River Calder,
from Wakefield to Brooksmouth, and from thence to
Salter-Hebble Bridge, near Halifax, taken in the Months
of October and November 1757, by John Smeaton, it
appears* . . .

n.p.

Signed: J. Smeaton. Halifax, 26 Nov 1757.

Single sheet   32 × 19 cm

BM, RS (Smeaton), BDL

1296   *A Plan of the River Calder from Wakefield to
Brooksmouth and thence to Salter Hebble Bridge, laid
down from a Survey taken in October and November
1757; with a Projection for continuing the Navigation
from Wakefield to Salter Hebble Bridge near Halifax in
the County of York.*

By John Smeaton. R. W. Seale sc.

Engraved plan   23 × 82 cm   (scale 2 inches to
1 mile).

BM, RS (Smeaton), MCL, BDL, Sutro

See WEY, River Navigation, 1758.

1297   *A Plan of the River Clyde in the County of
Clydesdale from Dumbarton to Rose-Bank, laid down
from a Survey 1758, by James Barry, with a Projection of
the Navigation.*

By John Smeaton. R. W. Seale sc.

Engraved plan   36 × 75 cm   (scale $1\frac{3}{4}$ inches to
1 mile).

RS (Smeaton), MCL, Sutro

Smeaton's MS. report on the Clyde is dated 12
Dec 1758.

1298   *A Plan of the River Wear, from Biddick Ford to
the City of Durham, with a Projection of a Navigation
thereon.*

By J. Smeaton. R. W. Seale sc.

Engraved plan   23 × 39 cm   (scale 2 inches to
1 mile).

RS (Smeaton), MCL, BDL (Gough)

Smeaton gave evidence on this scheme to the
House of Lords in April 1759.

1299   *An Experimental Enquiry concerning the
Natural Powers of Water and Wind to turn Mills, and
other Machines, depending on a circular Motion.*

By J. Smeaton, F.R.S.

London: Printed in the Year 1760.

Dated: 3 May–14 June 1759.

4to   23 cm   77 pp.   3 folding engraved plates
(J. Mynde).

BM, NLS, CE, BDL

Offprint, with new title-page and resetting of
pp. 3 and 4, from *Phil. Trans.* Vol. 51 (1760 for
1759). For subsequent reprints *see* SMEATON
1794 and 1796 (with French translations 1810
and 1827), SMEATON 1814, TREDGOLD
1826 and 1836.

1300   *Mr Smeaton's Answer to the Misrepresentations
of his Plan for Black-Friars Bridge, contained in a late
anonymous Pamphlet, addressed to the Gentlemen of the
Committee for building a Bridge at Black-Friars.*

[London, 1760].

Signed: J. Smeaton. Furnivals-Inn-Court, 9 Feb
1760.

fol   33 cm   4pp.

BM, IC

For the anonymous pamphlet *see* BLACK-
FRIARS BRIDGE, Observations, 1760.

See BRINDLEY. Plan for a canal from Longbridge to Wilden, revised by Smeaton, 1760.

1301   *The Report of John Smeaton, Engineer, concern-
ing the Practicability of a Scheme of Navigation, from
Tetney Haven to Louth, in the County of Lincoln, from a
View taken thereof, in August 1760; As projected by Mr
John Grundy of Spalding, Engineer.*

Signed: John Smeaton. Austhorpe, 14 July 1761.

pp. 16–24.

[Printed with]

GRUNDY's report [q.v.].

[In]

A Scheme for executing a Navigation from Tetney-Haven to Louth; and for Draining the low Grounds and Marshes adjoining thereto. By John Grundy. To which is added the Report of John Smeaton [etc].
Nottingham: Printed by Samuel Creswell ... 1761.
8vo   22 cm   (i) + 24 pp.
LCL, Bristol County Library

See GRUNDY, EDWARDS and SMEATON. Report on the River Witham, 1761.

1302   *The Report of John Smeaton, upon the Harbour of Rye, in the County of Sussex.*
n.p.
Signed: J. Smeaton. London, 16 Feb 1763.
fol   31 cm   10 pp. Folding engraved plan.
A General Plan of the Bay & of the old & new Harbours of Rye ... in part extracted from former Plans, but viewed, revised, corrected and improved.
   By John Smeaton F.R.S. Anno 1763. E. Bowen sc.
BM, NLS, CE (plan only)

1303   *The Report of Mr John Smeaton, upon the Questions proposed to him by the Committee for improving, widening, and enlarging London Bridge.*
Signed: J. Smeaton. London, 18 March 1763.
With 3 folding engraved plans (Laurie & Whittle).
1. *Section of the Waterway at London Bridge.*
2. *Plan of the Foundations.*
3. *Plan and Section of the proposed Water Way under the great Arch of London Bridge.*
[In]
Second Report from the Select Committee upon the Improvement of the Port of London, 1799. pp. 25–35 and pls. 1–3.

1304   Second printing
In *Reports from Committees of the House of Commons* Vol. 14 (1803) pp. 470–474 and pls. 19–21.

See GRUNDY and SMEATON. Plan of part of Holderness with a scheme for draining the same, 1764.

1305   *Proposal for laying the Ships at the Key at Bristol Constantly Afloat, and for Enlarging Part of the Harbour by a New Canal through Cannon's Marsh.*
   By John Smeaton.
[Bristol]: Printed by S. Farley.
Signed: J. Smeaton. Bristol, 26 Jan 1765.
Single sheet   54 × 38 cm   including engraved plan.
The Plan of Bristol Key with the Projection of the Sluices and Canal for Floating the Shipping and Severn Trows.
Teasdale sc.
BM, RS (Smeaton), Bristol Library

1306   *The Report of John Smeaton, Engineer, upon the New-making and completing the Navigation of the River Lee, from the River Thames, through Stanstead and Ware, to the Town of Hertford.*
[with Estimates by Smeaton and an Appendix on Limehouse Cut by Thomas Yeoman].
Signed: J. Smeaton [Austhorpe, 24 Sept 1766].
[London, 1766].
fol   32 cm   12 pp.   Folding engraved plan.
A Plan of the River Lea from Hartford to Bow Bridge with a Profile of the Fall [and inset] *The River Lea from Bow Bridge to the Thames.*
P, B, BM (plan only), BDL (Gough, plan only)
   Lee Navigation Minute Book, 30 Sept 1766: Ordered that 1,000 copies of Mr Smeaton's report and plan be printed. In *Reports* Vol. 2, pp. 155–159 the report is dated but does not include the estimates or the Appendix by Yeoman. The plan is WHITTENBURY's of *c*. 1740 [q.v.] with addition of the inset and proposed cuts.

1307   *A Plan of the River Lee from Hertford to the River Thames, with a Profile of the Fall* [and an addition engraved by T. Jefferys 1767].
Engraved plan   23 × 95 cm   (scale 1½ inches to 1 mile).
Essex R.O., MCL, BDL (Gough)
   Third state of the plate showing a revised scheme the estimate for which (signed: J. Smeaton, London 26 Feb 1767) is printed in *Reports* Vol. 2, pp. 159–163.

1308   [Report relative to the Petition of the Proprietors of the London-Bridge Water-

Works] *To the Right Honourable the Lord Mayor, Aldermen, and Commons, in Common-Council Assembled.*
Signed: J. Smeaton. London, 5 Feb 1767.
pp. 2–4.

[In]

Kite, Mayor. A Common Council holden . . . on Wednesday the 25th Day of February 1767.
   Corporation of London, 1767.
fol   33 cm   7 pp.
CLRO

1309   *A Report upon the Harbour of King's Lynn, in Norfolk*
   By John Smeaton, Engineer, and F.R.S.
n.p.
Signed: J. Smeaton. Austhorpe, 14 Sept 1767.
4to   26 cm   27 pp.   Folding engraved plan.
A Plan of the Harbour of Lynn, extracted from a large plan Survey'd and Drawn by Mr Bell.
   T. Jeffreys sc.
CUL, RS (Smeaton, plan only)

1310   Second Edition
Title as No. 1309 with the date 1767.
n.p.
4to   24 cm   14 pp.   [no plan]
ULG
   Perhaps published *c.* 1793 in relation to the Eau Brink Cut project.

1311   *A View of the Bridge over Tweed at Coldstream. Finished in December 1766.*
   J. Smeaton, Engineer. R. Reid, Mason. T. Morris, Engraver.
Engraved elevation   21 × 74 cm   (scale 1 inch to 20 feet).
CE (Rennie)

1312   *The Report of John Smeaton Engineer, and F.R.S. concerning the Practicability and Expence of joining the Rivers Forth and Clyde by a Navigable Canal, and thereby to join the East Sea and the West.*
Edinburgh: Printed by Balfour, Auld and Smellie . . . 1767.
Signed: J. Smeaton. Austhorpe, 22 Dec 1764.
4to   26 cm   (i) + 35 + (4) pp.   2 folding engraved plans, the lines entered in colours (A. Bell sc.).
[1]   *A General Map of the Country betwixt the Forth & Clyde . . . showing the Course of the intended Canal.*
[2] *A Plan of the Tract of Country betwixt the Forth & Clyde proper for a Canal of Communication by way of the Rivers Carron Kilven &c together with a projection thereof.*
NLS, BDL (Gough)

1313   Second edition
Title as No. 1312.
Edinburgh: Printed by Balfour, Auld and Smellie . . . 1768.
4to   26 cm   (i) + 39 pp.   2 folding engraved plans (as in 1st edn).
BM (lacking pl. l), CE
   Forth & Clyde Canal Minute Book: 250 copies ordered to be printed 15 Jan 1768. Balfour's bill submitted 8 Feb 1768.

1314   *The Second Report of John Smeaton, Engineer, and F.R.S. touching the Practicability and Expence of making a Navigable Canal from the river Forth to the river Clyde . . . for vessels of greater burden, and draught of water, than those which were the subject of his First Report.*
Edinburgh: Printed by Sands, Murray, and Cochran. 1767.
Signed: J. Smeaton. Austhorpe, 8 Oct 1767.
4to   26 cm   36 pp.   Folding engraved plan, the line entered in colour.
Plan of the Great Canal from Forth to Clyde with the Extension at both Ends.
A. Bell sc.
NLS, ULG, BDL (Gough)

1315   Second issue
As No. 1314 with same typesetting, but note on p. 36 omitted and no plan.
CE
   Stitched as issued in grey wrappers.

1316   *A Review of Several Matters relative to the Forth and Clyde Navigation, as now settled by Act of Parliament: with Some Observations on the Reports of Mess. Brindley, Yeoman, and Golburne.*

By John Smeaton, Civil Engineer, and F.R.S.
[Edinburgh]: Printed by R. Fleming and A. Neill.
1768.
Signed: J. Smeaton, 28 Oct 1768.
4to   25 cm   (i) + 34 pp.
BM, NLS, SML, ULG
Forth & Clyde Canal Minute Book, 28 Nov
1768: bill for printing 500 copies. [The first
printed document with 'civil engineer' on
title-page].

1317   *Mr Smeaton's Report on Lewes Laughton Level.*
Lewes: Printed by William Lee. 1768.
Signed: J. Smeaton. Austhorpe, 27 July 1768.
fol   33 cm   19 pp.   Folding engraved plan.
*Projection of Works for the Drainage of Lewes Laughton
Level. Lying upon the Rivers Ouse and Glynd in the
County of Sussex.*
By John Smeaton Engineer 1768. T. Marchant,
Survey'd & delin.
BM, RS (Smeaton, plan only), CE, ULG

1318   *The Report of John Smeaton, Engineer, concern-
ing the Drainage of the North Level of the Fens and the
Outfall of Wisbeach River.*
n.p.
Signed: John Smeaton. Austhorpe, 22 Aug 1768.
4to   26 cm   24 pp.   Folding engraved plate.
*A Chain and Scale of Levels along Wisbeach River and
Channel from Peterborough Bridge down to the Eye at
Sea.*
Taken in 1767 by William Elstobb.
BM, CRO, BDL (Gough), CUL

1319   Second Edition
Title as No. 1318 with *Printed in the Year
MDCCLXVIII.*
n.p.
4to   24 cm   20 pp.   [no plate]
CE

1320   *The Report of John Smeaton, Engineer, upon the
Harbour of Dover.*
London: Printed by J. Hughs . . . 1769.
Signed: J. Smeaton. Austhorpe, 17 June 1769.
fol   36 cm   17 pp.   Folding engraved plan.
Plan of the Pier Heads of Dover Harbour.

Andrews sc.
BM, CE, BDL (Gough)

See SMEATON, ADAM and BAXTER. Report
on the new bridge at Edinburgh, 1769.

1321   *The Report of John Smeaton, Engineer, upon the
Harbour of the City of Aberdeen.*
[Aberdeen: Printed by Jas. Chalmers & Co . . .
1772].
Signed: J. Smeaton. Austhorpe, 19 Feb 1770.
fol   33 cm   6 pp.   Docket title on p. (8).
NLS
Aberdeen Shore Works Accounts, September
1772: Paid Jas. Chalmers & Co for printing Mr
Smeaton's report. Stitched as issued.

1322   Reprint
See ABERDEEN HARBOUR. Reports, 1834.
pp. 3–11.

1323   *The Report of John Smeaton, Engineer, upon the
Harbour of Port Patrick* [and] *Explanation of the Plan
for completing the Interior Harbour.*
Signed: J. Smeaton, Austhorpe, 18 May 1770
[and] 6 Jan 1774.
[In]
*Report from the Committee on the Communication be-
tween England and Ireland.*
House of Commons, 1809. pp. 62–67.

See MACKELL. Revised plan of Forth & Clyde
Canal, 1770.

1324   *The Report of John Smeaton, Engineer, upon the
means of improving the navigation of the rivers Aire and
Calder, from the free and open tides-way to the towns of
Leeds and Wakefield respectively.*
[Published January 1772].
Signed: J. Smeaton. Austhorpe, 28 Dec 1771.
Not seen, but reprinted in *Reports* Vol. 2
pp. 131–140.

1325   *Plan of the Rivers Air and Calder, from
the Towns of Leeds & Wakefield to Snaith.*
[1772]
Engraved plan   30 × 73 cm   (scale $1\frac{1}{4}$ inches to
1 mile).

RS (Smeaton)

Includes the navigation cuts and canal proposed in the report of 28 Dec 1771.

1326   Aire & Calder Navigation

A set of seven engraved plans (scale 8 inches to 1 mile) giving details of the proposed new cuts and canal. [1772]

(a) *The Leeds Cutt**. 21 × 25 cm

MCL, BDL

(b) *The Knostrop Cutt**. 23 × 31 cm

BM, MCL, BDL

(c) *The Woodlesford Cutt**. 28 × 40 cm

MCL, BDL

(d) *Methley Cutt.* 20 × 25 cm

MCL, BDL

(e) *The Castleford Cutt.* 17 × 23 cm

MCL, BDL

(f) *The Brotherton Cutt**. 28 × 46 cm

BM, Humberside R.O., BDL

(g) *Canal from Haddlesey to Gowdale.* 28 × 73 cm

BM, RS (Smeaton)

Smeaton's original drawings for these exist in the Royal Society collection.

1327   Another issue of No. 1326(f)

A Plan of the River Air &c. over Brotherton Marsh.

Engraved by T. Jefferys . . . 1772.

Engraved plan   32 × 42 cm   (scale 8 inches to 1 mile).

ULG

See SMEATON and WOOLER. Report on Tyne Bridge, 1772.

1328   *The Report of John Smeaton, Engineer, upon the Plan and Projection of a Canal, upon the North-Side of the River Air from Haddlesey to . . . Brier Lane End in the Township of Newlands; as laid down by Mr Jessop, Engineer.*

Signed: J. Smeaton. Austhorpe, 5 Dec 1772.

pp. 5–6.

[Printed with]

The Report of W. Jessop . . . 25 Nov 1772. See JESSOP, W.

n.p.

fol   31 cm   6 pp. Folding engraved plan.

ULG

This replaced the earlier plan for a canal from Haddlesey to Gowdale.

See SMEATON and JESSOP. Plan of the Rivers Air and Calder, 1773.

1329   *Letters between Redmond Morres, Esq; one of the Subscribers to the Grand Canal, and John Smeaton, Esq; Engineer, and F.R.S. in 1771, and 1772, relative to the Manner of Carrying on that Navigation* [etc].

Dublin: Printed by Oli. Nelson . . . 1773.

4to   25 cm   39 pp.

National Library Dublin

Includes five letters from Smeaton written between Aug 1771 and Dec 1772.

1330   *The Report of John Smeaton, Engineer, upon his view of the Country through which the Grand Canal is proposed to pass, and in answer to the several Matters contained in the Queries . . . agreed to on the 3d of August, 1773.*

Signed: J. Smeaton. Dublin, 6 Oct 1773.

pp. 10–20.

[In]

The Queries proposed by the Company of Undertakers of the Grand Canal, to John Smeaton, Esq; Engineer, and F.R.S. Together with his General Report thereon.

Dublin: Printed by Oliver Nelson . . . 1773.

4to   25 cm   20 pp.

National Library Dublin, CUL

1331   *The Report of John Smeaton, Engineer, upon the Means of improving the Drainage of the Level of Hatfield Chace.*

n.p.

Signed: J. Smeaton. Austhorpe, 7 Oct 1776.

fol   33 cm   9 pp.   Docket title on p. (12)

Folding engraved plan, lines entered in colours.

A Sketch of the River Torne from its Entry into the Level of Hatfield Chace at the Crooked Dyke End, to the River Trent at Althorp. Also of the Course of Drainage from Tunnel Pit by the New Idle to Durtness Bridge, together with Course of a proposed new Outfall Drain to the Trent near Waterton.

By direction of Mr Smeaton surveyed by Henry Eastburn 1776. Flyn sc.

Doncaster Archives, Nottingham University
  Library
  Stitched as issued.

1332 *Estimate for the Execution of the New Works
proposed for the Improvement of the Drainage of the Level
of Hatfield Chace . . .*
  By John Smeaton, Engineer.
n.p.
Signed: J. Smeaton. Austhorpe, 7 Oct 1776.
fol   33 cm   3 pp.   Docket title on p. (4).
Nottingham University Library.

1333 *A Report of John Smeaton, Engineer, upon the
State of the Bridlington Piers.*
n.p.
Signed: J. Smeaton. Austhorpe, 15 May 1778.
fol   32 cm   8 pp.
CE

1334 *The Report of John Smeaton, Engineer, upon Mr
James Shout's Plan for rebuilding and extending the old
pier of the harbour of Sunderland, in the County of
Durham.*
n.p.
Signed: J. Smeaton. Gateshead, 16 Jan 1780.
4to   27 cm   7 pp.
CE

1335 *The Report of John Smeaton, Engineer, upon the
State and Condition of Wells Harbour in the County of
Norfolk.*
n.p.
Signed: J. Smeaton. London, 4 May 1782.
fol   33 cm   15 pp.   Docket title on p. (16).
BM, ULG, Sutro

1336 *Mr Smeaton's Memorial Concerning Hexham
Bridge.*
n.p.
[Dated: Gray's Inn, 16 May 1783].
8vo   23 cm   (i) + 29 pp.
Northumberland R.O., Newcastle University
  Library
  Date known from Smeaton's diary.

1337 *Second Report by John Smeaton, Engineer, upon
the Inrun of the Seas into the Harbour of Aberdeen, in
Easterly Winds.*
pp. 13–18 with lithograph plan.
*Plan of the Harbour of Aberdeen with its Alteration as
proposed 1787 by Mr Smeaton.*
[Printed with]
*Letter from Mr Smeaton respecting the Harbour of Aber-
deen.*
pp. 19–20.
Both dated: Austhorpe, 22 March 1788.
[In]
*The Reports by Smeaton, Rennie and Telford, upon the
Harbour of Aberdeen.* See ABERDEEN HAR-
  BOUR. Reports, 1834.

1338 *A Narrative of the Building and a Description of
the Construction of the Edystone Lighthouse with Stone:
to which is subjoined, an Appendix giving some Account
of the Lighthouse on the Spurn Point, built upon a Sand.*
  By John Smeaton, Civil Engineer, F.R.S.
London: Printed for the Author, by H. Hughs:
  Sold by G. Nicol . . . 1791.
fol   55 cm   xiv + 198 pp.   23 engraved plates
  (Edwd Rooker, Hen. Roberts, J. Record, W.
  Faden) + vignette on title-page (M. Dixon del.
  A. Birrel sc.).
BM, NLS, RS, CE, BDL, CUL
  The RS copy presented by Smeaton in January
  1791.

1339   Second edition
Title as No. 1338 but *Second Edition, Corrected.*
London: Printed for G. Nicol . . . 1793.
fol   55 cm   xiv + 198 pp.   23 engraved
  plates + vignette (as in 1st edn.).
BM, SML, CE, RIBA, ULG
  New setting of type, typographical corrections
  including paragraph numbering, omission of
  note on p. vii. Price £3 3s. in boards.

1340   Third edition
Title as No. 1338 but *Second Edition* [i.e. Third].
London: Printed by T. Davidson . . . for Longman, Hurst . . . 1813.
fol   54 cm   xiv + 198 pp.   23   engraved
  plates + vignette (as in 1st edn.).

NLS, RS, CE, ISE, IC, BDL
> New setting of type. The original copper plates used for third time. 300 copies printed, price £6 6s. in half leather.

1341 *An Historical Report on Ramsgate Harbour: written by order of, and addressed to the Trustees.*
By John Smeaton, Civil Engineer, F.R.S. and Engineer to Ramsgate Harbour.
London: Printed in the Year 1791.
Signed: J. Smeaton, March 1791.
8vo   24 cm   vi + (2) + 85 pp.   2 folding engraved plans.
[1] *A Map of the Downs.*
[2] *Plan of Ramsgate Harbour and Principal Works thereof.*
BM, SML, CE, IC, BDL (Gough), CUL

1342   Second edition
As No. 1341 but *Second Edition.*
London: Printed in the Year 1791.
8vo   24 cm   vi + (2) + 86 pp.   2 folding engraved plans (as in 1st edn.)
BM, NLS, RS, CE, BDL (Gough)
> New setting of type, with additional note on p. 86. The RS copy presented by Alexander Aubert in November 1791.

1343   *Experimental Enquiry concerning the Natural Powers of Wind and Water to turn Mills and other Machines depending on a Circular Motion. And an Experimental Examination of the Quantity and Proportion of Mechanic Power necessary to be employed in giving different degrees of Velocity to Heavy Bodies from a State of Rest. Also New Fundamental Experiments upon the Collision of Bodies.*
By the late Mr John Smeaton, F.R.S.
London: Printed for I. and J. Taylor . . . 1794.
8vo   23 cm   iv + 110 pp.   5 folding engraved plates.
BM, NLS, SML, CE, BDL, CUL
> Reprints of Smeaton's three papers on mechanics originally published in *Phil. Trans.* Vol. 51 (1760 for 1759), Vol. 66 (1776) and Vol. 72 (1782). The plates re-engraved.

1344   Second edition
Title as No. 1343 but *Second Edition.*

London: Printed for I. and J. Taylor . . . 1796.
8vo   25 cm   viii + 110 pp.   5 folding engraved plates (as in 1st edn.).
SML, CE, ULG, BDL
> New setting of type and addition of *Contents.* Price 5s. in boards. French translation published in Paris, 1810 and 1827. For further reprints of these three papers *see* TREDGOLD 1826 and 1836.

1345   *Reports of the late Mr John Smeaton, F.R.S. made on Various Occasions in the Course of his Employment of an Engineer.*
Printed for a Select Committee of Civil Engineers, and Sold by Mr Faden. Vol. I.
London: Printed by S. Brooke . . . 1797.
4to   27 cm   xxx + [xxv–xxxii] + 412 pp. Engraved portrait frontispiece (W. Bromley after Mather Brown).
BM, NLS, BDL
> Approximately 850 copies printed. Published March 1798, price 18s. in boards. The prelims include a *Preface* [by Robert Mylne] and *Some Account of the Life . . . of Mr John Smeaton* by Charles Hutton with a letter from Mrs Mary Dixon giving a sketch of her father's character.

1346   *Reports of the late John Smeaton, F.R.S. made on Various Occasions, in the Course of his Employment as a Civil Engineer.*
In Three Volumes.
London: Printed for Longman, Hurst . . . 1812.
3 vols.   4to   27 cm   **1,** (i) + xxx + [xxv–xxxii] + 412 pp.   Portrait frontispiece + 33 plates. **2,** xi + 440 pp.   23 plates. **3,** vii + 420 pp.   16 plates. The plates engraved by Wilson Lowry from drawings by John and Joseph Farey reduced from Smeaton's originals.
BM, NLS, SML, RS, CE, ULG, BDL, CUL
> Five hundred sets published August 1812, price £7 7s. in boards. Volume 1 made up from remaining sheets of the 1797 printing with new title-page (and half-title) and the addition of thirty-three plates. The three volumes contain more than 200 reports, an almost complete collection of those written between October 1760 and January 1792.

1347 Second edition
Title as No. 1346 but *Second Edition. In Two Volumes.*
London: Printed by M. Taylor . . . 1837.
2 vols.   4to   28 cm   **1**, xxxv + 423 pp.   Portrait frontispiece + 38 plates.   **2**, xii + 423 pp.
34 plates.
BM, CE, IC, CUL
> Unabridged text and prelims of the 1812 edition, using the same copper plates, with a new introduction by W. M. Higgins.

1348 *The Miscellaneous Papers of John Smeaton, Civil Engineer, F.R.S. Comprising his Communications to the Royal Society, printed in the Philosophical Transactions.*
London: Printed for Longman, Hurst . . . 1814.
4to   27 cm   viii + 208 pp.   12 engraved plates.
BM, SML, CE, ULG, BDL, CUL
> Five hundred copies printed. Published September 1814, price £1 11s. 6d. in boards. Reprints of Smeaton's eighteen papers in *Phil. Trans.* 1750–1788 using eleven of the original copper plates and one re-engraved by Lowry.

SMEATON, John, ADAM, John and BAXTER, John

1349 *Report of Messrs Smeaton, Adam, and Baxter* [on the North Bridge].
Signed: J. Smeaton, John Adam, John Baxter. Edinburgh, 22 Aug 1769.
pp. 9–16.
                    [In]
Acts and Proceedings of the Town Council of Edinburgh . . . so far as it concerns the New Bridge. With the Report of Messrs. Smeaton, Adam, and Baxter.
Printed by Order of the Council
Edinburgh: Printed for L. Hunter. 1769.
8vo   28 pp.
NLS

SMEATON, John and JESSOP, William

1350 *Plan of the Rivers Air and Calder, from the Towns of Leeds & Wakefield* [*to Armin*].
Engraved plan, the cuts and canal entered in colour   30 × 91 cm   (scale 1¼ inches to 1 mile).
BDL
> Endorsed 'Plan of the Rivers Air & Calder with the Intended Improvements 1773'. The plate of SMEATON's 1772 plan [q.v.] with an addition showing Jessop's Haddlesey–Brier Lane canal. Copies with MS. alterations are in MCL and BDL (Gough).

SMEATON, John and WOOLER, John

1351 *A Report relative to Tyne Bridge.*
Newcastle: Printed by T. Saint . . . 1772.
Signed: J. Smeaton, John Wooler. 4 Jan 1772.
8vo   21 cm   16 pp.
BM, CE, BDL (Gough)

SMITH, David [d. 1854]

See TELFORD. Plan of the Crinan Canal, surveyed by David Smith, 1823.

1352 *Plan of the River Cart from Clyde to Paisley, shewing intended Improvements.*
Surveyed by David Smith 1834. T. Atkinson sc.
Engraved plan   33 × 62 cm   (scale 5 inches to 1 mile).
HLRO

SMITH, Edward

1353 *A Description of the Patent Perpendicular Lift, erected on the Worcester and Birmingham Canal, at Tardebig, near Bromsgrove.*

By Edward Smith.

Birmingham: Printed by J. Orton Smith . . . Price One Shilling.

Signed: Edward Smith. Birmingham, 6 March 1810.

8vo   22 cm   16 pp.   3 engraved plates.

CE

The canal lift by John Woodhouse.

SMITH, Humphry [*c.* 1671–1743]

1354 *Mr Humphry Smith's Scheme for the Draining of the South and Middle Levels of the Fens.*

n.p.

Signed: Humphry Smith. 28 Aug 1729.

4to   27 cm   8 pp.

CE, CRO, BDL (Gough), Sutro

SMITH, John *Engineer* [?1725–1783] M.Soc.C.E.

1355 *A Plan of the Intended Drainage & Navigation from the River Humber to Market Weighton in the East Riding of the County of York, with the Intended Drainage & Navigation up Long Dike . . . Foulney, Everingham & Harswell Beck, to Shipton Common.*

Survey'd by Smithson Dawson in Feb.ʸ 1772. John Smith, Engineer.

Engraved plan   45 × 30 cm   (scale 1⅓ inches to 1 mile).

Humberside R.O., Hull University Library.

1356 *The Report of John Smith, Engineer, Pointing out the Method of making a Navigable Canal from Wakefield to the Head of the River Went, and from thence, by the Course of that River, into the Dun, at a Place called Went-Mouth, and also of Draining certain Low Grounds . . . together with an Estimate of the Expences thereof* [etc].

n.p.

Signed: John Smith, Engineer. Wakefield, 21 Oct 1772.

fol   32 cm   17 pp.   2 folding engraved plans (Forrest sc.).

[1] *A Plan of the Intended Navigation, and Drainage, from Wakefield to the River Dun. By the Course of the River Went.*

Taken by William Pape, Surveyor. John Smith Engineer.

[2] *A Section of the Canal and Navigation from the River Dun, to the River Calder.*

By John Smith, Engineer. Anno 1772.

ULG

1357 *The Report and Opinion of John Smith, Engineer, concerning the present State of the Drainage of the Low Lands on both Sides of the River Witham, from the City of Lincoln, through Boston to the Sea . . . together with Observations on the Plan and Estimate drawn by Mr Creassy, for effecting the purposes of a General Drainage of this Extensive Country.*

Boston: Printed by C. Preston, 1776.

4to   32 pp.

Not seen; details from bookseller's catalogue.

1358 *Plan of the Intended [Erewash] Canal from the River Trent to Langley Bridge, in the Counties of Derby and Nottingham.*

John Smith Eng.ʳ Ella & Fletcher Survey.ˢ Engraved by W.ᵐ Faden, 1777.

Engraved plan, the line entered in colour   37 × 109 cm   (scale 3¼ inches to 1 mile).

BDL (Gough)

SMITH, John *Surveyor*
[*c. 1747–1820*]

1359 *A State of the Depths of Water upon the Shallows, in the River Trent; between Cavendish-Bridge and Gainsbro': As taken by W. Jessop, Engineer, in . . . 1782; and also those taken by J. Smith, Surveyor, on the 17th, 18th, and 19th of July, 1786.*
Nottingham: Printed by George Burbage. 1786.
Signed: John Smith.
fol 34 cm 15 pp.
PRO

1360 *A State of the Depths of Water upon the Shallows in the River Trent, between Cavendish Bridge and Gainsbro'; as taken by W. Jessop in . . . 1782; and those taken by J. Smith, Surveyor, in July, 1786; and also those taken by the said surveyor . . . in July 1792.*
Nottingham: Printed by G. Burbage. 1792.
Signed: J. Smith.
fol 32 cm 16 pp.
PRO

See WHITWORTH. Plan of intended Canal from Ashby de la Zouch to Griff, surveyed by John Smith, 1792.

See WHITWORTH and JESSOP. Plan of the Ashby de la Zouch Canal, surveyed by J. Smith, 1794.

SMITH, William [1769–1839]

1361 *Observations on the Utility, Form and Management of Water Meadows, and the Draining and Irrigating of Peat Bogs, with an Account of Prisley Bog.*
By William Smith, Engineer and Mineralogist.
Norwich: Printed by R. M. Bacon . . . 1806.
8vo 22 cm (iv) + 121 pp 2 folding engraved plates.
BM, ULG

1362 *Report on the Plan for draining the Low Ground north and south of the River Went, between the Rivers* Aire and Dun, *in conjunction with the proposed Aire and Dun Canal, and Went Branch.*
By William Smith, Engineer.
Pontefract: Printed by William Hunt.
Dated: 12 Sept 1818.
4to 20 cm 10 pp.
Not seen.

1363 *The Proposed Aire & Dun Junction Canal to drain the contiguous Lands and to shorten and connect the present Navigations.*
William Smith, Engineer 1819. Drawn on Stone by John Phillips.
Litho plan 21 × 32 cm (scale 1 inch to 4 miles).
[In]
Aire and Dun Junction Canal, and Extensive Drainage in Yorkshire.
n.p.
fol 34 cm 3 pp. The plan on p. 3, docket title on p. (4).
BDL

SNAPE, John [*c. 1738–1816*]
M.Soc.C.E.

1364 *A Plan of the Intended Extension of the Dudley Canal into the Birmingham Canal near Tipton Green in the County of Stafford: And also of the Dudley and Stourbridge Canals.*
Surveyed in the Year 1785 by John Snape. J. Russel sc.
Engraved plan 30 × 50 cm (scale 2½ inches to 1 mile).
RS (Smeaton)

1365 *A Plan of the intended Navigable Canal from the Town of Birmingham into the River Severn near the City of Worcester.*
Surveyed in the Year 1789 by John Snape. Ross sc.
Engraved plan 34 × 73 cm (scale 1 inch to 1 mile).
BM

1366 *A Plan of the Intended Navigable Canal, from the Worcester & Birmingham Canal at Kings Norton, in the County of Worcester; to Stratford upon Avon, in the County of Warwick; also the collateral Branches to Grafton &c.*

Surveyed by John Snape 1792.
Engraved plan   29 × 70 cm   (scale 1¼ inches to 1 mile).
BM

1367 *A Plan of the intended Navigable Canal from the present Dudley Canal Navigation at Netherton in the Parish of Dudley . . . to the Worcester and Birmingham Canal at Selly Oak . . . with the collateral Branches to communicate therewith.*

Surveyed by John Snape in 1792.
Engraved plan   30 × 55 cm   (scale 1 inch to 3 miles).
BM, RGS

SOUTHERN, John [c. 1758–1815]

1368 *Answers [to the Questions respecting the Construction of a Cast Iron Bridge, of a Single Arch, 600 Feet in the Span, and 65 Feet Rise].*

By Mr John Southern, Engineer.
Signed: John Southern. Soho, Birmingham, 23 April 1801.
With folding engraved plate.
Elementary Elevation of London New Bridge, proposed by John Southern.
April 1801.

[In]

Report from the Select Committee upon Improvement of the Port of London, 1801. pp. 64–70 and pl. 2.

1369 Second printing
In *Reports from Committees of the House of Commons*
Vol. 14 (1803) pp. 627–630 and pl. 55.

SPECIFICATIONS

1370 *Specification of the Locks proposed to be built on the Royal Canal of Ireland, on the Lands of Porterstown, Riverstown [etc].*
Dublin: Printed by T. M'Donnel.
Signed: John Rennie. London, 29 Nov 1803.
Single sheet   40 × 25 cm
ULG

1371 *Specification (and Tender) of the Masonry, Brick Work, Piling, Planking, &c. of a New Entrance to the Basin at the Royal Dock Yard at Deptford, and also for the Building of a River Wall from the said entrance to the Second Slip on the South.*
[John Rennie, Engineer].
[London]: Printed by W. Clowes . . . for His Majesty's Stationery Office.
fol   34 cm   4 + 3 pp.
CE

A private collection copy is endorsed 'Works at Deptford Feby 17th 1815'.

1372 *Specification of the manner of quarrying Rock . . .for the Works of Dunleary Harbour [and] Specification of Horse Work proposed to be done by Contract in drawing the Waggons on the Rail-road for Dunleary Harbour.*
[John Rennie, Engineer]
Dublin. 19 May 1817.
fol   37 cm   2 pp.
NLS

1373 *Specifications of the Earthwork of the New proposed Cut for the river Ouze, between Eau Brink and Lynn.*
London: Printed by Luke Hansard & Sons.
Signed: John Rennie. London, 7 Jan 1818.
fol   33 cm   3 pp. Docket title on p. (4).
B

1374 *Edinburgh & Glasgow Union Canal. Conditions under which the Proposals for contracting for the Different Lots of Work are to be given in.*
[Hugh Baird, Engineer].
Edinburgh: 26 Dec 1817.

fol   34 cm   2 pp.
CE (Telford)

1375   *Edinburgh & Glasgow Union Canal. Specification No. 1. For Cutting, Embanking, Lining, Puddling, Fencing, &c.*
[Hugh Baird, Engineer].
Edinburgh: [1818].
fol   34 cm   3 pp.   Docket title on p. (4).
CE (Telford)

1376   *Edinburgh & Glasgow Union Canal. Specification No. VI. For an Aqueduct intended over the River Avon.*
[Hugh Baird, Engineer].
Edinburgh: 12 March 1818.
fol   34 cm   3 pp.   Docket title on p. (4).
CE (Telford)
   In this collection there are another nine specifications for works on the Union Canal.

1377   *Particulars and Specification of the proposed New West Pier of Scarborough.*
[William Chapman, Engineer].
Newcastle: E. Walker, Printer. December 1818.
fol   32 cm   2 pp.
CE

1378   *Holyhead Road. Braunston Hill and Stow Hill, near Coventry. Specifications.*
Signed: Thos. Telford, 3 July 1821.
[In] TELFORD. *Life of Thomas Telford.* London, 1838 [q.v.] pp. 525–528.

1379   *Holyhead Road. St. Alban's and South Mims Trust. Specification.*
Signed: John Easton. London, 9 July 1823.
[In] TELFORD. *Life of Thomas Telford.* London, 1838 [q.v.] pp. 529–536.

1380   *Specification, &c. of a Proposed Jetty at Scarborough.*
[William Chapman, Engineer].
Newcastle: Printed by Edward Walker. September 1823.

fol   32 cm   3 pp.   Docket title on p. (4).
CE

1381   *Bedford Level Corporation. Specification for Building a Sluice with Four Openings, at Wellmore Lake, adjoining the Hundred Feet River near Welney.*
[John Dyson, Engineer].
Dated: Fen Office, 28 Dec 1824.
fol   33 cm   2 pp.
CRO

1382   *Specification and Description of Mason Work, to be done in erecting the projected Bridge of Suspension, over the River Tyne, between North and South Shields, according to Plans, Elevations, and Sections hereunto referred, together with the Conditions of Contract.*
Newcastle: Printed by T. & J. Hodgson.
Signed: John Green. [Sept 1825].
fol   35 cm   4 pp.   2 engraved plans.
Plan. John Bell, Surveyor. M. Lambert sc.
Plan [i.e. Elevation] *of the Suspension Bridge proposed to be Erected over the River Tyne between North and South Shields.*
Designed by Cap.! S. Brown, R.N.   I. Green del.   M. Lambert sc.
IC
   Dated by local newspaper notices.

1383   *Specification (and Tender) of the mode of Constructing the Sea Boundary Wall . . . [and] the West Pier Proposed for the Improvement of the Harbour of Leith.*
[Sir John Rennie, Engineer].
n.p.
fol   33 cm   8 pp.
CE
   With MS. entries of tenders from five contractors, endorsed '16 Nov. 1826'.

1384   *Newcastle-upon-Tyne and Carlisle Railway. Specification of the Work of Building a Bridge over Corby Beck . . . [and] a Bridge over the River Eden to consist of 5 Semicircular Arches, each of 80 Feet Span.*
n.p.
Signed: Francis Giles, 21 Aug 1830.
fol   34 cm   2 + 2 pp.
CE (Telford)

1385 *Specification of a Timber Bridge, for the River Clyde, at Glasgow.*

By Robert Stevenson, Civil Engineer.
[Glasgow]: Edward Khull, Printer. 1831.
8vo   20 cm   15 pp.
NLS

1386 *Glasgow Bridge Specification and Tender.*
Signed: Thomas Telford. London, 6 Nov 1832.
[In] TELFORD. *Life of Thomas Telford.* London, 1838 [q.v.] pp. 507–525.

1387 *Specification (and Tender) of the Repairs of Blackfriars Bridge.*
[James Walker, Engineer].
London: Printed by Arthur Taylor . . . 1833.
fol   33 cm   19 + 9 pp.
CE

1388 *Specification of the Harbour Works at Perth, including Tide-Basin, Canal, and Dock, agreeably to plans and sections by Robert Stevenson & Son, of Edinburgh, Civil Engineers.*
Perth: Printed by C. G. Sidey.
Dated: Edinburgh, 28 April 1835.
fol   34 cm   7 pp.
NLS

1389 *Great Western Railway. Specification of Rails.*
[I. K. Brunel, Engineer].
n.p.
fol   33 cm   3 pp.   with section on p. 3.
CE
Endorsed 'May 1837'.

1390 *Manchester and Leeds Railway. Contract. Tender for Works.*
[Thomas Gooch, Engineer].
n.p. [1837].
fol   39 cm   6 pp.   Docket title on p. (8).
CE

With MS. 'Prices at which the Ludden, Sowerby, Copley, Elland and Rastrick contracts were executed'.

STAVELEY, Christopher [1759–1827]

See JESSOP and STAVELEY. Plan of intended navigation from Loughborough to Leicester and Melton Mowbray, and railways to Swannington, etc. 1790.

See JESSOP and STAVELEY. Plan of intended navigation from Melton Mowbray to Oakham, 1792.

See VARLEY and STAVELEY. Plan of intended Union Canal from Leicester to the Grand Junction Canal, 1792.

STEEDMAN, John

See STEVENSON, Robert. Surveys by John Steedman, 1814–1826.

1391 *To the Subscribers to the Survey of the proposed Railway from the Forth and Clyde Canal to Stirling . . . and to Alloa Ferry, near Kersie Nook, the Report of John Steedman, Engineer.*
[Edinburgh, 1827].
Signed: John Steedman. Edinburgh, 25 August 1827.
4to   25 cm   18 pp.   Hand-coloured folding litho plan.
Reduced Plan & Section of a proposed Railway from the Forth & Clyde Canal to the Town of Stirling & Callander with a Branch to Alloa Ferry.
By John Steedman, Engineer, 1826.
ULG (Rastrick)

STEERS, Thomas [*c.* 1672–1750]

1392 *A Map of the Rivers Mersey and Irwell from Bank-Key to Manchester with an Account of the rising of* the *Water & how many locks it will require to make it Navigable*.

Survey'd by order of the Gentlemen at Manchester by Tho. Steers. 1712. I. Senex sc.

Engraved map   19 × 29 cm   (scale 1 inch to 1½ miles).

Chetham's Library Manchester, BDL (Gough)

See STEERS and EYES. Plan of the River Calder, 1741.

STEERS, Thomas and EYES, John

1393 *A Plan of the River Calder, from Wakefield to Ealand, and thence Continued to Salter hebble Bridge, in the County of York*.

Survey'd in 1740 & 1741 by John Eyes.

Eman: Bowen sc.   [with inset] *The Projection of the intended Navigation on this River*.

By Tho.ˢ Steers & Jn.º Eyes.

Engraved plan   24 × 68 cm   (scale 1¾ inches to 1 mile).

BDL

Shows the proposed locks and cuts.

STEPHENSON, George [1781–1848] M.Soc.C.E.

1394 *Plan and Section of the intended Railway or Tramroad from Stockton by Darlington to the Collieries near West Auckland in the County of Durham and several Branches therefrom, and of the Variations and Alterations intended to be made therein respectively & also of the Additional Branch of Railway or Tramroad proposed to be made*. 1822. W. Miller sc.

Engraved map, the lines entered in colours   38 × 87 cm   (scale 1¼ inches to 1 mile).

BM

The large-scale MS. deposited plan has exactly the same title and is signed: Geo. Stephenson, Engineer.

1395 *A Plan and Section of an intended Railway or Tram-Road from Liverpool to Manchester, in the County Palatine of Lancaster*.

Surveyed by George Stephenson, Engineer. 20th day of Nov. 1824.

Engraved by J. & A. Walker.

Signed: Geo. Stephenson, Engineer Tho.ˢ O. Blackett, Surveyor.

Engraved map, hand-coloured   66 × 97 cm (scale 1 inch to 1 mile).

BM, SML, Inst. Mech. Eng.

For the route finally adopted *see* RENNIE, George and John, 1826.

1396 *Report and Estimate, of an intended Railway, from Bolton le Moors to Eccles; where it is proposed to join the intended railway from Liverpool to Manchester*.

By George Stephenson, Civil Engineer.

Newcastle: Printed by J. Clark . . . 1825.

Signed: George Stephenson. Newcastle, 12 Dec 1825.

8vo   22 cm   8 pp.

SML

See STEPHENSON, Robert and LOCKE. Observations on comparative merits of locomotive and fixed engines, 1830.

1397 *Report by George Stephenson, Esquire, Civil Engineer, on Line of Edinburgh, Glasgow and Leith Railway*.

Signed: Geo. Stephenson. Liverpool, 30 Dec 1830.

pp. (1)–5.

[Printed with]

Report by GRAINGER and MILLER, Dec 1830 [q.v.].

[In]

Edinburgh, Glasgow & Leith Railway. Reports by Messrs Grainger & Miller of Edinburgh, and Mr George Stephenson of Liverpool, Civil Engineers. January 1831.

[Edinburgh, 1831]

4to   29 cm   (iii) + 2 + 5 + 18 + 6 pp.   Folding engraved plan.

NLS, CE

Stitched as issued in printed wrappers with engraving of *Locomotive Steam Engine with a Train of Railway Carriages*. (Lizars sc.).

1398   Second, revised edition

Report by George Stephenson, Esq. Civil Engineer, Liverpool. To the Committee of the Edinburgh, Glasgow, and Leith Railway.

Signed: Geo. Stephenson. Liverpool, 6 June 1831. pp. 11–16.

[Printed with]

Report by GRAINGER and MILLER, 23 Nov 1831 [q.v.].

[In]

Edinburgh and Glasgow Railway. Reports by Mr George Stephenson, of Liverpool, and Messrs Grainger and Miller, of Edinburgh, Civil Engineers. November 1831.

Edinburgh: Printed by J. & C. Muirhead, 1831.

4to   28 cm   28 pp.   Folding engraved plan.

BM, NLS

Issued in printed wrappers similar to the January 1831 publication. The reports differ substantially from the earlier versions.

1399   *Plan and Section of an Intended Railway, commencing at or near to Ranelagh Street and Bold Street in Liverpool, in the County of Lancaster, and terminating in the Township of Chorlton . . . in the County of Chester, at or near to the Public Road leading to Nantwich, and intended to form a Portion of a Projected Railway, from Liverpool aforesaid to Birmingham, in the County of Warwick.*

Surveyed under the Direction of Geo. Stephenson, Esq.ʳ by Joseph Locke, Civil Engineer. [1831].

Signed: Geo. Stephenson, Engineer.

Engraved map, hand-coloured   52 × 186 cm (scale 2 inches to 1 mile).

BM

An early version of the northern part of the Grand Junction Railway. The MS. deposited plan of the revised line [1832], from Warrington to Chorlton, is also 'Surveyed under the direction of George Stephenson, Esq.ʳ by Joseph Locke Civil Engineer'.

1400   *To the Committee of the Sheffield and Manchester Railway Company. Report upon the Practicability of making the Line, and the Mode and Cost of working it.*

By Geo. Stephenson, Esq.

Signed: Geo. Stephenson. Liverpool, 20 Sept 1831.

pp. (1)–22.

[In]

Appendix to the Report of the Provisional Committee of the Sheffield & Manchester Railway to the Company of Proprietors, at their First General Assembly, to be held October 20, 1831 . . . at Manchester.

Liverpool: Printed by Wales and Baines . . . [1831].

8vo   21 cm   (iii) + 52 pp.   Folding engraved plan.

Plan of the Sheffield and Manchester Railway.

H. Austen, Surveyor. Liverpool, Sept 1830. T. Smith sc.

ULG, Elton

See STEPHENSON, George and Robert. London and Birmingham Railway map, 1833.

1401   *The York & North Midlands Railway, uniting York with . . . the Manufacturing Districts of Yorkshire, Lancashire, and the Midland Counties.*

Surveyed under the Direction of Geo. Stephenson Esq. Engineer, by Fred.ᵏ Swanwick, Assistant Engineer. 1835.

Litho map   48 × 33 cm   (scale 1 inch to 12 miles).

BM

1402   *Plan and Section of an Intended Railway to commence at or near to the Town of Stockport, and to terminate at . . . Manchester.*

Surveyed under the Direction of George Stephenson, Esq., Engineer, by Frederick Swanwick, Assistant Engineer. November 1835.

Litho plan   59 × 76 cm   (scale 4 inches to 1 mile and 1 inch to 100 feet).

MCL

1403   *London & Blackwall Railway, with Branches to the East & West India Docks, and proposed Steam Navigation Depot.*

George Stephenson Esq. Engineer. W. K. Morland, del. Waterloo & Morland, Lithog.

Litho plan   37 × 50 cm   (scale 4 inches to 1 mile).

[In]

Prospectus of the London and Blackwall Rail-Way Company. [1836].

[London]: Waterlow and Morland, Printers.

fol 42 cm (4) pp. including plan (as above).
BM, CE

1404 *Plan and Section of the Proposed Chester and Crewe Railway. 1836.*
Geo. Stephenson, Engineer. Murray Gladstone, Assistant Engineer. John Palin, Surveyor. On Stone by C. F. Cheffins.
Litho map, hand-coloured 35 × 115 cm (scale 2 inches to 1 mile).
ULG (Rastrick)

See STEPHENSON, George and Robert. Map of proposed railway between London and Brighton, 1836.

1405 *Report to the Provisional Committee of the Maryport and Carlisle Railway.*
Signed: George Stephenson. Duke-street, Westminster, 3 May 1836.
p. (2).
[In]
Maryport and Carlisle Railway. [Prospectus].
Maryport: Robert Adair, Printer.
fol 42 cm (4) pp Docket title on p. (4).
ULG

1406 *Map of the Maryport and Carlisle Railway, shewing its Junction with the Newcastle and Carlisle Railway.*
George Stephenson Esq.ʳ Engineer.
Drawn under the Direction of Geo. Stephenson by Mr W. S. Hall. C. F. Cheffins lithog.
C. F. Cheffins lithog.
Litho map, partly hand-coloured 36 × 53 cm (scale 1 inch to 5 miles).
[In]
Maryport and Carlisle Railway. [Prospectus].
Maryport: Robert Adair, Printer.
fol 42 cm (4) pp. including plan (as above).
CE, ULG
Another edition of the Prospectus.

See GRAINGER and MILLER. Map of proposed Edinburgh, Haddington & Dunbar Railway, 1836.

1407 *Reports on the Formation of a Railway between Lancaster and Carlisle, (via Ulverston and Whitehaven) with Observations on the Mode of Crossing Morecambe Bay.*
By George Stephenson, Esquire, Civil Engineer. And other information by the Grand Caledonian Junction Railway Committee.
Whitehaven: Printed by Robert Gibson . . . 1837.
8vo 20 cm 20 pp.
Contains (pp. 5–16) three reports by Stephenson, dated Alton Grange 12 Oct 1836, 16 Aug 1836 and London 13 March 1837, and Observations on Morecombe Bay dated Alton Grange 16 Aug 1837. Also: Observations by the Committee (pp. 3–5) and Remarks (pp. 16–20).
SML

See STEPHENSON, George and BIDDER. London and Blackwall Railway report, 1838.

1408 [Report] *To the Committee appointed for the Promotion of a Railway from Newcastle to Edinburgh.*
Signed: George Stephenson. Tapton House, near Chesterfield, 13 Sept 1838.
pp. 4–7.
[In]
Prospectus of the Great North-British Railway between Edinburgh and Newcastle-upon-Tyne. 1838.
Edinburgh: Printed by Thomas Allan & Co.
Dated: Edinburgh, 10 Dec 1838.
fol 37 cm 7 + (1) pp. Docket title on p. (8).
BM

1409 *Report of George Stephenson, Esquire, on the comparative merits of the Railway from Chester to Holyhead, and that from Wolverhampton to Porthdynllaen.*
Chester: Printed by R. H. Spence . . . 29 April 1839.
Signed: George Stephenson.
8vo 21 cm (i) + 8 pp.
BM, ULG

STEPHENSON, George and BIDDER, George Parker

1410 *London & Blackwall Commercial Railway.*
Report of Messrs. Geo Stephenson, & G. P. Bidder.
[London]: F. Mansell, Printer.
Dated: 6 Jan 1838.
8vo   21 cm   22 pp.
SML, ULG

STEPHENSON, George and Robert

1411 *London and Birmingham Railway.*
Geo. Stephenson & Son, Engineers. [1833].
Drawn by G. Hennet. Engraved by J. Dower.
Includes *Section of the Line of Railway showing the Rise and Fall* [and a Table of] *Comparative Distance from the Metropolis to the Cities & Towns ...*
Engraved map   55 × 43 cm   (scale 1 inch to 8 miles).
RGS, ULG
> The RGS copy is printed in a Prospectus dated 1833. For a later version *see* STEPHENSON, Robert, 1835. In the 1833 MS. deposited plans the section is signed: George Hennet, Surveyor.

1412 Deposited plan
Plan of a Proposed Railway, to join the intended Railway from London to Southampton and to form a Communication between London & Brighton, by way of Shoreham.
Messrs George and Robert Stephenson—Consulting Engineers. G. P. Bidder—Acting Engineer. 1836.
Litho map in 10 sheets each 24 × 64 cm   (scale 4 inches to 1 mile).
CE

STEPHENSON, Robert [1803–1859] F.R.S., M.Soc.C.E., M.Inst.C.E.

See STEPHENSON and LOCKE. Observations on comparative merits of locomotive and fixed engines, 1830.

See STEPHENSON, George and Robert. London and Birmingham Railway map, 1833.

1413 *London and Birmingham Railway. 1835.*
Robt. Stephenson, Engineer.
Engraved by I. Dower.
Includes *Section of the Line of Railway showing the Rise and Fall and the Inclinations in feet per mile* [and a Table of] *Comparative Distance from the Metropolis to the Cities & Towns ...*
Engraved map   55 × 43 cm   (scale 1 inch to 8 miles).
CE
A later version of the 1833 map.

1414 *London & Birmingham Railway. Plan of the Line and Adjacent Country. 1835.*
Robert Stephenson, Engineer.
Engraved by I. Dower.
Includes *Section showing the Inclinations of the Railway.*
Engraved map, hand-coloured   72 × 152 cm (scale 1 inch to 2 miles).
ULG

1415 Later edition
London & Birmingham Railway. Plan of the Line and Adjacent Country. 1835.
Robert Stephenson, Engineer.
Published by Chaʃ F. Cheffins, Surveyor, Engineering Draughtsman & Lithographer.
London: Published 10 Sept 1835 by C. F. Cheffins.
Includes *Section showing the Inclinations of the Railway*, and inset plans of *Birmingham Depot* and *London Depot at Chalk Farm and Euston Square.*
Engraved map, hand-coloured   72 × 152 cm (scale 1 inch to 2 miles).
BM, ULG (Rastrick), BDL

See STEPHENSON, George and Robert. Map of proposed railway between London and Brighton, 1836.

1416 *Report from Mr Stephenson to the Directors of the Western London and Brighton Railway Company.*
London: Printed by James & Luke G. Hansard . . . 1837.
Signed: Robert Stephenson. Great George Street, Westminster, 15 Aug 1837.
4to   25 cm   7 pp.
CE

1417 *London and Brighton Railway. Mr Robt. Stephenson's Reply to Capt. Alderson.*
London: John Weale . . . 1837.
Signed: Robert Stephenson. Great George Street, 24 Oct 1837.
fol   33 cm   16 pp.   Litho map.
The map, untitled, shows in colours Rennie's or the Direct Line and Stephenson's or the Western Line. (C. F. Cheffins Lithog).
Inst. Mech. Eng., CE (lacks wrappers)
    Issued in yellow printed wrappers with title and publication line on front cover only. The report is headed: *London and Brighton Railway. To the original Shareholders of Stephenson's Brighton Railway*. A full-length copy of Captain Alderson's report is included.

1418 *Northern and Eastern Railway. Mr Stephenson's Report.*
[London]: R. Clay, Printer.
Signed: Robert Stephenson. Great George Street, Westminster, 25 Sept 1839.
fol   45 cm   (4) pp.   including litho map.
Map (and Section) of the Northern & Eastern Railway.
CE

1419 *London and Westminster Water Company. Report by Robert Stephenson, Esq. Civil Engineer.*
[London]: Office of the Company . . . 1840.
Signed: Robert Stephenson. London, 16 Dec 1840.
8vo   21 cm   16 pp.   Hand-coloured litho frontispiece (geological section).
BM, CE

1420 *London, Westminster, and Metropolitan Water Company. Mr Stephenson's Second Report to the Directors.*
Westminster: Printed by J. Bigg and Son.
Signed: Robert Stephenson. London, 24 Aug 1841.
8vo   21 cm   16 pp.
BM, CE

STEPHENSON, Robert and LOCKE, Joseph

1421 *Observations on the comparative merits of Locomotive & Fixed Engines, as applied to Railways; being a Reply to the Report of Mr James Walker, to the Directors of the Liverpool and Manchester Railway, compiled from the Reports of Mr George Stephenson. With an Account of the Competition of Locomotive Engines at Rainhill in October, 1829, and of subsequent experiments.*
    By Robert Stephenson and Joseph Locke, Civil Engineers.
Liverpool: Printed by Wales and Baines . . . [1830].
Dated: Liverpool, February 1830.
8vo   25 cm   viii + 83 pp.   Folding litho plate.
NLS, SML, CE, ULG

STEVENSON, Alan [1807–1865] F.R.S.E., M.Soc.C.E., M.Inst.C.E.

1422 *The British Pharos: or a list of the Light-Houses on the coasts of Great Britain and Ireland, descriptive of the Appearance of the Lights at Night.*
Leith: Published and Sold by W. Reid & Son . . . 1828.
Signed: Alan Stevenson.
12mo   15 cm   xiv + 111 pp.
NLS

Issued in publisher's cloth with printed label on front cover. Price 2s. 6d.

1423   Second edition
As No. 1422 but *Second Edition*.
Leith: Published and Sold by W. Reid & Son . . . 1831.
Signed: Alan Stevenson.
8vo   17 cm   xvi + 118 pp.
BM, NLS

See STEVENSON, Robert and Alan. Reports and plans, 1832–1837.

1424   *Letter to the Author of an Article on the British Light-House System, in Number CXV of the Edinburgh Review.*
By Alan Stevenson, Civil Engineer.
Edinburgh: William Blackwood . . . 1833.
Signed: Alan Stevenson. Edinburgh, 1 July 1833.
8vo   22 cm   32 pp.
BM, NLS, CE

1425   *Report to the Committee of the Commissioners of Northern Lights appointed to take into consideration the subject of Illuminating the Lighthouses by Means of Lenses.*
By Alan Stevenson, M.A. Civil Engineer.
Edinburgh: Printed by Neill & Company . . . 1835.
Signed: Alan Stevenson. Edinburgh, 10 Dec 1834.
4to   28 cm   (i) + 41 + (1) pp.   6 engraved plates.
BM, NLS, SML, CE
The first of four reports on this subject (all in CE).

See STEVENSON, Robert, Alan and David. Chart of the River Lune, 1838 and Report on the River Dee, 1839.

1426   *Lighthouses: being the article Sea-Lights, in the Seventh Edition of the Encyclopaedia Britannica.*
By Alan Stevenson, LL.B., F.R.S.E. Civil Engineer.
Edinburgh: Printed for Adam and Charles Black. 1840.

4to   29 cm   (19) pp.   with woodcuts in text. 2 engraved plates.
CE
Offprint, with new title-page, from *Encyclopaedia Britannica*. 7th edn. Vol. 20 (1840).

STEVENSON, David [1815–1886] F.R.S.E., M.Soc.C.E., M.Inst.C.E.

1427   *Observations on the Liverpool and Manchester Railway: with Remarks on the Dublin and Kingstown Railway.*
By David Stevenson, Civil Engineer.
[Edinburgh, 1836].
Dated: Edinburgh, 21 Feb 1835 [and] 9 March 1836.
8vo   23 cm   (i) + 16 pp.   3 engraved plates (E. Mitchell).
CE
Offprints, with a new title page, from *Edinburgh New Philosophical Journal* for April 1835 and April 1836.

1428   *Sketch of the Civil Engineering of North America; comprising remarks on the harbours, river and lake navigation, lighthouses, steam-navigation, water-works, canals, roads, railways, bridges, and other works in that country.*
By David Stevenson, Civil Engineer.
London: John Weale . . . 1838.
8vo   22 cm   xv + errata leaf + 320 pp.   Folding engraved map + 14 engraved plates, mostly folding (Geo. Aikman sc.)
BM, NLS, SML, CE, ULG (Rastrick)

See STEVENSON, Robert, Alan and David. Chart of the River Lune, 1838 and Report on the River Dee, 1839.

STEVENSON, Robert [1772–1850] F.R.S.E., M.Soc.C.E., M.Inst.C.E.

1429 *Address received from Mr Robert Stevenson.*
Dated: Edinburgh, 23 Dec 1800.
pp. 13–38.

[In]

Memorial and State relative to the Light-Houses erected on the Northern Parts of Great Britain; and relative to a Proposal for Erecting a Light-House upon the Bell or Cape Rock.
Published by Direction of the Commissioners.
Edinburgh: Printed by Alex. Smellie. 1803.
8vo   23 cm   38 pp.
NLS

1430 *Report on the Improvement of the Harbour of Dundee.*
By Robert Stevenson, Esq.
[Dundee]: 1814.
Signed: Robert Stevenson. Edinburgh, 2 Feb 1814.
4to   28 cm   (i) + 13 pp.   Folding engraved plan, hand coloured.
Reduced Plan for the enlargement of the Harbour of Dundee
By Robert Stevenson Esq. Civil Engineer 1814.   John Steedman del.   Thos. Ivory sc.
NLS, CE

1431 *Observations upon the Alveus or General Bed of the German Ocean and British Channel.*
By Robert Stevenson, Civil Engineer, F.R.S.E.
Edinburgh: Printed by Neill & Co. 1817.
Dated: 2 March 1816.
8vo   20 cm   (i) + 27 pp.
NLS, BDL

1432 *Information relative to the Caledonian Canal, communicated to the Editor of the Supplement to the Encyclopaedia Britannica.*
By Robert Stevenson, F.R.S.E. Civil Engineer.
[Edinburgh]: Printed by George Ramsay & Co.
4to   26 cm   10 pp.
CE
Reprint, with new setting of type, from *Supplement* to the Fourth, Fifth and Sixth Editions of the *Encyclopaedia Britannica* Vol. 2 (1817).

1433 *Report relative to a Line of Canal upon One Level, between the Cities of Edinburgh and Glasgow, to form a junction with the Forth and Clyde Canal at Lock No. 20.*
By Robert Stevenson, Civil Engineer, F.R.S.Edin. Engineer for the Commissioners of the Northern Light-Houses.
Edinburgh: Printed by Neill and Company. 1817.
Signed: Robert Stevenson. Edinburgh, 17 Feb 1817.
4to   (i) + 29 + 7 pp.   Folding engraved plan, hand-coloured.
Reduced Survey & Section of a Line of Canal between Edinburgh & Glasgow upon One Level, connecting Leith and the Broomielaw, made in the years 1814 & 1815 . . .
By Robert Stevenson F.R.S.Edin. Civil Engineer. 1817. John Steedman Surveyor. J. & G. Menzies sc.
NLS, CE, ULG, BDL

1434 *Report relative to the Strathmore Canal, or Inland Navigation, between the Royal Boroughs of Forfar and Aberbrothwick . . .*
By Robert Stevenson, Civil Engineer [etc].
Edinburgh: Printed by Neill and Company. 1817.
4to   27 cm   (i) + 17 + (2) pp.   Folding engraved plan, hand coloured.
Reduced Survey & Section of the proposed Canal into the Vale of Strathmore between . . . Forfar & Aberbrothwick.
Surveyed . . . by Robert Stevenson . . . 1817. John Steedman, Surveyor. J. & G. Menzies sc.
BM, NLS, CE

1435 *The Report of Robert Stevenson, Civil Engineer, relative to the Improvement of the Communication by the Ferries betwixt Fife and Forfar.*
[Edinburgh]: P. Neill, Printer.
Signed: Robert Stevenson. Edinburgh, 4 Aug 1818.
4to   28 cm   25 pp.   2 folding engraved plans, hand coloured.
[1] *Plan with Sections shewing the proposed improvements at the Craig Rock and the mode of converting the*

Western Pier of the New Harbour of Dundee into a landing slip for Ferry boats . . . by Robert Stevenson . . . 1818.

[2] *Reduced Survey of part of the Frith of Tay as seen at low water of Spring tides referring to a Report relative to the improvement of the Ferry . . .* by Robert Stevenson . . . 1818. John Steedman, Surveyor. J. & G. Menzies sc.

NLS, CE

Issued in printed wrappers with title: Report relative to the Ferries across the Tay . . . 1818.

1436 *Report relative to various Lines of Railway, from the Coal-Field of Mid-Lothian to the City of Edinburgh and Port of Leith; with plans and sections, showing the practicability of extending these lines of railway to Dalkeith, Musselburgh, Haddington and Dunbar.*

By Robert Stevenson, Civil Engineer [etc].

Edinburgh: Printed by P. Neill. 1819.

Signed: Robert Stevenson. Edinburgh, 28 Dec 1818.

4to   27 cm   (i) + 47 pp.   2 folding engraved plans, hand coloured.

[1] *Reduced Plan of part of the Shires of Edinburgh & Haddington shewing the Lines of the Proposed Railways . . .* By Robert Stevenson . . . 1818. J. Steedman, Surveyor. J. & G. Menzies sc.

[2] *Sections of the Ground Surveyed for the several Lines of the Edinburgh Railway.* 1818.

NLS, CE, ULG

1437 *Report relative to an Iron Railway, between the Port of Montrose and the Town of Brechin.*

By Robert Stevenson, Civil Engineer [etc].

Edinburgh: Printed by P. Neill. 1819.

Signed: Robert Stevenson, Edinburgh, 31 March 1819.

4to   27 cm   (i) + 25 pp.   Folding engraved plan, hand coloured.

Reduced Survey & Sections of several Lines of Railway from the Port of Montrose to the Town of Brechin calculated to be extended in various directions into the vale of Strathmore.

By Robert Stevenson . . . 1819. J. Steedman, Surveyor. T. Clerk sc.

CE

1438   Revised scheme

Signed: Robert Stevenson. Edinburgh, 30 Nov 1819.

4to   27 cm   3+ (1) pp.   Folding engraved plan, title etc as in No. 1437. No title-page.

CE

Second state of the plan with new line and additional topography.

1439 *Report relative to the Compensation Reservoir for the Mills on the Water of Leith and Bevelaw Burn.*

By Robert Stevenson, Civil Engineer [etc].

Edinburgh: Printed by P. Neill. 1819.

Signed: Robert Stevenson. Edinburgh, 17 April 1819.

4to   28 cm   (i) + 19 pp.

CE

1440 *Reduced Plan of Part of the Parishes of Penicuk, Currie & Kirknewton shewing the Situation of Two Compensation Reservoirs for the use of the Mills on Bevelaw Burn & the Water of Leith.*

Referred to in a Report by Robert Stevenson, Civil Engineer, 1819, Engr. by T. Clerk.

Engraved plan, hand coloured   28 × 44 cm (scale 1 inch to 2,000 feet).

P

1441 *Copy of Mr Stevenson's Observations* [on Tay Ferries at Dundee].

n.p.

Dated: Edinburgh, 6 Jan 1820.

4to   25 cm   4 pp.   Folding engraved plan, hand coloured.

Plan by Mr Stevenson [and] *Plan by Mr Telford.*

NLS

See also STEVENSON's report, 4 Aug 1818.

1442 *To the Right Honourable the Chairman and the Directors of the Edinburgh Joint Stock Water Company, the Report of Robert Stevenson, Civil Engineer, relative to conveying the Crawley Aqueduct or Pipe Track to the New Town by the Lothian Road.*

n.p.

Signed: Robert Stevenson. Edinburgh, 26 Jan 1820.

4to   27 cm   7 pp.

NLS

1443 *Memorial relative to Opening the Great Valleys of Strathmore and Strathearn, by means of a Railway or Canal, with Branches to the Sea from Perth, Arbroath, Montrose, Stonehaven and Aberdeen; together with Observations on Interior Communication in General.*

By Robert Stevenson, F.R.S.E., Civil Engineer.
Edinburgh: Printed for A. Constable & Co. 1821.
Signed: Robert Stevenson. Edinburgh, 1 July 1820.
4to   27 cm   (i) + 13 pp.   Folding engraved plan, hand coloured.
Sketch [map] of the Country from Stirling to Aberdeen referring to a Memorial regarding a communication by Canal or Railway through the Great Vallies of Strathearn & Strathmore.
By Robert Stevenson, Civil Engineer. 1819.
CE

1444   Another edition
4to   27 cm   13 pp.   Folding engraved plan (as in No. 1443). No title-page.
NLS, CE
   Some changes in text and different setting of type throughout.

1445   *To His Grace the Duke of Roxburgh . . . and other Noblemen and Gentlemen of the Shires of Mid-Lothian, Roxburgh, Selkirk, and Berwick. The Report of Robert Stevenson, Civil Engineer.*

n.p.
Signed: Robert Stevenson. Edinburgh, 22 May 1821.
4to   28 cm   12 pp.   Folding litho plan.
Reduced Plan of part of the Shires of Edinburgh, Haddington, Berwick, Roxburgh & Selkirk shewing Proposed Lines of Railway from the Coal Fields of Mid Lothian to the Rivers Tweed & Leeder.
By Robert Stevenson, Civil Engineer. 1821.
J. Robertson Lithog.
NLS, CE, ULG (Rastrick)
   Issued in printed wrappers with title: Roxburgh, Selkirk, &c Railway Report.

1446   *To the Proprietors and Tenants of the Mill Property, situate on the Course of the Water of Leith, in the County of Mid-Lothian. The Report of Robert Stevenson, Civil Engineer.*

n.p.

Signed: Robert Stevenson. Edinburgh, 25 Jan 1822.
4to   28 cm   8 pp.   Folding litho plan.
Reduced Survey with Sections shewing the Position of Reservoirs for the Supply of the Mills on the Water of Leith.
By Robert Stevenson, Civil Engineer 1821.
C.G. Scott del.
J. Robertson Lithog.
NLS, CE

1447   *Report relative to the Improvement of the Communication across the Bristol Channel.*

By Robert Stevenson, Civil Engineer.
[Edinburgh]: P. Neill, Printer.
Signed: Robert Stevenson. London, 29 April 1822.
4to   28 cm   20 pp.   3 folding litho plans, hand coloured (all by Robertsons Lithogr. Press Edinburgh).
[1] *Reduced Survey & Section of Part of the River Severn shewing the Proposed Improvement upon New Passage Ferry* by Robert Stevenson, Civil Engineer. Surveyed by W. H. Townsend of Bristol & J. Steedman of Edinburgh, 1822.
2 [and] 3.   *Design for the Improvement of New Passage Ferry on the Gloucester [and] on the Monmouth Side.* By Robert Stevenson . . . 1822.
NLS, CE

1448   *An Account of the Bell Rock Light-House, including the Details of the Erection and Peculiar Structure of that Edifice. To which is prefixed a Historical View of the Institution and Progress of the Northern Light-Houses. Drawn up by desire of the Commissioners . . .*

By Robert Stevenson, Civil Engineer; F.R.S.E., [etc].
Edinburgh: Printed for Archibald Constable & Co. . . . 1824.
4to   31 cm   xix + 533 + (2) pp.   Frontispiece engraved by J. Horsburgh after J. M. W. Turner + vignette after Miss Stevenson + 21 engraved plates (drawn by W. Lorimer, J. Slight, J. Steedman, G. C. Scott, D. Logan; engraved by W. H. Lizars, E. Mitchell, R. Scott, W. Miller, J. Moffat).

BM, NLS, SML, CE, ULG, BDL
  Published June 1824, price £5 5s. (in boards).
  300 copies printed. The last plate is incorrectly
  numbered in the list of contents.

1449 *Notes by Mr Stevenson, in reference to the Essays
on Railways presented to the Highland Society.*
[Edinburgh].
8vo   20 cm   17 pp.
CE
  Offprint from *Prize Essays and Transactions of the
  Highland Society of Scotland* Vol. 6 (1824)

1450 *Roads and Highways: communicated to the
Editor of the Edinburgh Encyclopaedia.*
By Robert Stevenson, Civil Engineer.
Edinburgh: Printed by A. Balfour and Co. 1824.
4to   27 cm   (i) + 12 pp.   Engraved plate
(drawn by G. C. Scott, engraved by J. Moffat).
NLS, CE
  Offprint, with new title-page, from *Edinburgh
  Encyclopaedia* Vol. 17 (1824).

1451 *Sketch Plan of a Design for Obtaining New
Access to the Cross of Edinburgh, prepared at the desire of
the Proprietors of the Grass Market and its vicinity.*
By Robert Stevenson, Civil Engineer Dec.ʳ
1824. Reduced & Sketched at Robertson & Bal-
lantine's lithog.ʳ . . . Edinburgh.
Litho plan, hand coloured   27 × 70 cm
NLS, CE

1452 *Report of a Survey for the East Lothian Rail-
way.*
By Robert Stevenson, Civil Engineer.
Edinburgh: Printed by P. Neill. 1826.
Signed: Robert Stevenson. Edinburgh, 22 Dec
1825.
4to   27 cm   (i) + 16 pp.   Folding litho plan,
  hand coloured.
*Plan of part of the Shires of Edinburgh & Haddington
with a Section of the proposed East Lothian Railway
from Carnie . . . to Haddington and port of Dunbar.*
  By Robt. Stevenson . . . 1825.
Robertson & Ballantine lith.
NLS (plan only), CE

1453 *Report relative to Lines of Railway, surveyed
from the Ports of Perth, Arbroath, and Montrose, into the
Valley of Strathmore; with an Appendix, containing
Estimates of Expence and Revenue . . . and a Book of
Reference . . .*
  By Robert Stevenson, F.R.S.E. Mem.Inst.
  Civ.Eng.
Edinburgh: Printed by P. Neill. 1827.
Signed: Robert Stevenson. Edinburgh, 14 Aug
  1826.
4to   27 cm   (i) + ii + 79 pp.   2   folding
engraved plans, hand coloured.
1. *Reduced Plan & Sections of the Valley of Strathmore
shewing a Line of Railway from the Ports of Montrose
& Arbroath to Perth.* Surveyed under the direc-
tion of Robert Stevenson Civil Engineer 1826.
Wm. Blackadder, John Steedman Surveyors.
John Sherar del.   T. Clerk sc.
2. *Sketch of the Country from Stirling to Aberdeen referring to a . . . Canal or Railway through the Great
Vallies of Strathearn & Strathmore . . .* 1819.
J. & G. Menzies sc.
NLS, CE

1454 *Exerpt from a Memoir on British Harbours,
drawn up in the year 1824.*
  By R. Stevenson, Esq. F.R.S.E. . . . Civil
  Engineer.
[Edinburgh, 1828].
8vo   22 cm   6 pp.   Engraved plate. (E. Mit-
  chell).
Sketch Plan in reference to Leith Harbour.
  Alan Stevenson del.
NLS
  Offprint from *Edinburgh New Philosophical Journal*
  for January 1828.

1455 *To the Hon. the Provost, Magistrates, and
Council, of the Royal Burgh of Stirling.*
  The Report of Robert Stevenson, Civil Engin-
  eer.
Stirling: Printed by E. Johnston.
Signed: Robert Stevenson. Edinburgh, 26 Nov
  1828. With prefatory note dated 10 Dec 1838.
4to   25 cm   (i) + 8 pp.   Folding litho plan,
  hand coloured.
Reduced Plan and Section of the Forth from Alloa Pier to

Stirling Shore, shewing the Means of Improving the Navigation;

Made under the Direction of Robert Stevenson, Civil Engineer, in the years 1826–27. Forrester & Nichol, lithographed to accompany revised Report 1838. Thomas Stevenson del.

NLS

1456 *To the Commissioners for Preserving and Improving the Port and Harbour of Sunderland.*

The Report of Robert Stevenson, F.R.S.E., Civil Engineer.

n.p.

Signed: Robert Stevenson. Edinburgh, 28 Sept 1829.

fol   33 cm   9 pp.   Folding engraved plan, hand coloured.

Reduced Plan of the Harbour of Sunderland shewing the Practicability of its Extension by means of Wet Docks referred to in a Report by Robert Stevenson . . .

James Ritson del.   W. H. Lizars sc.

NLS

See SPECIFICATION of a timber bridge at Glasgow, 1831.

See STEVENSON, Robert and Alan. Reports and plans, 1832–1837.

See STEVENSON, Robert, Alan and David. Chart, 1838 and report, 1839.

STEVENSON, Robert and Alan

1457   *A Chart of the Coast of Scotland with part of England & Ireland shewing the positions of the several Lighthouses, the principal Anchorages, Rocks, Shoals & Soundings, together with the . . . rise of Spring & Neap Tides.*

Made by Order of the Board of Commissioners of the Northern Lighthouses . . . by Robert Stevenson, F.R.S.E. Engineer and Alan Stevenson M.A. Clerk of Works to the Board. Edinburgh 1832.

J. Ritson del.   W. H. Lizars sc.

Engraved map, hand coloured   173 × 104 cm (scale 1 inch to 8 miles).

BM, NLS

1458   *To the Most Noble the Marquis of Ely, and . . . the Committee for the Improvement of Ballyshannon Harbour, and its Connexion with Lough Erne, the Report of Robert Stevenson and Son, Civil Engineers.*

n.p.

Signed: Robert Stevenson, Alan Stevenson. Edinburgh 15 June 1832.

fol   33 cm   10 pp.   2 folding engraved plans.

[1] *Plan of the Bar & Entrance of the Harbour of Ballyshannon . . .* by Robert & Alan Stevenson. 1832.

J. Menzies sc.

[2] *Plan of the Harbour of Ballyshannon and of the Country between it and Lough Erne . . .* By Robert Stevenson F.R.S.E. & Alan Stevenson M.A. Civil Engineers . . . 1832.

James Ritson del.   J. Menzies sc.

CE, Elton

Issued in printed wrappers with title: Report relative to the Improvement of Ballyshannon Harbour.

1459   *To the Lord Provost, Magistrates, and Town-Council of the City of Perth, the Report of Robert Stevenson and Son, Civil Engineers [on Perth Harbour].*

n.p.

Signed: Robert Stevenson, Alan Stevenson. Edinburgh, 22 Jan 1834.

fol   32 cm   8 pp.   2 folding engraved plans, No. 2 hand-coloured.

1. *Survey of the River Tay from Perth to Invergowrie and Balmerino in reference to the Improvement of its Navigation and the Extension of the Harbour of Perth by means of a Wet Dock.*

Surveyed under the direction of Robert Stevenson and Alan Stevenson Civil Engineers Edinburgh 1833. James Ritson Surveyor. W. H. Lizars sc.

2. *Plan & Section of the Wet Dock, Ship Canal and Tide Harbours at Perth.*

By Robert Stevenson & Son . . . 1834. James Ritson Surveyor. Lizars sc.

NLS, CE
> Issued in printed wrappers with title: Report on the Navigation of the Tay, and Extension of Perth Harbour.

1460 *To His Grace the Duke of Buccleugh and Queensbery, the Report of Robert Stevenson and Son, Civil Engineers [on Granton Harbour].*
n.p.
Signed: Robert Stevenson, Alan Stevenson. Edinburgh, 22 May 1834.
fol 33 cm 7 pp. 2 folding plans, hand coloured.
[1] *Chart of the Frith of Forth from Queensferry to Inchkeith shewing the relative position of the Proposed Harbour at Granton.*
> By Robert and Alan Stevenson . . . 1834. James Ritson Surveyor. W. H. Lizars sc.

[2] *Plan no. II of Granton Harbour . . .*
> By Robert Stevenson & Son . . . 1834. J. Ritson Surveyor. [litho].

NLS, CE, ULG
> Issued in printed wrappers with title: Report relative to Granton Harbour. 1834.

See SPECIFICATION of the Harbour Works at Perth, 1835.

1461 *Plan of the Edinburgh & Glasgow Railway.*
> From a Survey by Robert Stevenson & Son Civil Engineers.

Edinburgh 1835.
Litho plan, the line entered in colour 20 × 51 cm (scale 1 inch to $2\frac{1}{2}$ miles).
[In]

Prospectus of a Company to be called the Edinburgh, Leith & Glasgow Railway Co.
Dated: Edinburgh, 26 Dec 1835.
fol 45 cm (3) pp.
BM

1462 *Chart of Skerryvore Rocks lying off the Coast of Argyle-shire.*
> From a Survey made by Robert & Alan Stevenson Civil Engineers, by Order of the Commissioners of the Northern Lighthouses. 1836. W. H. Lizars sc.

Engraved plan 61 × 147 cm (scale 3 inches to 1 nautical mile).
BM

1463 *Plan of the Edinburgh and Dundee Railway.*
> From a Survey by Robert Stevenson & Son. Civil Engineers. Edinburgh 1836.

Forrester & Nichol litho.
Litho plan 32 × 67 cm (scale 1 inch to $1\frac{1}{2}$ miles).
NLS
> The line runs from Burntisland to Newport. Plan probably issued with a prospectus.

1464 *To the Mayor and Council of the Borough of Preston. The Report of Robert Stevenson and Son, Civil Engineers [on the River Ribble].*
n.p.
Signed: Robert Stevenson & Son. Edinburgh, 16 March 1837.
4to 25 cm 8 pp. 2 folding litho plans, hand coloured.
[1] *Chart of the Ribble from the Admiralty Survey by Captain Belcher R.N. with a Section of the River from the North Union Railway Bridge, to the Striped Buoy off Lytham; shewing the Improvement of the Navigation,* proposed by Robert Stevenson and Son . . . 1837. Lithographed by Binns & Clifford.
3. *Plan of the River Ribble from Preston to the Naze Point with a Longitudinal Section of the River from the Quarry on Ashton Marsh to the Naze.*
> By Robert Stevenson & Son . . . 1837.

Lithographed by Binns & Clifford.
NLS
> Plan No. 2 has not been found.

STEVENSON, Robert, Alan and David

1465 *Chart of the River Lune from Lancaster to Glasson shewing the Proposed Improvements in the Navigation.*
> By Robert Stevenson & Sons Civil Engineers 1838. James Andrews del. Geo. Aikman sc.

Published by A. & C. Black, Edinburgh.
Engraved plan, hand coloured   21 × 42 cm
NLS

1466   *Reports of the Committee appointed to consider the best means of Promoting the Improvement of the Port and Harbour of Chester; and of Robert Stevenson and Sons, Civil Engineers, relative to the Improvement of the Navigation of the River Dee.*
Chester: Printed by R. H. Spence.
Signed: Robert Stevenson and Sons. Edinburgh 21 Aug 1819.
8vo   22 cm   14 pp.   Folding litho plan, hand coloured.
Reduced Plan of the River Dee from Chester to Flint in reference to the Improvement of its Navigation.
   By Robert Stevenson & Sons Civil Engineers. Edinburgh, 1839.
James Andrews Surveyor and Del.
Lithographed by W. Nichol.
NLS, CE
   Stevenson's report is on pp. 5–14.

STICKNEY, Robert

1467   *A Plan of part of the River Witham with Brayford Mere and the several Watercourses communicating therewith through the City of Lincoln.*
   Rob.! Stickney, 1792.
Engraved plan   32 × 40 cm   (scale 10 inches to 1 mile).
RGS, Sutro

See STICKNEY and DICKINSON. Plan of the Horncastle Navigation, 1792.

See PILLEY and PICKERNELL. Plan of Grimsby navigation, surveyed by Stickney, 1795.

1468   *A Map of part of . . . Holderness in the East Riding . . . comprizing the Levels of Keyingham & Burstwick and the Drainages thereof and projected improve-*

ments. Also the Level of Winestead, the Sunk Island, and the new Land gained from the River Humber.
   Rob.! Stickney, Surveyor 1802.
Engraved by C. Smith.
Engraved plan   45 × 35 cm   (scale 1½ inches to 1 mile).
HRO

STICKNEY, Robert and DICKINSON, Samuel

1469   *A Plan shewing the Course of the Rivers Baine and Waring, and the Works proposed to be executed thereon, to open a Navigation from Horncastle to the River Witham in the County of Lincoln.*
   Surveyed by Rob.! Stickney and Sam.! Dickinson. 1792. W.F. 1792.
Engraved plan   57 × 32 cm   (scale 2½ inches to 1 mile).
BM, CE (Page), RGS, Sutro

STONE, Thomas [d. 1815]

1470   *A Letter on the Drainage of the East, West and Wildmore Fens, addressed to the Proprietors.*
   By Thomas Stone, Land-Surveyor.
London: Printed and Published by G. Cawthorne . . . 1800. Price One Shilling.
8vo   23 cm   20 pp.
ULG, BDL (Gough)

STOREY, Thomas [1789–1859] M.Inst.C.E.

1471   *Report on the Great North of England Railway, connecting Leeds and York, with Newcastle-upon-Tyne.*

By Thomas Storey, C.E.

Darlington: Printed by Coates and Farmer . . . 1836.

Dated: St. Helen's Auckland, 24 Feb 1836.

8vo   23 cm   25 pp.   Folding engraved plan + folding litho section.

Plan of the Great North of England Railway.

  Tho.ᵇ Storey Esq. Engineer. R. Otley & T. Sopwith, Surveyors. W. Collard sc.

Sections of the Great North of England Railway.

SML, ULG

1472   Deposited Plan

Plan & Section of an Intended Railway to be called the Great North of England Railway commencing at . . . Gateshead, in the County of Durham, and terminating at the River Tees near to Croft Bridge in the said County.

  Thomas Storey, Engineer. [1836]. R. Martin & Co. Lithog.

Title page + 7 sheets of plans + 3 sheets of sections, each 53 × 69 cm   (scale 4 inches to 1 mile).

RGS

1473   Deposited Plan

Plan (and Sections) of the Extension of the Great North of England Railway from the River Tees to the City of York.

  Thomas Storey, Engineer. [1837] R. Martin & Co. lith.

Title page × 8 sheets, each 53 × 75 cm   (scale 4 inches to 1 mile).

RGS

STRATFORD, Ferdinando [1719–1766]

1474   *An Extract from Mr Stratford's Plan for extending the Navigation from Bath to Chippenham.* [1765]. pp.16–20.

[In]

Observations on a Scheme for extending the Navigation of the rivers Kennett and Avon, so as to form a Direct Inland Communication between London, Bristol, and the West of England, by a Canal from Newbury to Bath.

Marlborough: Printed by E. Harold. 1788.

8vo   20 cm   20 pp.   Folding engraved plan.

CE (Page), ULG

STRICKLAND, William [c. 1787–1854]

1475   *Reports on Canals, Railways, Roads, and other Subjects, made to 'The Pennsylvania Society for the Promotion of Internal Improvement'.*

  By William Strickland, Architect and Engineer.

Philadelphia: H. C. Carey & I. Lea . . . 1826.

Oblong folio   27 × 45 cm   vi + 51 pp.   58 engraved plates, some double-page.

BM, SML, ULG, Elton

  Reports by Strickland made during his visit to Great Britain and Ireland in 1825. Sold in London by J. Taylor, price £3 13s. 6d. in boards. Some of the plates were reprinted in SIMMS's *Public Works of Great Britain*, 1838 [q.v.].

SUTCLIFFE, John

1476   *Report on the Proposed Line of Navigation from Stella to Hexham, on the South Side of the River Tyne; with an estimate of the expence of executing the Line.*

  By John Sutcliffe, Engineer.

Newcastle: Printed for John Bell . . . [1796]. Price One Shilling.

Dated: Newcastle upon Tyne, 5 October 1796.

8vo   20 cm   45 pp.

CE, ULG (Rastrick), BDL (Gough), Elton

1477   *Report on the Line of Navigation from Hexham to Haydon-Bridge, proposed as a continuation of the*

Stella and Hexham Canal . . . and a Report on the Line from Newcastle to Haydon-Bridge, [etc].

By John Sutcliffe, Engineer.

Newcastle: Printed for John Bell . . . [1797]. Price One Shilling.

Dated: Halifax, 3 Jan 1797.

8vo   20 cm   74 pp.

BM, CE, ULG (Rastrick)

1478   [Report] *To the Proprietors of the Somerset-shire Coal Canal.*

[With a summary of Benjamin Outram's observations].

n.p.

Signed: John Sutcliffe. Halifax, 26 May 1800.

fol   46 cm   4 pp.

CE (Page)

1479   *A Treatise on Canals and Reservoirs, and the best mode of designing and executing them; with Observations on the Rochdale, Leeds and Liverpool, and Huddersfield Canals . . . likewise Observations on the best Mode of Carding, Roving, Drawing, and Spinning all Kinds of Cotton Twist. Also instructions for designing and building a Corn Mill,* [etc].

By John Sutcliffe, Civil Engineer.

Rochdale: Printed for the Author, by J. Hartley . . . 1816.

8vo   21 cm   (iii) + xiv + 413 + (1) pp.

BM, NLS, ULG, BDL

SUTHERLAND, Alexander [d. 1802]

1480   *Reports, with Estimates, Plans, and Sections, &c. First, of the proposed Canal through the Weald of Kent, intended to form a Junction of the Rivers Medway and Rother, from near Yalding in Kent to the Tideway near the Port of Rye in Sussex; Secondly, of a Branch by the River Teise to the Town of Lamberhurst. Thirdly, of a Branch . . . to the Town of Hedcorn; and Fourthly, of a Branch . . . to the Town of Cranbrook.*

London: Printed by R. B. Scott . . . 1802.

Signed: Alexander Sutherland, 30 July 1801.

4to   25 cm   vi + 28 pp.   2 folding engraved plans.

[1]   *Map shewing the Line with a Profile or Section of the proposed Kent and Sussex Junction Canal thro' the Weald of Kent.* Survey'd 1801 by Alex.ʳ Sutherland, Engineer.

[2]   *Map shewing the Line of the propos'd River Teise Navigation.* Survey'd by A. Sutherland, Engineer 1801.

East Sussex R.O., ULG

Issued in blue printed wrappers. Price 3s. or 5s. with map of Kent.

SWANWICK, Frederick [1810–1885] M.Inst.C.E.

1481   *Plan and Section of an Intended Railway commencing . . . in the Township of Brightside Bierlow and Parish of Sheffield and terminating . . . in the Township of Kimberworth and Parish of Rotherham all in the West Riding of the County of York.*

Fred.ᵏ Swanwick, Engineer. Nov. 1834.

Henry Sanderson, Surveyor. On stone by Jobbins and Cheffins.

Litho map, hand-coloured   28 × 96 cm   (scale 1 inch to 12 chains, or $6\frac{2}{3}$ inches to 1 mile).

CE

See STEPHENSON, George. York & North Midlands Railway map, 1835, and plan of Stockport–Manchester railway, 1835.

SWINBURN, William   M.Inst.C.E.

1482   *Mr Swinburne's Reports* [*on the foundation of Westminster Bridge*].

Dated: Westminster Bridge Wharf, 15 July 1835 (and) 24 June 1836.

pp. 30–32 and 47–48.

[In]

Reports by Messrs Telford, Cubitt, and Swinburne, Civil Engineers, as to the State of the Foundations, &c. of Westminster Bridge. London, 1836. See WEST-MINSTER BRIDGE.

SWITZER, Stephen [c. 1682–1745]

1483 *An Introduction to a General System of Hydro-staticks and Hydraulicks, Philosophical and Practical.*
By Stephen Switzer.
London: Printed for T. Astley . . . 1729.
2 vols. 4to 25 cm **1**, engraved frontispiece, (vi) + xxxii + (4) + 133 + (15) + [129–274] + 10 pp. **2**, (viii) + [275–352] + (4) + [353–413] + 12 pp. In all, 60 folding engraved plates (Toms sc.).
BM, NLS, SML, RS, RIBA, IC

SYLVESTER, Charles   M.Inst.C.E.

1484 *Report on Rail-Roads and Locomotive Engines, addressed to the Chairman of the Committee of the Liver-pool and Manchester projected Rail-Road.*
By Charles Sylvester, Civil Engineer.
Liverpool: Printed by Thos. Kaye . . . 1825.
Signed: Charles Sylvester. Liverpool, 15 Dec 1824.
8vo 22 cm 39pp.
SML, CE, ULG (Rastrick)

1485 Second edition
As No. 1484 but *Second Edition* . . . 1825.
BM, SML, CE, ULG
With minor corrections including some page numbers.

TAYLOR, James and VAZIE, William

1486 [*Report*]
Signed: James Taylor, William Vazie. Edin-burgh, 6 June 1806. pp. 6–14.
[Printed with]
Letter and report by John GRIEVE, November 1805 [q.v.].
[In]
Reports of Surveys made for ascertaining the Practicabil-ity of making a Land-Communication by a Tunnel under the River Forth, at or near Queensferry.
1806.
[Edinburgh]: Mundell, Doig, and Stevenson, printers.
4to 27 cm (i) + 16 pp Engraved map.
BM, CE

TELFORD, Thomas [1757–1834] F.R.S.L. & E., M.Inst.C.E.

Note that Telford's parliamentary plans and reports on Highland Roads and Bridges, the Caledonian Canal and the Holyhead Road are listed separately (*see* items 1558–1596).

1487 *A Copy of a Letter to the Secretary of the British Society, from Thomas Telford, their Engineer, containing a Course of Experiments made by him on Mr James Parker's Cement.*
n.p. 1796.
12mo 17 cm 16 pp.
BM, ULG

1488 *Some Account of the Inland Navigation of the County of Salop.*
Signed: Thomas Telford. Shrewsbury, 13 Nov 1800.
pp. 284–333 with 4 folding engraved plates (Neele sc.) including:
No. 2. *Plan & Elevation of the Inclined Planes upon the Shropshire & Shrewsbury Canal.*

No. 3. *Perspective View of a part of the Iron Aqueduct which conveys the Shrewsbury Canal over the River Tern at Longden.*
[In]
General View of the Agriculture of Shropshire.
By Joseph Plymley.
London: Printed for Richard Phillips . . . 1803.
8vo   22 cm   xxiv + 366 pp.   5 engraved plates.
BM, SML, ULG

See TELFORD and DOUGLASS. Plans for iron bridges of three and five spans over the Thames, 1800.

See TELFORD and DOUGLASS. Plans and an account of a proposed single arch iron bridge of 600 feet span over the Thames, 1800–1802.

1489   *Mr Telford's Reports of Cromarty, Aberdeen and Wick, made to the Lords of the Treasury in 1801.*
pp. 5–6, 9–12 and 15–17.
[In]
Fourth Report from the Committee on the Survey of the Coasts &c. of Scotland: Naval Stations and Fisheries.
House of Commons, 1803.

1490   *Report by Thomas Telford on the Harbour of Aberdeen.*
Signed: Thomas Telford. London, 28 April 1802.
pp. 41–46   with folding hand-coloured litho plan.
Plan of the Harbour of Aberdeen with the proposed improvements in the year 1802 by Mr Telford.
[In]
Reports upon the Harbour of Aberdeen . . . 1834. See
ABERDEEN HARBOUR.

1491   *A Survey and Report of the Coasts and Central Highlands of Scotland, made by the Command of the . . . Treasury, in the Autumn of 1802.*
By Thomas Telford, Civil Engineer, F.R.S.-Edin.
[House of Commons] 1803.
Signed: Thomas Telford. London, 15 March 1803.
fol   34 cm   27 pp.
The introductory report on Highland roads, bridges and harbours, and the Caledonian Canal. *See also* in TELFORD. *Life of Thomas Telford.* London, 1838. pp. 290–301.

1492   *A Survey and Report of the proposed extension of the Union Canal; from Gumley-Wharf, in Leicestershire, to the Grand Junction Canal, near Buckby-Wharf in Northamptonshire.*
By Thomas Telford, Civil Engineer, F.R.S.-Edin.
London: Printed by B. McMillan . . . 1803.
Signed: Thomas Telford. London, 23 May 1803.
4to   26 cm   22 pp.
CE

1493   Second edition
Title as No. 1492.
Leicester: Printed by J. Throsby . . . 1804.
4to   27 cm   18 pp.
CE

1494   *Suggestions by Thomas Telford, Esq. Civil Engineer, relative to the Canal from Glasgow to the West Coast of the County of Air. In a Letter . . . to the . . . Earl of Eglinton.*
Air: Printed by J. & P. Wilson . . . 1804.
Signed: Thomas Telford. Saltcoats, 15 Oct 1804.
4to   27 cm   6 pp.
CE

1495   *Report by Thomas Telford, Esq. Engineer, relative to the Proposed Canal from Glasgow to the West Coast of the County of Air; and the Harbour at Ardrossan Bay, at the West End of the Canal.*
Glasgow: Printed by William Tait . . . 1805.
Signed: Thomas Telford, 7 Jan 1805.
4to   27 cm   15 pp.
NLS, CE
Includes (pp. 13–15) a report on the proposed harbour by DOWNIE, 5 Jan 1805 [q.v.]. Some minor errors in page numbering are here neglected.

1496   *Report relative to the proposed Canal from the City of Glasgow, to the Harbour of Ardrossan, on the West Coast of the County of Ayr, in Scotland.*
[London]: Strahan and Preston.
Signed: Thomas Telford, 8 June 1805.

fol   39 cm   14 pp.   Docket title on p. (16).
3 folding partly hand-coloured engraved plans
(J. Barlow sc.)

[1] *A Map of a Part of the Northern Coasts of Great
Britain & Ireland.*

[2] *Map of a Canal from the City of Glasgow through
Paisley to the Harbour of Ardrossan on the West Coast
of Ayrshire.* 1805. Surveyed by Jn.º Howell for
Thomas Telford.

[3] *Plan of a Harbour with Wet and Dry Docks
intended to be constructed in Ardrossan Bay . . .* By
Thos. Telford.

CE

Includes (pp. 13–14) the report by Murdoch
DOWNIE, 5 Jan 1805 [q.v.].

1497   *Mr Telford's Report of the General State of the
Grand Junction Canal.*
Signed: Thomas Telford. London, 3 June 1805.
pp. 5–33.

[In]

*Report of the General Committee of the Grand Junction
Canal Company, to the General Assembly of Proprietors
on the 4th of June, 1805.*
London: Printed by M. and S. Brooke . . . 1805.
8vo   22 cm   35 pp.
CE

See JESSOP and TELFORD. Report (and estimate) on Aberdeen Harbour, 1805 and 1810.

1498   *A Report respecting supplying the City of Glas-
gow and its Suburbs, with Water.*
Glasgow: Printed by S. Hunter & Co.
Signed: Thomas Telford. Glasgow, 21 Feb 1806.
4to   25 cm   4 pp.
NLS

1499   *Report respecting the Cumberland Canal.*
Signed: Thomas Telford. Carlisle, 6 Feb 1808.
pp. 1–10.

[Printed with]

CHAPMAN's further report, 24 Feb 1808 [q.v.].

[In]

*Mr Telford's Report on the Intended Cumberland Canal;
and Mr Chapman's Further Report or Observations
thereon.*
Carlisle: Printed by W. Hodgson and Co. 1808.

8vo   21 cm   (i) + 16 pp.
BM, CE

See TELFORD and McKERLIE. Report on
communication between the North of England
and Ireland, 1808.

1500   *Report of the Gotha Canal; being the result of a
Survey made in 1808.*
By Thomas Telford, Civil Engineer.
Dated: September 1808.

[In] TELFORD. *Life of Thomas Telford.* London,
1838 [q.v.]. pp. 348–363.

1501   *Report by Thomas Telford on the Harbour of
Aberdeen.*
Signed: Thos. Telford. Aberdeen, 11 Aug 1809.
pp. 51–55   with folding hand-coloured litho
plan.
*Plan of the Harbour of Aberdeen in the year 1810 with
Proposed Docks by Mr Telford.*

[In]

Reports upon the Harbour of Aberdeen . . . 1834. See
ABERDEEN HARBOUR.

1502   *Report respecting the Proposed Cast-Iron Railway from Glasgow to Berwick.*
Signed: Thos. Telford, London, 12 March 1810.
pp. 1–19.

[Printed with]

Jessop's opinion on Mr Telford's report, 31
March 1810.

[In]

*Report by Mr Telford relating to the proposed Railway
from Glasgow to Berwick-upon-Tweed; with Mr Jessop's Opinion thereon; and Minutes of a Meeting* [etc].
Edinburgh: Printed by A. Neill & Co . . . 1810.
4to   29 cm   (i) + 36 + (1) + 9 pp.
BM, NLS, CE

1503   *Plan and Section of the Track of a proposed Cast
Iron Railway, from the City of Glasgow, to Berwick-
upon-Tweed, passing through the Counties of Lanark,
Peebles, Selkirk, Roxburgh, Berwick, and Northumberland.*

Thos. Telford. London, March 1810. Engraved
by James Barlow.

Engraved plan   4 sheets each 60 × 82 cm
(scale 1 inch to 1 mile and 1 inch to 200 feet
vertical).
NLS

1504   *Report and Estimates relative to a proposed Road
in Scotland from Kyle-Rhea in Inverness-shire to Killin in
Perthshire, by Rannoch-Moor.*
By Thomas Telford.
London: Printed by Luke Hansard & Sons . . .
1810.
Signed: Thomas Telford, May 1810.
8vo   23 cm   15 pp.   Folding engraved plan,
the lines entered in colour.
*Proposed Road from Kyle Rhea . . . to Killin in Perth-
shire. 1810.*
BM, NLS, CE (Telford), BDL

1505   *Report of Mr Telford, respecting the Stamford
Junction Navigation*
Stamford: Printed by J. Drakard.
Signed: Thomas Telford. Stamford, 28 July 1810.
8vo   23 cm   27 pp.
BM, CE

1506   *Plan of the proposed Stamford Junction Navigation, from Oakham in the County of Rutland, to Stamford and Boston in the County of Lincoln, and from
Stamford to Peterborough in the County of Northampton.*
Surveyed under the direction of Thos. Telford,
by Hamilton Fulton, and Drawn by W. A. Provis 1810.
Engraved plan, the line entered in colour
69 × 94 cm   (scale 1 inch to 1 mile).
BM

1507   *Report of Thomas Telford, Esq. respecting the
Improvement of the Supply of Water to the City of Edin-
burgh.*
Signed: Thos. Telford. Edinburgh, 10 Aug 1810.
pp. 1–56.
[Printed with]
Report of Dr T. C. Hope, 30 July 1811.
[In]
*Reports on the Means of Improving the Supply of Water
for the City of Edinburgh.*
Edinburgh: Printed for Archibald Constable and
Company, 1813.

4to   28 cm   (iii) + ii + (i) + 76 pp.
NLS, CE, ULG

1508   *Mr Telford's Survey of New Galloway Bridge:
Glenluce Road; and Carlisle and Garistown Road.*
Signed: Thos. Telford. London, 16 March 1811.
pp. 8–9   with folding engraved plan, the lines of
road entered in colours.
*Plan of Improvements in the Road between Carlisle and
Portpatrick as proposed by Mr Telford and Mr Rennie
[and plans of nine proposed bridges]. W. A.
Provis del.   J. Basire sc.*
[In]
*Report from the Committee, upon the Roads between Car-
lisle and Port Patrick.*
House of Commons, 1811.

1509   *Report respecting the Harbour of Dundee in the
County of Forfar.*
Dundee: Printed by R. S. Rintoul.
Signed: Thos. Telford. 24 Aug 1814.
4to   28 cm   4 pp.   Folding engraved plan.
*Map of the Harbour of Dundee with the Proposed
Improvements August 1814.*
Reduced and Engraved from Mr Telford's Plan
by Thos. Ivory.
CE
The plan and MS. estimate, signed Thos. Tel-
ford, in HLRO.

1510   *Report of Mr Telford, Civil Engineer, on the
intended Edinburgh and Glasgow Union Canal.*
London: Printed by A. Strahan.
Signed: Thos. Telford. London, 5 April 1815.
fol   33 cm   3 pp.   Docket title on p. (4).
CE (Telford)
For a reprint of this report *see* No. 1513.

1511   *Report, by Mr Thomas Telford, respecting the
Road from Carlisle to Glasgow.*
Signed: Tho. Telford. London, 10 June 1815.
pp. 14–20   with folding engraved plan in 2
sheets, the lines of road entered in colours.
*Map of the Mail Road between Glasgow and Carlisle,
with the Proposed Improvements.*
Surveyed and Drawn under the Direction of
Thomas Telford Civil Engineer, by W. A. Pro-
vis. James Basire sc.

[In]
Report from Select Committee on Carlisle and Glasgow Road.
House of Commons, 1815.

1512 *Report of the State of the Crinan Canal; with an Estimate annexed.*
Signed: Thos. Telford. Inverary, 11 Jan 1817.
[In]
Fourteenth Report of the Commissioners for the Caledonian Canal.
House of Commons, 1817. pp. 47–48.

1513 *First (and Second) Report of Mr Telford, Civil Engineer, on the intended Edinburgh and Glasgow Union Canal.*
Signed: Thomas Telford. London, 5 April 1815 (and) Edinburgh, 29 Jan 1817. pp. Appendix 1–6.
[In]
Observations by the Union Canal Committee, on the Objections made by the Inhabitants of Leith to this Undertaking.
Edinburgh: Oliver & Boyd, Printers [1817].
8vo   23 cm   31 + 7 pp.
CE

1514 *Mr Telford's Report (and Supplementary Report) respecting Runcorn Bridge.*
Signed: Thomas Telford. Liverpool, 13 March 1817 (and) London, 22 July 1817.
pp. 9–22   with folding engraved plate.
Design for a Bridge over the River Mersey at Runcorn, to connect the Counties of Chester and Lancaster.
By Thos. Telford, F.R.S.E. 1814. Drawn by W. A. Provis, Edm.ᵈ Turrell sc.
[In]
Report of the Select Committee on the proposed Bridge at Runcorn.
House of Commons, 1817.

1515 *Map of a proposed line of Navigable Canal from near the town of Knaresbro' to the River Ouse at Ancaster Sailby with a line of Rail-Road from Knaresbro' . . . to Pately Bridge in the County of York.*
Surveyed for Thos. Telford by H. R. Palmer 1818.

Engraved plan   22 × 47 cm   (scale 1 inch to 2 miles).
[In]
Report of the Knaresbrough Rail-Way Committee.
Leeds: Edward Baines, Printer. [1819]
8vo   24 cm   vi + 36 pp.   Folding engraved plan (as above).
CE, ULG
Includes a summary of Telford's (undated) report.

See EDINBURGH, water supply. Plan of proposed aqueduct, 1818.

1516 *Northern Roads. Copy of a Report and Estimate, of two Proposed Lines of Roads: The one leading from Catterick Bridge, in the County of York, to Carter Fell, on the borders of Scotland: The other leading from Catterick Bridge aforesaid, to New Castleton, in the County of Roxburgh; made under the Instructions of the . . . Treasury.*
By Thomas Telford, Civil Engineer.
House of Commons, 1820.
Signed: Thos. Telford. London, 6 July 1820.
fol   33 cm   6 pp.   Docket title on p. (8).

1517 *Report by the Committee of Management, relative to the Plans of the New Ferry Harbours at Dundee and the opposite coast of Fife, proposed by Mr Telford.*
Dundee: Printed by R. S. Rintoul, 1821.
4to   27 cm   (i) + 13 + (3) pp.   2 folding engraved plans.
NLS
Includes (pp. 4–6) extracts from Telford's report, following instructions of 18 Nov 1820. *See also* Robert STEVENSON on the Tay Ferries, 1820.

1518 *Mr Telford's Report on the Points referred to him by the Corporation of Wisbech.*
Signed: Thomas Telford. London, 31 May 1821. pp. 22–24.

[Printed with]
Letter from John RENNIE, 30 Jan 1821 [q.v.].
[In]
Report of the Proceedings of the Committee for taking into Consideration Mr Rennie's Reports on the Improvement of the Outfall of the River Nene.

Wisbech: Printed by White and Leach. 1821.
8vo   22 cm   (i) + 24 + (1) pp.
BM, Cambs R.O., CE

See SPECIFICATIONS for Holyhead Road near
  Coventry, 1821.

1519   *Morpeth and Edinburgh Road. Mr Telford's
Report and Estimate.*
Signed: Thos. Telford. London, 17 Aug 1821.
pp. 19–22   with folding hand-coloured engraved
  plan, in two sheets.
*Map and Sections of a Proposed Mail Road between
London and Edinburgh by Morpeth, Wooler and Cold-
stream.*
  Surveyed . . . by Thos. Telford, Civil Engineer,
  1820 & 1821. A. Arrowsmith sc.
                    [In]
*Report from Select Committee on Morpeth and Edinburgh
Road.*
  House of Commons, 1822.
  The report, without the plan, was published by
  the House of Commons as a separate paper
  earlier in 1822.

See TELFORD and RENNIE. Report on the Eau
  Brink Cut, 1822.

1520   *Mr Telford's Reports [on the foundations of
Westminster Bridge].*
Dated: 12 May 1823, 14 May 1823, 9 June 1829, 1
  Aug 1829, and 12 Sept 1831.
pp. 5–8, 9–11, 14–17, 19–20 and 26–29.
                    [In]
*Reports by Messrs Telford, Cubitt, and Swinburne, Civil
Engineers, as to the State of the Foundations, &c of
Westminster Bridge.*
  London,   1836.   *See*   WESTMINSTER
  BRIDGE.

1521   *Report of Thomas Telford, Esq. On the Effects
which will be produced on the River Thames by the
rebuilding of London Bridge.*
London: Printed by A. Taylor.
Signed: Thomas Telford. Abingdon Street, 11
  June 1823.
fol   33 cm   7 pp.   Docket title on p. (8).
CE, RIBA

1522   Another edition
[In] *Heygate, Mayor . . . Report to Common Council
  . . . 13 June 1823.*
Corporation of London.
fol   32 cm   11 pp.
CLRO, CE

1523   Reprint
With two short reports by Telford dated 7 March
  and 8 April 1823. pp. 876–881.
                    [In]
*Minutes of Evidence taken before the Lords Committee
. . . on the Bill . . . for improving the Approaches to
London Bridge.*
  1829.

1524   *Eau-Brink Cut. Copy of Mr Telford's Report.*
Lynn: Mugridge. [1825]
Signed: T. Telford, 5 July 1823.
fol   32 cm   3 pp.   Docket title on p. (4).
CRO
  See also in TELFORD. *Life of Thomas Telford.*
  London, 1838. pp. 317–319.

1525   *Plan of the Crinan Canal.*
  Surveyed by David Smith, under the direction
  of Thos. Telford Civil Engineer Oct. 1823. A.
  Arrowsmith sc.
Engraved plan   29 × 16 cm   (scale 1 inch to 1
  mile).
                    [In]
*Twenty-First Report of the Commissioners for the
Caledonian Canal.*
  House of Commons, 1824.

1526   *Report from Mr Telford on Roads from Glas-
gow to Port-Patrick.*
Signed: Thos. Telford. London, 26 March 1824.
pp. 7–11   with folding engraved plan, the lines
  of road entered in colour.
*Map of the Roads between Girvan and Stranraer with
Proposed New improved Lines.*
  Surveyed by James Mills under the direction of
  Thomas Telford Esq. Civil Engineer 1824. A.
  Arrowsmith sc.
                    [In]
*Report from the Select Committee on Glasgow and Port-
Patrick Road.*
  House of Commons, 1824.

1527 *English and Bristol Channels Ship Canal. Prospectus and Mr Telford's Preliminary Report.*
London: Printed by S. Brooke . . . 1824.
Report signed: Thos. Telford. London, 2 Aug 1824.
8vo   23 cm   10 + 8 pp.   Small folding engraved plan (title as No. 1528).
ULG, Elton
   The report is reprinted in Nos. 1529 and 1530.

1528 *Plan of the Line of the Proposed English and Bristol Channels Ship Canal.*
   Surveyed under the direction of Thomas Telford Esq. Civil Engineer, F.R.S.E. By James Green, Civil Engineer, 1824. Ingrey & Madeley lithog.
Litho plan   39 × 97 cm   (scale 1 inch to 1 mile).
CE (Telford), Elton

1529 *Ship Canal, for the Junction of the English and Bristol Channels. Reports of Mr Telford and Captain Nicholls.*
London: Printed by S. Brooke . . . 1824.
Signed: Thomas Telford. London, 14 Dec 1824 (and) Geo. Nicholls. London, 3 Dec 1824.
fol   37 cm   (iii) + 41 pp.   Small engraved plan (title as in No. 1528) and folding engraved map.
Map of the South West Parts of England & Wales, shewing the English and Bristol Channels, also the course of the Proposed Ship Canal, between Beer Harbour and Bridgewater Bay.
   Made under the Direction of Thos. Telford, Civil Engineer, F.R.S.E. 1824. A. Arrowsmith sc.
CE, ULG, Elton
   See also in TELFORD. *Life of Thomas Telford.* London, 1838. pp. 586–620.

1530   Another edition.
Title as No. 1529.
London: Printed by S. Brooke . . . [1825].
8vo   23 cm   47 pp.   Folding engraved map.
BM

1531 *Plan of the Proposed St. Katharine's Docks*
   Designed by Thomas Telford, Civil Engineer,

F.R.S.E. Philip Hardwick, Architect. C. Hullmandel's Lithog.
Litho plan, hand coloured   42 × 58 cm   (scale 1 inch to 100 feet).
PLA
   This copy endorsed 'For Jas. Mountague Esq. 1 Nov 1824'.

1532 *Report respecting the Mail-Road between the City of Edinburgh and Town of Morpeth, by the Towns of Berwick & Alnwick.*
   Made under the direction of Thos. Telford, Civil Engineer, F.R.S.E.
London: Printed by Henry Teape . . . 1824.
Signed: Thos. Telford. London, 20 Aug 1824.
fol   32 cm   15 pp.
ULG

1533   Another edition, with plan
Berwick and Morpeth Road. Copy of a Report made by Mr Telford, to the . . . Treasury, respecting the Mail Road between the City of Edinburgh and the Town of Morpeth [etc].
   House of Commons, 1825.
fol   33 cm   8 pp.   Docket title on p. (10).
   Folding engraved plan, the lines of road entered in colours.
Map of the Present Mail Road between Edinburgh and Morpeth through Haddington, Dunbar, Berwick and Alnwick with Proposed New Lines and Variations.
   Surveyed . . . under the direction of Thos. Telford Civil Engineer. By James Mills. A. Arrowsmith sc.

1534 *Map or Plan & Sections describing the line of an intended Turnpike Road to be made from the Parish of West Wycombe in the County of Buckingham, through Thame to Chilworth in the County of Oxford.*
   As altered by Mr Telford, Nov. 15, 1824. H. Stokes lithog.
Litho plan and section, the lines entered in colours   34 × 80 cm   (scale $1\frac{3}{4}$ inches to 1 mile and 1 inch to 150 feet vertical).
RGS, BDL

1535 *Mr Telford's Report on his Survey of Mail Roads in South Wales.*
Signed: Thos. Telford. London, 17 June 1825.
   [In]

South Wales Roads. A Copy of the Postmaster General's Letter to the . . . Treasury . . . upon the subject of the Roads through South Wales; together with a Copy of the Report from Mr Telford upon those Roads.

House of Commons, 1826. pp. 5–22.

1536   *Extract from Report on the Mail Road through South Wales.*

Made under the direction of Thomas Telford.

pp. 89–92   with folding hand-coloured litho plan.

Map shewing the present Mail Roads from Milford and Pembroke to Caermarthen with the various New Lines which have been surveyed . . . also Milford Bay and adjacent Coast with the sundry proposed Landing Places.

Surveyed under the directions of Thomas Telford Esq. Civil Engineer, F.R.S.E. by Alexander Easton. 1824.

[In]

Report from Select Committee on Milford Haven Communication.

House of Commons, 1827.

1537   *Plans and Elevations of a Church and (two) Manses.*

Drawn under the direction of Thomas Telford Esq. F.R.S.E. by George May. James Basire sc.

Engraved plate   30 × 18 cm.

[In]

First Report of the Commissioners for Building Churches in the Highlands and Islands of Scotland.

House of Commons, 1825.

1538   *Liverpool and London Road. Copy of the Reports of Mr Telford, on the State of the Road from London to Liverpool.*

House of Commons, 1827.

Signed: Thomas Telford. London, 31 May (and) 17 June 1826.

fol   33 cm   8 pp.   Docket title on p. (10).

Reports on the roads from Weedon to Lichfield and from Lichfield by Runcorn to Liverpool. *See also* No. 1547.

1539   *Mr Telford's Report on the Improvement of Swansea Harbour.*

Signed: Thos. Telford. London, 5 Feb 1827.

pp. 23–26   with folding litho plan.

Plan of the Harbour of Swansea showing the proposed Improvements.

By Thos. Telford, Civil Engineer. T. Bedford's Lithog.

[In]

Reports on the Harbour of Swansea.

Swansea: Printed . . . by W. C. Murray and D. Rees. 1831.

8vo   20 cm   28 pp.

West Glamorgan County Library.

1540   *Report respecting the Mail Road between London and the Town of Morpeth, made under the direction of His Majesty's Postmaster General.*

By Thomas Telford, Esq. F.R.S.E.

London: Printed by J. Hartnell . . . for His Majesty's Stationery Office. 1827.

Signed: Thos. Telford. London, 16 May 1827.

fol   32 cm   32 pp.

CE

1541   *Map of the London and Edinburgh Mail Roads from London to East Retford, including the Two Lines to Alconbury Hill through Hatfield and Ware; delineating all the Variations and Improvements of which they are capable* [etc].

Surveyed under the direction of Thos. Telford Esq. by James Mills 1826. Drawn by L. Hebert. Printed at the Lithographic Establishment, Q.M.G. Office.

Litho plans, 5 sheets each   47 × 67 cm   (scale 1 inch to 1 mile)   the lines in colours.

BDL

1542   *Map of the London and Edinburgh Mail Roads from East Retford to Morpeth including Two Lines through Doncaster and Boroughbridge, and through Thorne and York; delineating all the Proposed New Lines and Improvements* [etc].

Surveyed under the direction of Thomas Telford Civil Engineer F.R.S.E. by Henry Welch 1826.

[Signed] Thomas Telford. London, May 1827.

Litho plans, 5 sheets each 46 × 60 cm and one sheet 46 × 34 cm (scale 1 inch to 1 mile) the lines in colours.
BM

1543 *London and Morpeth Mail Road. Index Map.* Surveyed by Thos. Telford. 1827. Drawn by L. Hebert. Printed at the Lithographic Establishment, Q.M.G. Office.
Litho plan, the lines entered in colours 7 sheets each 64 × 49 cm (scale 1 inch to 2 miles).
BM, SML, BDL

See ROGET, BRANDE and TELFORD. Report on the supply of water to the Metropolis, 1828.

1544 *Report of Thomas Telford, Esq. on the Roads from London to Holyhead, and from London to Liverpool.*
[London]: Printed by Order of the Commissioners for Carrying into Execution the Acts 4 Geo. IV c. 74 and 7 and 8 Geo. IV c. 35.
Signed: Thomas Telford. London, 15 May 1828.
8vo 21 cm 28 pp.
CE (Telford)

1545 *Report respecting the Lower Ferry between the Counties of Mid-Lothian and Fife.*
By Thomas Telford, Esq. Civil Engineer.
Edinburgh: Auchie, Printer.
Signed: Thos. Telford. London, 7 Oct 1828.
8vo 21 cm 22 pp. with table.
NLS

1546 *Plan of the St. Katharine Docks.*
Thomas Telford, F.R.S. Civil Engineer. Philip Hardwick, F.A.S. Architect. Printed by C. Hullmandel.
Litho plan, hand coloured 19 × 25 cm
PLA
Issued with Programme of Opening, 25 Oct 1828.

See PROVIS, W. A. Account of the Menai suspension bridge constructed from designs by, and under the direction of Thomas Telford. London, 1828.

1547 *Mr Telford's Reports, Estimates and Plans for improving the Road from London to Liverpool.*
House of Commons, 1829.
Signed: Thomas Telford. London, 31 May and 17 June 1826 (as in No. 1538), 3 Nov 1828 and 6 Feb 1829.
fol 32 cm 11 pp. 4 folding hand-coloured litho plans.
1. *Map of the Country between the Village of Weedon in the County of Northampton, and the City of Lichfield . . . showing the Several Lines in which the Liverpool Mail Road may be carried.* Surveyed under the direction of Thos. Telford Esq. F.R.S.L. & E. by Thomas Casebourne. J. Basire lithog.
2. *Sections of the Several Lines in which the Liverpool Mail Road may be carried between Weedon and Lichfield.* Thos. Telford. J. Basire lithog.
3. *Map of the Country between the Towns of Liverpool and Talk on the Hill, on the borders of Staffordshire and Cheshire, Shewing the Comparative Merits of sundry distinct Lines of Road.* 1828. Surveyed under the direction of Thos. Telford, Esq. F.R.S.L. & E. by Thos. Casebourne. A. Arrowsmith lithog.
4. *Sections of Sundry Distinct Lines of Mail Road between Liverpool and Talk on the Hill.* Thos. Telford. A. Arrowsmith lithog.

1548 *Liverpool and Manchester Rail-Way. Mr Telford's Report to the Commissioners for the Loan of Exchequer Bills. With Observations in Reply, by the Directors of the said Rail-way.*
Liverpool: Printed by Thos. Kaye . . . 1829.
Report signed: Thos. Telford. London, 4 Feb 1829.
8vo 21 cm (i) + 16 pp. 2 folding hand-coloured litho plans.
[1] *Section of the Broad-Green Embankment.*
[2] *Section of part of the Embankment between the Sankey Valley and Newton Bridge* [and] *Section of the Sankey Embankment.*
SML, CE, ULG (Rastrick), Elton
Field investigations by James Mills.

1549 *Report of the proposed New Mail Road, from Carlisle to Edinburgh, by the Town of Langholm, in the County of Dumfries;* made under the direction of His Majesty's Postmaster General.

By Thomas Telford, Esq. F.R.S.E.
London: Printed by J. Hartnell ... for His
   Majesty's Stationery Office. 1829.
Signed: Thos. Telford. London 14 Aug 1829.
fol   33 cm   8 pp.
NLS

1550   Second printing
[In] *Report from the Select Committee on the State of the Northern Roads.*
House of Commons, 1820. pp. 26–28.

1551   *Carlisle and Edinburgh Road. Map shewing Portions of the Present Mail and other Roads, and Proposed New Lines of Road, between the Cities of Carlisle and Edinburgh, in the Counties of Cumberland, Dumfries, Selkirk, Pebbles and Edinburgh, also Roxburgh.*
Surveyed under the direction of Thomas Telford, Esq. Civil Engineer by Henry Welch, 1828. Drawn in Lithograph by L. J. Hebert.
Litho plan, the lines entered in colours
   42 × 188 cm   (scale 1 inch to 1 mile).
NLS

1552   [Report on] *Glasgow Railway.*
Signed: Thomas Telford. London, 14 Nov 1829.
5 pp.
[In]
Glasgow Railway and Tunnel. See GLASGOW, Railway and tunnel, 1829.

1553   *The Report of Mr Telford, on the Road from Ketley Iron Works, in the County of Salop, to Chirk in North Wales.*
House of Commons, 1830.
Signed: Thos. Telford. Abingdon Street, 10 Feb 1830.
fol   33 cm   4 pp.   2 folding litho plans (A. Arrowsmith).
[1] *Map shewing the Lines of the Present & Proposed Mail Coach Roads between Chirk and Ketley.* 1830. Surveyed by J. Macneil. Road Superintendent, under the direction of Thomas Telford, Civil Engineer.
[2] *Section of the Proposed and Present Roads.* Levels for Sections taken by J. Macneil.

1554   *Plan for improving the Harbour of Aberdeen in the year 1831*
By Thomas Telford F.R.S.L. & E.
Litho plan, hand coloured   34 × 59 cm   (scale 1 inch to 300 feet).
[In]
Reports upon the Harbour of Aberdeen ... 1834. See
   ABERDEEN HARBOUR.

See TELFORD and WALKER. Report on the present state of the New London Bridge, 1831.

See SPECIFICATION for Glasgow Bridge, 1832.

1555   *Metropolis Water Supply. Report of Thomas Telford, Civil Engineer, on the Means of supplying the Metropolis with Pure Water.*
House of Commons, 1834.
Signed: Thomas Telford. London, 17 Feb 1834.
fol   33 cm   12 pp.   3 folding hand-coloured litho plans, all signed Thos. Telford; S. Arrowsmith lithog.
1. *Plan and Section of a Line of Aqueduct from the River Verulam above Watford to Primrose Hill.*
2. *Plan and Section of a Line of Aqueduct from the River Wandle at Beddington to Clapham Common.*
3. *Map of the New River from its Source near the Town of Ware to London, and part of the River Lea.*

1556   *Life of Thomas Telford, Civil Engineer, written by himself; containing a Descriptive Narrative of his Professional Labours: with a Folio Atlas of Copper Plates. Edited by John Rickman ... With a Preface, Supplement, Annotations, and Index.*
London: Printed by James and Luke G. Hansard and Sons ... 1838.
Preface signed: J. R. April 1838.
4to   38 cm   xiv + 719 pp.   Woodcuts in text and 1 engraved plate.
[issued with]
Atlas to the Life of Thomas Telford, Civil Engineer, containing Eighty-Three Copper Plates, illustrative of his Professional Labours.
[London]: Sold by Payne and Foss ... 1838.
fol   57 cm   (iv) pp.   Portrait frontispiece (W. Raddon after Samuel Lane) + 82 engraved plates numbered 1–83 but No. 28 omitted; mostly drawn by George Turnbull, others by

Thomas Casebourne and H. R. Palmer; engraved by Edmund Turrell, chiefly, with some by W. A. Beever, T. Bradley and F. J. Havell.

Text and Atlas: BM, NLS, SML, CE, ULG, IC, BDL

Price £8 8s.

Highland Roads and Bridges

See Telford's report of 15 March 1803 (no. 1491).

1557 *Reports of Mr Thomas Telford respecting the Loch-na-Gaul (and) the Glengarry Road.*
Signed: Thos. Telford, 16 Dec [and] 21 Dec 1803.
[In]
First Report of the Commissioners for Highland Roads and Bridges.
House of Commons, 1804. pp. 28–31.

1558 *Map of Intended Roads & Bridges in the Highlands of Scotland.*
Engraved by J. Barlow.
Engraved map, the lines entered in colours 63 × 54 cm (scale 1 inch to 12 miles) with accompanying 3 pp. of references.
[In]
First Report of the Commissioners for Highland Roads and Bridges.
House of Commons, 1804.

1559 Revised map
Title as No. 1558 but dated 1805.
[In]
Second Report of the Commissioners for Highland Roads and Bridges.
House of Commons, 1805.

1560 *Map of Scotland from original materials obtained by the Parliamentary Commissioners for Highland Roads and Bridges and exhibiting the Intended Roads and Bridges 1807.*
Drawn by A. Arrowsmith.
Engraved map, the lines entered in colours 59 × 50 cm (scale 1 inch to 12 miles).
[In]

Third Report of the Commissioners for Highland Roads and Bridges.
House of Commons, 1807.

1561 Revised map
Title as No. 1560 but . . . *exhibiting the Roads and Bridges made, Contracted for, or under consideration. 1809.*
[In]
Fourth Report of the Commissioners for Highland Roads and Bridges.
House of Commons, 1809.

1562 New plate
Title as No. 1560 but . . . *exhibiting the Roads, Bridges and Harbours, Made, Contracted for, or under Consideration. 1811.*
[In]
Fifth Report of the Commissioners for Highland Roads and Bridges.
House of Commons, 1811.
Updated versions are in the Sixth Report (1813) and Seventh Report (1815).

1563 New plate
Title as No. 1562 but dated 1817.
[In]
Eighth Report of the Commissioners for Highland Roads and Bridges.
House of Commons, 1817.

1564 Final version
Map of Scotland, from Original Materials obtained by the Parliamentary Commissioners for Highland Roads and Bridges: and exhibiting the Roads, Bridges, and Harbours, Made, Improved, or Repaired by them. 1821.
Drawn by A. Arrowsmith.
[In]
Ninth Report of the Commissioners for Highland Roads and Bridges.
House of Commons, 1821.

1565 *Report and Estimates, relative to Rannoch Road.*
By Thomas Telford.
Signed: Thomas Telford, May 1810.
[In]
Fifth Report of the Commissioners for Highland Roads and Bridges.
House of Commons, 1811. pp. 39–43.

1566 *Mr T. Telford's Report and Estimate of Bonar Bridge.*
Signed: Thomas Telford, 3 Jan 1811.
[In]
Fifth Report of the Commissioners for Highland Roads and Bridges.
    House of Commons, 1811. p. 46.

1567 *Structure of the Iron Arch at Bonar.*
Signed: Tho. Telford. Feb 1813.
pp. 37–38   with engraved plate
Bonar Bridge. General Elevation (and Structural Details).
Drawn by W. A. Provis, E. Turrell sc.
[In]
Sixth Report of the Commissioners for Highland Roads and Bridges.
    House of Commons, 1813.

1568 *Report concerning Craig-Ellachie Bridge.*
    By Mr Telford.
Signed: Thomas Telford. Sept 1812.
[In]
Sixth Report of the Commissioners for Highland Roads and Bridges.
    House of Commons, 1813. p. 36.

1569 *General Inspection of the Roads and Bridges in Scotland maintained under the direction of the Parliamentary Commissioners; May, June, and July 1828.*
    By Thomas Telford, F.R.S.L. & E.
Signed: Thos. Telford. London; 12 Feb 1829.
[In]
Fifteenth Report of Commissioners for Repair of Roads and Bridges in Scotland.
    House of Commons, 1829. pp. 6–13.

Caledonian Canal

1570 *Extracts from Mr Telford's Reports of 1801 [on the Caledonian Canal].*
[In]
Third Report from the Committee of Survey of the Coasts &c of Scotland; Caledonian Canal.
    House of Commons, 1803. pp. 5–15.

See TELFORD's report of 15 March 1803 (No. 1491).

1571 *Report on the intended Inland Navigation from the Eastern to the Western Seas, by Inverness and Fort William.*
    By Thomas Telford.
Signed: Thos. Telford. London, 2 Feb 1804.
[In]
First Report of the Commissioners for the Caledonian Canal.
    House of Commons, 1804. pp. 14–17.

1572 *General Plan of the Intended Inland Navigation, from the Eastern to the Western Sea, by Inverness and Fort William.*
    By Messrs Telford & Downie.
Reduced from the original by John Howell. J. Barlow sc.
Engraved plan, hand coloured   30 × 123 cm (scale 3 inches to 5 miles).
[In]
First Report of the Commissioners for the Caledonian Canal.
    House of Commons, 1804.
    The survey delivered 16 Dec 1803.

1573 Second state
General Map of the Intended Caledonian Canal or Inland Navigation . . . by Inverness and Fort William [1805] etc.
[In]
Second Report of the Commissioners for the Caledonian Canal.
    House of Commons, 1805.
    Shows progress of work to April 1805.

See JESSOP and TELFORD. Reports on the Caledonian Canal, 1804.

1574 *A Description of the Ground along the Line of the Caledonian Canal, from North-East to South-West; as ascertained by Trial-Pits and Borings.*
Signed: Thomas Telford. Clachnacharry, 3 May 1805.
[In]
Second Report of the Commissioners for the Caledonian Canal.
    House of Commons, 1805. pp. 47–52.

1575 *General Map of the Intended Caledonian Canal or Inland Navigation, from the Eastern to the Western Sea, by Inverness and Fort William.*
Drawn by A. Arrowsmith.
Engraved plan, hand coloured   30 × 95 cm
(scale 1 inch to 2 miles).
[In]
Third Report of the Commissioners for the Caledonian Canal.
House of Commons, 1806.

See JESSOP and TELFORD. Annual reports on the Caledonian Canal, 1805–1812.

1576 *Comparative Expence of the Locks finished at Corpath and Clachnacharry, on the Caledonian Canal.*
Signed: Thos. Telford.
[In]
Fourth Report of the Commissioners for the Caledonian Canal.
House of Commons, 1807. pp. 25–26.

1577 *Plan of the Caledonian Canal between Loch Beauley and Loch Ness (and) between Loch Eil and Loch Lochie.*
Drawn by A. Arrowsmith.
Engraved plan, the lines entered in colours
30 × 59 cm   (scale 3 inches to 1 mile).
[In]
Fourth Report of the Commissioners for the Caledonian Canal.
House of Commons, 1807.
Updated versions are in the 6th Report (1809) and 7th Report (1810).

1578 *Plan of the Caledonian Canal between Loch Eil and Loch Lochie (and) between Loch Oich and Loch Ness [and] between Loch Ness and Loch Beauley.*
Engraved plan, the lines entered in colour
30 × 46 cm   (scale 2 inches to 1 mile).
[In]
Tenth Report of the Commissioners for the Caledonian Canal.
House of Commons, 1813.

1579 Revised plate
Title as No. 1578 but with A. Arrowsmith sc.
[In]

Twelth Report of the Commissioners for the Caledonian Canal.
House of Commons, 1815.
Updated versions are in the 14th Report (1817) and 16th Report (1819).

1580 *Report and Estimate by Mr Telford.*
Signed: Thos. Telford. October 1813.
[In]
Eleventh Report of the Commissioners for the Caledonian Canal.
House of Commons, 1814. pp. 32–38.
Telford's annual reports on the canal continue in the 12th to 20th Reports of the Commissioners, 1815–1823.

1581 *Caledonian Canal or Inland Navigation from the Eastern to the Western Sea.*
Drawn by A. Arrowsmith.
Engraved plan, hand coloured   30 × 96 cm
(scale 1 inch to 2 miles).
[In]
Eighteenth Report of the Commissioners for the Caledonian Canal.
House of Commons, 1821.

Holyhead Road

1582 *Mr Telford's Report to the ... Treasury, respecting the Great Roads, from Holyhead through North Wales.*
Signed: Thos. Telford. London, 22 April 1811.
pp. 13–24   with 2 folding engraved plans and 3 engraved plates.
[1] *Map of Mail Road from Shrewsbury to Holyhead.* Thos. Telford 1811. A. Arrowsmith sc.
[2] *Map of the Menai near the present Ferry and proposed Bridges.*
[3] *Design for a Bridge proposed to be erected over the Menai upon the Swilley Rocks.* Thos. Telford, Engineer. Drawn by W. A. Provis. W. Lowry sc.
[4] *Design for a Bridge proposed to be erected over the Menai at Ynns-y-moch* [and] *Design for the Centering.* T. Telford, Engineer. Drawn by W. A. Provis. W. Lowry sc.

[5] *Design for a Bridge and Embankment for the River Conway.* Thos. Telford, Engineer. W. A. Provis del. Js. Basire sc.
[In]
Report from Committee on Holyhead Roads.
House of Commons, 1811.

1583 *Report, Plan, and Estimate, for building a Bridge over the Menai Strait, near Bangor Ferry.*
Signed: Thomas Telford. London 5 May 1818.
pp. 3–4 with engraved plate.
Design for a Bridge over the Menai Straits at Ynns-y-moch.
By Thomas Telford, F.R.S.E. Drawn by W. A. Provis. Edm. Turrell sc.
[In]
Papers relating to the building a Bridge over the Menai Straits near Bangor Ferry.
House of Commons, 1819.
Preliminary design of the Menai suspension bridge.

1584 *Report and Estimate for Improving the Navigation of the Menai Straits, at and near the Swilley Rocks.*
Signed: Thomas Telford. London, 16 Feb 1819.
p. 22 with folding engraved plan.
Map of that Part of the Menai Strait which includes the Site of the New Bridge and the Swilly, Cribinniau and Britannia Rocks.
Thos. Telford. A. Arrowsmith sc.
[In]
Papers relating to the building a Bridge over the Menai Straits near Bangor Ferry.
House of Commons, 1819.

1585 *Report of Mr Telford respecting the Roads across the Island of Anglesey.*
Signed: Thomas Telford, 16 Feb 1819.
p. 15 with folding engraved map, the lines entered in colours.
Map of the Present & Proposed Mail Roads between Holyhead & Bangor Ferry in the County of Anglesey.
Thos. Telford. A. Arrowsmith sc.
[In]
Second Report on the Road from London to Holyhead.
House of Commons, 1819.

1586 *Mr Telford's Reports on the English Part of the Holyhead Road [and] on Holyhead Road through North Wales.*
Signed: Thomas Telford. London, 30 June 1819.
pp. 105–133 [and] 175–178.
[In]
Sixth Report from the Select Committee on the Road from London to Holyhead.
House of Commons, 1819.

1587 *Reports of Mr Telford to the Commissioners for the Improvement of the Holyhead Road, upon the State of the Road between London and Shrewsbury.*
House of Commons, 1820.
Signed: Thomas Telford. London, 30 June 1819 [and] 5 June 1820.
fol 33 cm 44 pp. 9 folding engraved plans, the lines entered in colours, and 1 engraved plate.
I–IX. *Holyhead Road. Maps and Sections of the present Road, where Variations are Proposed, also of the Proposed Variations.*
By Thomas Telford. A. Arrowsmith sc.
X. *Holyhead Road. Tools for Making and Repairing Roads.* I. Easton del. E. Turrell sc.
The report of 30 June 1819 is a reprint of the 'English Part' in the previous item. Annexed to the report of 5 June 1820 are *General Rules for Repairing Roads* later published separately: *see* ROADS.

1588 *Report of the Progress and Present State of the Anglesea Road and Menai Bridge [and] Mr Telford's Reports to the Commissioners for Improving the Holyhead Road, upon the State of that Road.*
House of Commons, 1821.
fol 33 cm 23 pp.
Includes: Instructions to Mr Easton, 14 Oct 1820. General Report, 22 May 1821. Progress reports on road from Bangor to Shrewsbury and Shrewsbury to London, February and May 1821; all signed by Telford.

1589 *Report of Thomas Telford to the Commissioners for Improvements of the Holyhead Road [on roads in North Wales].*
Signed: Thomas Telford. Bangor, 8 Feb 1822.
pp. 5–6.

[In]
First Report of the Select Committee on the Roads from London to Holyhead, and from Chester to Holyhead.
House of Commons, 1822.

1590 *Report of Mr Thomas Telford to the Commissioners for the Improvement of the Holyhead Road . . . upon the State of the Road from Shrewsbury by Coventry to London.*
House of Commons, 1822.
Signed: Thos. Telford, 19 March 1822.
fol   33 cm   7 pp.   Docket title on p. (8).

1591 *Holyhead Road. Map of the Present and Proposed Lines between Holywell and Northope on the Chester Road.*
A. Arrowsmith sc.
Engraved plan, the lines entered in colours
   20 × 35 cm   (scale 2 inches to 1 mile).
[and]
Sections of the Road between Bangor and Chester.
Engraved plan   20 × 53 cm   (scale 2 inches to 1 mile and 1 inch to 400 feet vertical).
[In]
Third Report of the Select Committee on the Roads from London to Holyhead, and from Chester to Holyhead.
House of Commons, 1822.

1592 *Mr Telford's Third Report, by order of the Lords of the Treasury.*
Signed: Thomas Telford. London, 16 May 1822.
[In]
Fourth Report of the Select Committee on the Roads from London to Holyhead.
House of Commons, 1822. pp. 76–77.
Also in this volume is a reprint of TELFORD's report of 19 march 1822 [q.v.].

1593 *Report of Thomas Telford, Esq. upon the State of the Mail Road from London to Holyhead.*
House of Commons, 1823.
Signed: Thos. Telford. London, 7 April 1823.
fol   33 cm   5 pp.   Docket title on p. (6).

1594 *Report of Mr Telford to the Commissioners . . . for the further Improvement of the Road from London to Holyhead.*
Signed: Thomas Telford. London, 6 May 1824.

pp. 6–39   with engraved plate.
Section of a part of the Old Holyhead Road Improved. Section of an Embankment [and] of a Cutting as executed on the Holyhead Road Improvements.
J. Mitchell del.   A. Arrowsmith sc.
[In]
First Report of the Commissioners appointed under the Act of 4 Geo. IV c. 74 . . . for the Further Improvement of the Road from London to Holyhead.
House of Commons, 1824.

1595 Telford's annual reports on the Holyhead Road works continue in the 2nd to 11th Reports of the Commissioners, 1825–1834. Some of those include notes on the Weedon–Liverpool Road. His report of 15 May 1828 was also published separately, without plans: *see* No. 1544. Plans of Holyhead Harbour and Howth Harbour are given in the 5th (1828) and 9th (1832) Reports.

1596 *Statement of the Works performed between the Years 1815 and 1830.*
Signed: Thomas Telford, [April 1830].
[In]
Report of the Select Committee on the Holyhead and Liverpool Roads.
House of Commons, 1830. pp. 4–9.
The Committee report includes (pp. 31–47) an account of income and expenditure 1815–1830.

TELFORD, Thomas and DOUGLASS, James

1597 *Estimates and Observations by Messrs Telford and Douglass* [for iron bridges of three and five spans over the Thames].
Signed: Thos. Telford, Jas. Douglass [1800].
pp. 57–73   with 4 folding engraved plates (drawn by Wm. Jones).
[In]
Third Report from the Select Committee upon the Improvement of the Port of London.

House of Commons, 1800.
The plates (Nos. 9–12) issued in a separate volume.

1598   Second printing
In *Reports from Committees of the House of Commons* Vol. 14 (1803) pp. 564–573 and pls. 38–41.

1599   *Plan and Estimate for a Bridge, as proposed by Messrs Telford and Douglass, consisting of a single Arch of Cast Iron, 600 Feet in the Span.*
Signed: Thos. Telford, Jas. Douglass. [autumn 1800].
With folding engraved plate (issued separately).
Messrs Telford & Douglass's Design of a Cast Iron Bridge of a Single Arch proposed to be erected over the River Thames . . . Span of the Arch 600 Feet.
   Lowry sc.
                    [In]
Supplemental Appendix to the Third Report from the
   Select Committee. 1800. pp. 148–149 and pls.
   24.
   The only known copy of the text is in the
   Science Museum (Goodrich Collection).

1600   *Messrs Telford & Douglass's Explanatory Drawings. Plan of Framing. Shewing how the Ribs may be put together. Elevation and Plan*
[of a revised design of the 600 ft span bridge].
Folding engraved plate (J. Barlow sc).
                    [In]
Report from the Select Committee upon Improvement of the Port of London.
   House of Commons, 1801. p. 1

1601   Second printing
Nos. 1599 and 1600 with an additional folding
   engraved plate.
A Portion of Messrs Telford and Douglass's designed Cast Iron Bridge of a Single Arch, on an enlarged Scale, shewing the Structure of the Work & the Stile of the Ornaments.
                    [In]
Reports from Committees of the House of Commons
   Vol. 14 (1803) p. 603b and pls. 52–54.

1602   *An Account of the Improvements of the Port of London, and more particularly of the intended Iron*

Bridge, consisting of One Arch, of Six Hundred Feet Span.
London: Printed by E. Spragg . . . 1801.
8vo   20 cm   20 pp.
BM, ULG
   Essentially a reprint of Telford's paper in *Phil. Mag.* Vol. 10 (1801) pp. 59–67. The pamphlet was issued in the names of Telford and Douglass to subscribers to the large aquatint engraving (by Wilson Lowry and Thomas Malton) published October 1801.

1603   Second printing
Text as No. 1602 but with new setting of title
   page and dated 1802.
CE (Telford)
   Issued with the second state of the engraving.

TELFORD, Thomas and McKERLIE, John

1604   *Report respecting the Line of Communication between the North of England and Ireland.*
Signed: Thos. Telford, Jno. McKerlie. London,
   16 March 1808.
pp. 13–52   [with 16 folding plans issued in separate Atlas: *see* next item]
                    [In]
Report from the Committee on the Communication be-tween England and Ireland.
   House of Commons, 1809.

1605   *The Charts and Plans referred to in Mr Telford's Report and Survey on the Communication between England and Ireland by the North-West [i.e. South West] of Scotland.*
   House of Commons, 1809.
fol   68 × 53 cm   16 large folding plans, all
   engraved by James Basire.
These include:
Pl. 1 *Sketch of the Counties, Shores and Channel between the North of England and Ireland, shewing the Old, and New Roads; also the Present and Proposed Harbours.* Thos. Telford, Jno. McKerlie.

Pl. 2   *A Chart of Port Patrick.*

Pl. 7   *A Map of Donaghadee in the County of Down, Ireland.* Thos. Telford, Jn. McKerlie, March 1808.

Pl. 8   *A Plan of that part of Galloway in the vicinity of Newton Stewart, shewing the . . . road from Port Patrick to Dumfries; also a New Line . . .* by John Gillone, County Surveyor. 26 June 1807.

Pl. 10   *A Plan of Dee River and Roads, in the vicinity of Dee Bridge Village . . .* By John Gillone, 8 Feb 1808.

And 6 plans and elevations of proposed new bridges by Thos. Telford, March 1808.

TELFORD, Thomas and RENNIE, *Sir* John

1606   *Appendix A (and B). Messrs Telford and Rennie's Report* [on the Eau Brink Cut].
Signed: Thomas Telford, John Rennie. London 15 April 1822 [and] 15 May 1822.
pp. 25–31.
[In]
A Letter from Lord William Bentinck to the Eau Brink Commissioners.
London, 1822. *See* BENTINCK.
CE, CRO (Appendices only)

TELFORD, Thomas and WALKER, James

1607   *Report (and Second Report) on the present State of the New London Bridge.*
By Thomas Telford and James Walker, Civil Engineers.
Signed: Thomas Telford, J. Walker. London, 17 Oct 1831 [and] 7 Nov 1831.
[In]
Copy of the Reports presented to the Corporation of Lon-
don . . . relative to the Stability of the New London Bridge.
House of Commons, 1832. pp. 3–7 and 10–11.

THAMES NAVIGATION COMMISSIONERS

1608   *Report of the Committee of the Thames Navigation Commissioners, of the State of the Navigation below Great Marlow, June 24th, 1789. With Mr Brindley's Estimates for making a Navigable Canal from Monkey Island and Boulter's Lock to Isleworth.*
[Ordered to be printed 5 October 1791].
Dated: Great Marlow, 24 June 1789. Signed by John Call, William Vanderstegen and others.
8vo   22 cm   20 pp.
CE (Page), ULG
400 copies ordered to be printed. With BRINDLEY's estimates of 15 June 1770 [q.v.]. pp. 17–20.

1609   *Copies and Extracts of Proceedings of the Commissioners of the Thames and Isis Navigation, relative to a Proposal for making a Navigable Canal from Boulter's Lock to Isleworth.*
[Ordered to be printed 5 November 1791].
8vo   22 cm   16 pp.
CE (Page), ULG

1610   *A Report of the Committee of Commissioners of the Navigation of the Thames and Isis, appointed to Survey the Rivers from Lechlade to Whitchurch, by the General Meeting held 31st of May, 1791.*
Printed at Oxford, 1791.
Dated: Henley upon Thames, 23 July 1791. Signed by John Call, E. L. Loveden, William Vanderstegen and others.
8vo   22 cm   35 pp.
BM, CE (Page), ULG

1611   *Report from the Committee of the Commissioners of the Navigation of the Rivers Thames and Isis appointed to inquire into and report on the best method of* *improving and amending the said Navigation; from Eton Wharf to South Hope and also what part of Mr Mylne's Plan should be adopted.*

Ordered to be Printed 28 August 1793.
8vo   22 cm   8 pp.
BM, CE (Page), ULG

1612   *Two Reports of the Commissioners of the Thames Navigation, on the Objects and Consequences of the several projected Canals, which interfere with the interests of that River; and on the present sufficient and still improving State of its Navigation.*
Oxford: Printed by J. Munday. 1811.
Reports dated 22 October 1810, by E. L. Loveden; and 31 December 1810, by the Commissioners.
8vo   22 cm   50 pp.
BM, CE (Page)

1613   *Report of a Survey of the River Thames, from Lechlade to the City Stone near Staines, by a Committee of the Commissioners of the Upper Districts of the Navigation, read at a General Meeting held at Reading, August 23, 1811.*
Reading: Snare and Man. [1811].
8vo   22 cm   50 pp.
CE (Page)
    Written by Henry Allnutt, General Clerk to the Thames Commissioners. Members of the Committee include Sir Charles Palmer, Frederick Page, E. L. Loveden, Dr W. F. Mavor and Stephen Leach.

THAMES TUNNEL

1614   *The Thames Tunnel. Incorporated by Act 5 Geo. IV, dated June 24, 1824.*
London: Printed by Henry Teape . . . 1825.
Dated: Thames Tunnel Office, 12 April 1825.
8vo   21 cm   15 pp.
BM, SML, CE

1615   *Thames Tunnel Company Incorporated 1824.*
London: Teape and Son, Printers . . . 1827.

4to   24 cm   8 pp.   Folding litho plate, drawn by Wm. Westall, A.R.A.
Printed by Englemann Graf Coindet.
SML, ULG (Rastrick)

1616   Another edition
As No. 1615 but with 'Report of the Court of Directors' 6 March 1827.
4to   24 cm   8 + 5 pp.   Folding litho plate.
CE

1617   *The Origin, Progress, and Present State of the Thames Tunnel.*
London: Effingham Wilson . . . 1827.
8vo   23 cm   26 pp.
ULG
    Three further issues in 1827 (not seen).

1618   Fifth edition
Title as No. 1617 with *Fifth Edition.*
London: Effingham Wilson . . . 1827.
8vo   23 cm   28 pp.   Folding litho plate, drawn by Wm. Westall (as in No. 1615).
BM, CE, ULG (lacking plate)

1619   *Sketches and Memoranda of the Works for the Tunnel under the Thames, from Rotherhithe to Wapping.*
Published and sold at the Tunnel Works, Rotherhithe, and by Messrs Harvey and Darton . . . 1827.
Dated: September 1827.
Oblong 12mo   10 × 14 cm   (29) pp.
    including double-page hand-coloured section of strata   12 plates.
CE
    Primarily a visitor's guide to the tunnel works. Later editions have been seen with minor variations, dated Dec 1827 (P) and Jan 1828 (CE, ULG). Marbled boards with printed label. Price 2s.

1620   *Sketches of the Works for the Tunnel under the Thames, from Rotherhithe to Wapping.*
Published by Messrs. Harvey and Darton . . . 1828.
Dated: May 1828.

Oblong 12 mo 10 × 14 cm (34) pp. 14
plates including aquatint view of the West
Archway.
B

Variants and later editions exist dated: as
above (CE), April 1829 (SML), August 1829
(CE), and Oct 1830 (BM). Marbled boards
with printed label. Price 2s. 6d.

1621 *The Thames Tunnel. Report of the Court of
Directors and of M. J. Brunel, Esq.*
London: H. Teape and Son, Printers . . . 1828.
8vo 21 cm 15 + (1) pp.
BM, SML, ULG

1622 *Documents relating to the Thames Tunnel. Prefatory Remarks. Proceedings and Resolutions . . . Report
of the Directors . . . Report of the Engineer . . . Resolutions* [etc].
London: Printed by Arthur Taylor . . . 1829.
8vo 22 cm 31 + (1) pp.
BM, CE

1623 *A Letter to the Proprietors of the Thames Tunnel.*
[London]: Printed by A. J. Valpy.
Dated: 1 March 1832 (unsigned).
8vo 21 cm 12 pp. Folding litho plate.
*Section of the Tunnel as originally proposed and exhibited
in 1823 & 1824* [and] *Section of the Driftway . . .
abandoned in 1808* [and] *Section of the Tunnel as it has
been executed to the extent of 600 Feet* [and geological
strata beneath the river bed].
C. Ingrey lithog.
ULG (Rastrick)

See BRUNEL, Sir Marc. *An Explanation of the
Works of the Tunnel under the Thames*, 1836. The
first of a new series of guide books following
resumption of work.

1624 *Copies of Treasury Minutes and Correspondence
relating to . . . the Thames Tunnel Company.*
House of Commons, June 1837
fol 33 cm 8 pp.
Includes a report by James WALKER dated 22
April 1837 [q.v.].

1625 *Report from the Select Committee on the Thames
Tunnel; with the Minutes of Evidence.*
House of Commons, July 1837
fol 33 cm 28 pp.
Evidence by Marc Brunel, Thomas Page
(resident engineer), James Walker.

See WALKER. Further report to the Treasury,
published June 1838.

THOM, Robert [1774–1847]
M.Inst.C.E.

1626 *A Brief Account of the Shaws Water Scheme,
and present state of the Works: (and a Report on Supplying Greenock with Water) with . . . a Letter . . . on the
Principles of Filtration, as applicable for the Supply
of Populous Towns and Cities with Pure Water.*
Greenock: Printed at the Columbian Press, 1829.
Report signed: Robt. Thom, 22 June 1824. Letter
signed: Robt. Thom. Rothsay, 20 March 1829.
8vo 22 cm 88 pp. Engraved plate + folding
hand-coloured engraved plan.
*Reduced Plan of Lands drained into the Reservoirs, and
into the Aqueduct near Greenock, with Sections of the
Mill Seats, 1827.*
Surveyed by J⁵ Flint & Jⁿ Linn as directed by
R. Thom.
BM, CE

1627 *Report on Supplying Glasgow with Water (and
Extracts from Report . . . upon the Plentiful and Economi-
cal Supply of the City of Edinburgh with Pure Water,
and the Construction of Reservoirs).*
By R. Thom, C. E.
Glasgow: From the Steam Press of Edward Khull
. . . 1837.
Edinburgh report signed: Robt. Thom. Rothesay,
16 April 1829. Glasgow report signed: R.
Thom, C.E. [n.d.].

8vo 21 cm   (iii) + 27 pp.
BM, NLS

THOMAS, Richard [1779–1858]

1628 *Chart of the Severn.*
Surveyed in 1815 by R.ᵈ Thomas, Falmouth.
Engraved by H. Mutlow. Pub.ᵈ for the Author
by W. Faden.
Engraved map   54 × 180 cm   (scale 2 inches to
1 mile).
Includes: Section showing the Rise and fall of the
Tides between King-Road and Gloucester.
BM

1629 *Hints for the Improvement of the Navigation of
the Severn.*
By Richard Thomas, of Falmouth.
Falmouth: Printed by M. Brougham . . . 1816.
Signed: Richard Thomas, July 1816.
8vo   23 cm   35 pp.
CE

See WHISHAW and THOMAS. Report on proposed railway between Perranporth and Truro,
1831.

THOMAS, William

1630 *Observations on Canals and Rail-Ways, illustrative of the agricultural and commercial advantages to
be derived from an iron rail-way . . . between Newcastle,
Hexham, and Carlisle; with Estimates of the presumed
Expense, Tonnage, and Revenue.*
[By William Thomas, Esq. 1805].
pp. (9)–30.
[Printed with]
Report by B. R. DODD on proposed canal between Newcastle and Hexham, 22 Oct 1810
[q.v.].

[In]
*Observations on Canals and Rail-Ways . . . by the late
William Thomas, Esq. Also, second edition, Report of
Barrodall Robert Dodd, Esq . . . with Appendix.*
Newcastle: Printed by G. Angus . . . 1825.
8vo   22 cm   (i) + 52 pp.
BM, SML, ULG
A paper read to the Newcastle Lit. & Phil.
Society in 1805 by William Thomas of Denton
and here printed for the first time from a MS.
copy found in London.

THOMPSON, Benjamin [1776–1867]

1631 *A Plan & Section of an intended Railway or
Tram Road from the Town and County of Newcastle
upon Tyne to the City of Carlisle in the County of Cum-
berland with a branch therefrom.*
Surveyed under the Direction of Benjamin
Thompson by T. O. Blackett & I. Studholme.
August 1828.
Printed by Englemann, Graf, Coindet & Co.
Hand-coloured litho map in 3 sheets each
28 × 53 cm   (scale 1 inch to 1 mile).
BM

1632 Deposited Plan
Title, date etc as No. 1631.
Englemann, Graf, Coindet & Co. lithog.
Hand-coloured litho map in 12 sheets each
70 × 53 cm   (scale 4 inches to 1 mile).
BM, HLRO
An early example of lithography used for
deposited plans.

See THOMPSON *et al.* Reports on Newcastle &
Carlisle Railway, 1834–1841.

THOMPSON, Benjamin, JOHNSON, George and WOOD, Nicholas

1633   *The Managing Committee's Report to the Directors of the Newcastle and Carlisle Railway Company.*
Newcastle: Printed by J. Blackwell.
Signed: Benjamin Thompson, George Johnson, Nicholas Wood. 17 March 1834.
8vo   22 cm   6 pp.
BM

> Further reports, all with same title (as above), signed by Thompson, Johnson and Wood, and printed at Newcastle, were issued annually from 1835 to 1841.

THOMPSON, Isaac [*c.* 1703–1776]

See BURLEIGH and THOMPSON. Plan of the River Wear and Sunderland harbour, 1737.

See LABELYE. Report on Sunderland harbour, 1748. Abstract by Thompson.

THOMPSON, Jona

1634   *Observations on the most advantageous Line of Country through which a Canal Navigation may be carried, from Newcastle upon Tyne, or North Shields, towards Cumberland, &c. With a Proposal to Extend Collateral Branches . . . to Morpeth, the Port of Blyth, &c.*
By Jona Thompson.
Newcastle: Printed for R. Sands . . . 1795.
Signed: Jona Thompson. Sheepwash, 28 Feb 1795.
8vo   20 cm   24 pp.
BM, CE (Vaughan), BDL

THOMSON, Richard [1794–1865]

1635   *Chronicles of London Bridge.*
By an Antiquary.
London: Smith, Elder, and Co . . . 1827.
Dated: 15 June 1827.
8vo   20 cm   xv + (1) + 687 pp.   Frontispiece, and 55 wood-engraved illustrations in text (mostly by G. W. Bonner, W. Hughes, G. W. Moore and H. White).
BM, NLS, CE

> Known to be by Richard Thomson.

1636   *Second edition*
Title as No. 1635, but *Second Edition.*
London: Printed for Thomas Tegg . . . 1839.
8vo   15 cm   xiv + 518 pp.   Frontispiece, and 55 wood-engraved illustrations in text (as in first edition).
BM, NLS

> Issued in publisher's printed cloth; No. LXVI in The Family Library. Price 5*s.*

THOMSON, William

1637   *Report on the State of the Works on the Crinan Canal.*
Signed: William Thomson, Resident Engineer. Crinan Canal Office, 30 Nov 1838.
                    [In]
Report from the Select Committee on the Caledonian and Crinan Canals.
House of Commons, 1839. pp. 174–188.

TOFIELD, Thomas [1730–1779]

1638   *Mr Tofield's Report, on a Survey of Deeping Fen, for the more effectual Drainage thereof.*

n.p.   1768.
Signed: Thomas Tofield. Balby, 1 Nov 1768.
4to   28 cm   5 pp.   Docket title on p. (8).
LAO, Sutro

1639   *A Report on the Practicability of making a Navigable Canal, from the River Dun at Stainforth-Cut, to the River Trent at Althorpe.*
   From Levels taken by Mr John Thompson, and a View taken by Thomas Tofield.
n.p.
Signed: Thomas Tofield. Wilsick, 28 Oct 1772.
fol   30 cm   4 pp.
Sutro

1640   *A Scheme, for effectually securing the Level of Hatfield Chace, from being injured by the River Torn.*
   Proposed by Thomas Tofield.
n.p.   Printed in the Year MDCCLXXII.
Signed: Thomas Tofield. Wilsick, 20 Sept 1773.
fol   31 cm   4 pp.
Nottingham University Library, Sutro
   A second edition, printed at Doncaster by Sheardown, was issued in 1792.

1641   *A Report of the present State of a large Tract of Low Ground, extending from Muston to Malton: to which is added, a Scheme for the more effectual Drainage and Preservation therof.*
   Proposed by Thomas Tofield.
n.p.
Signed: Thomas Tofield. Wilsick, 20 Sept 1773.
fol   33 cm   4 pp.
Hull University Library
   Refers to the survey and plan by Isaac MIL-BOURN [q.v.].

TOMLINSON, John

1642   *Roads leading from Coleshill Guide-post to the Guide-post on Meredon Heath in the County of Warwick.*
   Survey'd by John Tomlinson 1760.
Engraved plan   40 × 46 cm   (scale 5 inches to 1 mile).
BM

TOMPSON, John [d. 1795]

1643   *A Plan of the intended Dearne and Dove Canal, describing the Branches from it with the adjacent Rivers & Brooks.*
   John Tompson, Engineer. Wᵐ Fairbank, Surveyor. 1793.
Engraved plan, the line entered in colour
   36 × 77 cm   (scale 2 inches to 1 mile).
BM, BDL

TOWNSHEND, Thomas [*c.* 1771–1846]

1644   *A Plan shewing the proposed Deviations upon that part of the Rochdale Canal situated between Manchester and Rochdale.*
   Survey'd by T. Townshend 1799. Engraved by B. Baker.
Engraved plan, the lines entered in colours
   21 × 40 cm   (scale 1½ inches to 1 mile)
BM

1645   *The Report of Mr Thomas Townshend, upon the District No. 6* [of the Shannon].
Dated: Balna Carig, 26 Feb 1811.
pp. 149–171   with 2 folding engraved plans.
[In]
Second Report of the Commissioners on the practicability of draining and cultivating the Bogs in Ireland.
   House of Commons, 1812.

1646   *The Report of Mr Thomas Townshend, on a great District which comprises a number of Bogs in the Counties of Armagh, Tyrone, Down, Antrim, and Londonderry, also Lough Neagh.*
Signed: Thomas Townshend. Dublin, 6 March 1813.
pp. 153–166   with 10 folding hand-coloured engraved plans.
[In]

Third Report of the Commissioners on the practicability of draining and cultivating the Bogs in Ireland.
House of Commons, 1814.

TRAFFORD, Sigismund [d. 1741]

1647 *An Essay on Draining: more Particularly with Regard to the North Division of the Great Level of the Fenns, called Bedford Level.*
London: Printed for J. Roberts . . . 1729.
Dated: 16 Dec 1728.
8vo   21 cm   23 pp.   Folding engraved plan, as in John PERRY's report of 25 Feb 1727 [q.v.].
BM, BDL (Gough), Sutro
Known to be by Sigismund Trafford.

TREACHER, John [*c.* 1735–1802]

1648 *A Report of John Treacher, Surveyor of the Upper Districts. To the Commissioners of the Thames Navigation.*
Dated: 5 May 1792.
pp. 66–67.

[In]
Report from the Committee appointed to enquire into . . . the Navigation of the Thames.
House of Commons, 1793.

TREDGOLD, Thomas [1788–1829] M.Inst.C.E.

1649 *Elementary Principles of Carpentry; a Treatise on the Pressure and Equilibrium of Beams and Timber Frames; the Resistance of Timber; and the Construction of Floors, Roofs, Centres, Bridges, &c. With Practical*

Rules and Examples. To which is added, an Essay on the Nature and Properties of Timber [etc].
By Thomas Tredgold.
London: Printed for J. Taylor . . . 1820.
Dated: London, 20 April 1820.
4to   28 cm   xx + 250 pp. + errata leaf   22 engraved plates, 3 double-page (James Davis sc.).
BM, NLS, SML, BDL, CUL
Price £1 4s. in boards.

1650   Second edition
Elementary Principles of Carpentry; a Treatise on the Pressure and Equilibrium of Timber Framing; the Resistance of Timber [etc].
By Thomas Tredgold, Civil Engineer. Second Edition; corrected and considerably enlarged.
London: Printed for J. Taylor . . . 1828.
Dated: October 1828.
4to   28 cm   xx + 280 pp.   22 engraved plates (as in 1st edn.).
BM, IC
An American edition was published in Philadelphia, 1837.

1651   Third edition
As No. 1650 but *Third Edition, corrected and considerably enlarged. With an Appendix, containing specimens of various Ancient and Modern Roofs.*
By Peter Barlow, F.R.S. . . . Hon.Mem.Inst. Civ.Eng.
London: John Weale . . . 1840.
4to   28 cm   (i) + xxii + 312 pp.   Portrait frontispiece + 50 engraved plates; 1–22 as in 1st edn., 23–50 mostly folding (W. A. Beever and J. Le Keux sc.).
BM, ISE

1652   *Tredgold's Elementary Principles of Carpentry. Appendix to the Second Edition.*
With interesting Additions by Sydney Smirke . . . John Shaw . . . Jos. Glynn . . . and other contributors.
London: John Weale . . . 1840.
4to   28 cm   22 pp.   Portrait frontispiece + 28 engraved plates.
BM, SML
The Appendix and pls. 23–30 of No. 1651 separately issued.

1653 *A Practical Essay on the Strength of Cast Iron
. . . containing Practical Rules, Tables, and Examples;
also an Account of some New Experiments, with an
Extensive Table of the Properties of Materials.*

By Thomas Tredgold, Civil Engineer. Mem.
Inst.Civ.Eng.

London: Printed for J. Taylor . . . 1822.

Dated: Lisson Grove, London, 16 March 1822.

8vo   22 cm   xvi + 175 pp. + errata leaf   4
engraved plates.

BM, IC

1654   Second edition

As No. 1653 but *Second Edition, improved and
enlarged.*

London: Printed for J. Taylor . . . 1824.

Dated: Lisson Grove, London, [November] 1823.

8vo   22 cm   xix + 305 pp + errata leaf   4
engraved plates.

BM, NLS, SML, IC, BDL, CUL

Price 15s. in boards. French and German translations were published in Paris, 1826 and Leip-
zig, 1827.

1655   Third edition

As No. 1653 but *Third Edition, improved and
enlarged.*

London: Printed for J. Taylor . . . 1831.

Dated: August 1831.

8vo   23 cm   xix + 307 pp.   4 engraved plates.

BM

1656   *Stone-Masonry and Stone-Cutting.*

Signed: H.H.H. [= Thomas Tredgold].

4to   28 cm   [552–570] pp.   3 engraved plates.

CE (from the author)

Offprint, in plain wrappers, from *Encyclopaedia
Britannica.* Supplement to the 4th, 5th and 6th
Editions. Vol. 6. Edinburgh . . . 1824.

1657   *A Practical Treatise on Rail-Roads and Carriages, shewing the Principles of Estimating their
Strength, Proportions, Expense, and Annual Produce
[etc].*

By Thomas Tredgold, Civil Engineer, Mem.
Inst.Civ.Eng.

London: Printed for Josiah Taylor . . . 1825.

8vo   22 cm   xi + errata leaf + 184 pp.   4 engraved plates, 1 folding (J. Dadley sc).

BM, NLS, SML, CE (Page), ULG (Rastrick),
BDL, CUL

Price 12s. in boards. An American edition was
published in New York, 1825 and French and
Spanish translations in Paris, 1826 and Mad-
rid, 1831.

1658   Second edition

As No. 1657 but *The Second Edition.*

London: Printed by and for J. B. Nichols and Son
. . . 1835.

8vo   22 cm   xi + 184 pp.   4 engraved plates
(as in 1st edn.).

SML, CE, ULG (Rastrick)

1659   *Tracts on Hydraulics.*

Edited by Thomas Tredgold, Civil Engineer.
Viz. Smeaton's Experimental Papers on the
Power of Water and Wind to turn Mills &c.
Venturi's Experiments on the Motion of Fluids.
Dr Young's Summary of Practical Hydraulics,
chiefly from the German of Eytelwein.

London: Printed for Josiah Taylor . . . 1826.

8vo   24 cm   xi + (1) + 219 pp.   7 engraved
plates.

SML, RS, ISE, IC, CUL

1660   Second edition

As No. 1659 but *Second Edition.*

London: Printed for M. Taylor . . . 1836.

8vo   24 cm   ix + (1) + 219 pp.   7 engraved
plates.

CE

TROTTER, William Edward

1661   *The Croydon Railway, and its adjacent Scenery.
Illustrated with six Views, Elevations of all the Bridges,
Plans of the Stations, and a Map and Section of the Line.*

By the Editor of 'Illustrated Topography of
Thirty Miles round London'.

London: Published by R. Tyas . . . [1839].

12mo 17 cm (v) + 57 pp. Large folding map etc, vignette and 5 plates (L. J. Wood del, J. R. Jobbins lith.).
CE

TURNBULL, William

1662 *A Treatise on the Strength and Dimensions of Cast Iron Beams, when exposed to Transverse Strains, from Pressure or Weight. With Tables of Constants to be used for calculating the strength and dimensions of similar beams of wrought iron, and several sorts of wood ... To which is added, the Theory of Bramah's Hydro-Mechanical Press.*
By William Turnbull.
London: Josiah Taylor ... 1831.
Dated: Fitzroy Square, June 1831.
8vo 22 cm viii + 86 pp. Woodcuts in text.
CE

1663 *A Treatise on the Strength, Flexure, and Stiffness of Cast Iron Beams and Columns, shewing their fitness to resist Transverse Strains, Torsion, Compression, Tension, and Impulsion; with Tables of Constants [etc].*
By William Turnbull.
London: Josiah Taylor ... 1832.
Dated: Fitzroy Square, November 1831.
8vo 22 cm viii + 194 pp. Woodcuts in text.
BM, SML
Price 10s. 6d. in boards.

TYNE, RIVER

1664 *An Account of the Great Flood in the River Tyne ... Dec. 30, 1815. To which is added a Narrative of the Great Flood in the Rivers Tyne, Tease, and Wear on the 16th and 17th Nov. 1771.*
Newcastle: Printed for John Bell ... 1816.
8vo 20 cm 16 pp.
BM

1665 *Papers relating to the River Tyne, ordered to be printed by the River Committee.*
Newcastle: Printed by T. and J. Hodgson ... 1836.
8vo 22 cm (i) + 48 pp.
CE

TYRWHITT, *Sir* Thomas

1666 *Substance of a Statement ... concerning the Formation of a Rail Road, from the Forest of Dartmoor to the Plymouth Lime-Quarries, with Additional Observations, and a Plan of the Intended Line.*
By Sir Thomas Tyrwhitt.
Plymouth-Dock: Printed by Congdon and Hearle ... and published by Harding ... London.
Dated: 1 Jan 1819.
8vo 21 cm 29 pp. Folding engraved plan.
CE

UPTON, John

1667 *Observations on the Gloucester & Berkeley Canal. Comprising Remarks on the past, present, and future management of that Important Undertaking.*
By John Upton, Engineer.
Gloucester: Printed by D. Walker ... 1815.
Signed: J. Upton. Gloucester, 18 Feb 1815.
8vo 23 cm 43 pp.
CE

1668 *Plan of the Gloucester & Berkeley Canal, shewing the different lines proposed from the Locks at Gloucester, to the intended Harbour and lower Junction with the River Severn at Sharpness Point.*
By John Upton.
W.F. 1816.
Engraved plan 24 × 47 cm (scale 1 inch to 1 mile).
BM

VALLANCEY, *Colonel* Charles [1726–1812] F.R.S., M.R.I.A.

1669   *A Treatise on Inland Navigation, or, the Art of making Rivers navigable, of making Canals in all Sorts of Soils, and of constructing Locks and Sluices. Extracted from the Works of Guglielmini, Michelini, Castellus Belidor, and others, with Observations and Remarks.*
By Charles Vallancey, Engineer.
Dublin: Printed for George and Alexander Ewing. 1763.
4to   27 cm   (i) + xi + (1) + 179 pp.   24 folding engraved plates.
BM, NLS, CE, ULG

1670   *A Report on the Grand Canal, or, Southern Line.*
By Charles Vallancey, Director of Engineers.
Dublin: Printed by Timothy Dyton . . . 1771.
Signed: Charles Vallancey. Dublin, December 1770.
4to   25 cm   66 pp.   Folding engraved plan.
A Sketch [map] of the Country between the City of Dublin and the River Shannon shewing the Course of the Southern Line or Grand Canal . . . and its Junction with the Rivers Barrow & Boyne.
ULG

VANDERSTEGEN, William

1671   *The Present State of the Thames considered; and a comparative view of Canal and River Navigation.*
By William Vanderstegen, Esq.
London: Printed for G. G. and J. Robinson . . . 1794. Price 1s 6d.
8vo   20 cm   (iii) + 76 pp.
BM, CE (Page), BDL (Gough)

VARLEY, John

See BRINDLEY. Plan of the intended canal from Chesterfield to the Trent at Stockwith, surveyed by John Varley, 1769.

See VARLEY and STAVELEY. Plan of intended Union Canal from Leicester to the Grand Junction Canal, 1792.

VARLEY, John and STAVELEY, Christopher

1672   *A Plan of the intended Union Canal, from Leicester to join a Branch of the Grand Junction Canal in the Parish of Hardingstone in the County of Northampton; with a Collateral Branch to Market Harborough.*
Survey'd in 1792 by Jn.º Varley Sen.ʳ and Christ.ʳ Staveley Jn.ʳ   Jn.º Varley Jun.ʳ del.
Engraved plan   39 × 84 cm   (scale 1 inch to 1 mile)
BM

1673   Another edition
A Plan of the Proposed Union Canal, to join the Leicester Navigation with a Branch of the Grand Junction Canal, near Northampton: also of a Branch to Market Harborough.
Surveyed in 1792, by John Varley Sen.ʳ & Christ.ʳ Staveley Jn.ʳ
Engraved by W. Faden.
Engraved plan   39 × 78 cm   (scale 1 inch to 1 mile).
BM

VAUGHAN, William [1752–1850] F.R.S., Hon.M.Soc.C.E.

1674   *On Wet Docks, Quays, and Warehouses, for the Port of London; with Hints respecting Trade.*

London. 1793.
Dated: London, 14 Dec 1793.
8vo   22 cm   (iv) + 27 pp.
BM, CE (Vaughan), ULG

1675   Second issue
A Treatise on Wet Docks, Quays, and Warehouses, for the Port of London; with Hints respecting Trade.
London: Printed for J. Johnson ... 1794. Price One Shilling.
Dated: London, 14 Dec 1793.
8vo   22 cm   (iv) + 27 pp.
CE (Page)
  Identical to No. 1674 except for the title-page.

1676   *Plan of the London-Dock, with some Observations respecting the River immediately connected with Docks in General, and of the Improvement of Navigation.*
London. 1794.
Dated: London, 24 Aug 1794.
8vo   22 cm   (ii) + 12 pp. Folding engraved plan.
Plan of the River with the Proposed Docks.
BM, CE (Vaughan), ULG, BDL (Gough)
  The plan shows the first scheme as drawn up by John Powsey.

1677   *Reasons in Favour of the London-Docks.*
London: 1795.
Dated: London, 21 April 1795.
8vo   22 cm   8 pp.
CE

1678   Second edition
Reasons in Favour of the London-Docks.
London, 1796.
Dated: London, 26 Feb 1796.
8vo   22 cm   (i) + 9 pp.
CE, BDL (Gough)
  A third edition, dated 31 Jan 1792, was printed for *A Collection of Tracts. See* VAUGHAN, 1797.

1679   *A Letter to a Friend on Commerce and Free Ports, and London-Docks.*
London: Printed in the Year 1796.
Dated: London, 16 Aug 1795. Preface dated: London, 26 March 1796.

8vo   20 cm   (iv) + 24 pp.
BM, CE, ULG, BDL (Gough)

1680   *Answer to Objections against the London-Docks.*
London, 1796.
Undated.
8vo   21 cm   21 pp.
CE

1681   Second edition
Answer to Objections against the London-Docks.
London, March 31, 1796.
Dated: London, 29 March 1796.
8vo   21 cm   22 pp.
ULG

1682   *Examination of William Vaughan, Esq. in a Committee of the Hon. House of Commons, April 22, 1796, on the Commerce of the Port of London, and the Accommodations for Shipping.*
London: 1796.
8vo   22 cm   23 pp.   Folding hand-coloured engraved plan.
Section of the River, Locks, Bason and Docks in Wapping at Spring and Neap Tides.
W.V. inv.   Allen sc.   1796.
CE (Telford), ULG, BDL (Gough)
  Text and plan reprinted from *Report from the Committee . . . on the Port of London.*
House of Commons, 1976. pp. 202–208 and pl. (16).
See also Reports from Committees of the House of Commons Vol. 14 (1803).

1683   *A Collection of Tracts on Wet Docks for the Port of London, with Hints on Trade and Commerce and on Free-Ports.*
[London] 1797.
8vo   23 cm   (ii), (iv) + 27, (ii) + 12, (iv) + 24, 23, (i) + 9, 22, 7 pp. Four folding engraved plans.
[1] *Plan of the River with the Proposed Docks* [a modified version of John Powsey's first scheme].
[2] *The London Docks.* Dan.ˡ Alexander, J. Cary sc. [the second scheme].
[3] *The London Docks.* D. Alexander Nov.ʳ 1796, J. Cary sc. [the revised scheme].

[4] *Section of the River Locks, Bason and Dock at Spring and Neap Tides.* W.V. inv. Allen sc.
BM, ULG, BDL

  Contents: 1. On Wet Docks (reprint); 2. Plan of the London Docks (reprint); 3. Letter to a Friend; 4. Examination of William Vaughan; 5. Reasons in Favour (third edition); 6. Answers to Objections; 7. London Docks, General Meeting 5 Jan 1796 (reprint). Issued in grey boards.

1684 *A Comparative Statement of the Advantages and Disadvantages of the Docks in Wapping and the Docks in the Isle of Dogs.*
London: Printed by H. L. Galabin . . . 1799.
Dated: London, 28 May 1799.
8vo   21 cm   43 pp.
BM, CE (Vaughan), ULG

1685   Second edition
Title as No. 1684 but *The Second Edition*.
8vo   21 cm   45 pp.
BM

1686   *Memoir of William Vaughan, Esq. F.R.S. with miscellaneous pieces relative to Docks, Commerce, etc.*
London: Smith, Elder, & Co . . . 1839.
8vo   22 cm   viii + 134 + 9 pp.   Portrait frontispiece.
RS, CE, ULG, BDL

  Issued in publisher's cloth with lettering label on spine. Bound in is an original edition of *Reasons in favour of the London-Docks*.

1687   *Tracts on Docks and Commerce, printed between the years 1793 & 1800 . . . with an Introduction, Memoir, and Miscellaneous Pieces.*
By William Vaughan, Esq. F.R.S.
London: Smith, Elder, and Co . . . 1839.
8vo   22 cm   viii + 134, (4), (iv) + 27, (ii) + 12, (iv) + 24, 23, (i) + 9, 22, 7, 45, (iii) + 55 pp. Portrait frontispiece, 2 folding engraved plans (as Nos. 1 and 4 in *A Collection*, 1797).
BM, SML, ULG, BDL

  Contents: Memoir, etc; Tracts 1–7 as in *A Collection*, 1797 (No. 3 now being a reprint); 8. A Comparative Statement, Second Edition; Reasons for extending the Public Wharfs.

VAZIE, Robert

1688   *Observations on the intended Archway through Highgate Hill.*
  By Robert Vazie, Mining Engineer.
Dated: Cornhill, London, 7 Jan 1809.
With folding engraved plan (James Basire sc.).
Plan and Sections of the proposed Archways through Highgate Hill . . . Projected and Surveyed by Mr Rob ! *Vazie . . . Jan 19th 1809.*
[In]
Third Report from the Committee on Broad Wheels and the Preservation of the Turnpike Roads and Highways.
  House of Commons, 1809. pp. 119–21.

1689   *A Report on the intended Shoreham Harbour-Docks.*
n.p.
Signed: Robert Vazie. Greville Street, Hatton Garden, 1 Jan 1810.
Single sheet   30 × 19 cm   2 pp.
ULG

1690   *A Plan & Section of the intended Harbour-Docks at New Shoreham.*
Rob.! Vazie, 7 Aug 1810.
Engraved plan   17 × 29 cm
West Sussex R.O.

1691   *A Comparative Statement of the specific Differences between forming a Detached Harbour, near to New Shoreham, as delineated in the Plan annexed, laid before Parliament in the Year 1810; and making an Entrance into the dangerous Existing Port, near to Kingston-by-Sea, as by another Plan proposed by Mr William Chapman, and submitted to Parliament in this present Session.*
[London]: Printed by J. S. Jordan . . . [1816].
Signed: Robert Vazie. Kentish Town, 26 March 1816.
fol.   32 cm   2 pp. + engraved plate (a modified version of No. 1690 dated Jan 1816).
CE, West Sussex R.O.

VAZIE, William

See TAYLOR and VAZIE. Report of Surveys for a tunnel under the River Forth, 1806.

See MILLAR and VAZIE. Observations on the proposed tunnel under the Forth, 1807.

VERMUYDEN, *Sir* Cornelius [*c.* 1595–1677]

1692 *A Discourse touching the Drayning the Great Fennes, lying within the severall Counties of Lincolne, Northampton, Huntington, Norfolke, Suffolke, Cambridge, and the Isle of Ely, as it was presented to his Majestie.*
By Sir Cornelius Vermuiden, Knight.
London: Printed by Thomas Fawcett . . . 1642.
4to   19 cm   (ii) + 32 pp.   With woodcut Royal arms on p. (i).
Folding engraved map (untitied) of the Fens showing the main drains already executed and proposed.
BM, ULG, BDL (Gough), CUL
  p. 32: Ordered by the Committee for the Great Levell that the Designe offered by Sir Cornelius Vermuiden, together with the Mappe be Printed. 22 Feb 1641/2.

1693 *The designe in what manner the South Levell of the Fennes is dreyned*, humbly presented to the right honourable the lordes and other comm^rs nominated in the late acte of parliament, at Elie the 24th day of March 1652 [i.e. 1653], by S.r Cornelius Vermuyden, Knight, director of the workes.
Signed: Corn: Vermuyden.
pp 269–275.
[In]
WELLS. *History of the Drainage of the Great Level of the Fens.* Vol. 1, London, 1830 [q.v.].

VIGNOLES, Charles Blacker [1793–1875] F.R.S., M.R.I.A., M.Soc.C.E., M.Inst.C.E.

See RENNIE, George and John. Plan of the Liverpool and Manchester Railway, 1826.

1694 *Map of the Intended Improvements along part of the Line of the existing Oxford Canal between Longford in the County of the City of Coventry and Wolphamcote in the County of Warwick, together with the Sections thereof.*
Laid down from Surveys and Levels taken by and under the immediate Direction of Charles Vignoles, Engineer. R. Cartwright, lithog.
Dated: 8 Oct 1828.
Litho map,   hand-coloured   38 × 119 cm (scale 2 inches to 1 mile).
BM, CE

1695 *Report of the Engineer to the Directors of the Preston and Wigan Railway.*
Signed: Charles Vignoles, Engineer. Liverpool, 31 Oct 1831.
pp. 15–21 with folding litho plan.
Map and Sections of a certain intended line of Railway between . . . Preston and Wigan together with Several Intended Branches therefrom, all in the County Palatine of Lancaster.
From Surveys and Levels made under the immediate direction of C. B. Vignoles, Civil Engineer. Surveys and Levels taken by C. G. Forth. 27 Nov 1830.
[In]
Report of the Directors of the Preston and Wigan Railway Company . . . at the General Meeting, held November 1, 1831 . . . together with the Proceedings and Resolutions . . . the Engineer's Report, and the Treasurer's Statement of Account.
[Liverpool: 1831].
8vo   22 cm   24 pp.   Folding litho plan (as above).
CE

1696 [Report] *To the Directors of the Liverpool and Manchester Railway.*

Signed: Charles Vignoles. C.E. Trafalgar Square London, 21 June 1835.

pp. (3)–20.

[Printed with]

Report by Joseph LOCKE, 17 Jan 1835 [q.v.].

[In]

Two Reports addressed to the Liverpool & Manchester Railway Company, on the projected North Line of Railway from Liverpool to the Manchester, Bolton, and Bury Canal, near Manchester, exhibiting the extent of its cuttings and embankings, with estimates of the cost of completing the said Railway.

By Charles Vignoles, Esq. and Joseph Locke, Esq. Civil Engineers.

Liverpool: Printed by Wales and Baines . . . 1835.

8vo   22 cm   32 pp.

SML, CE, ULG (Rastrick)

1697   Deposited plan

Map of the Proposed Sheffield, Ashton under Lyme and Manchester Railway.

Laid down from Surveys made under the immediate directions and Superintendence of Charles Vignoles, Civil Engineer. November 30th, 1836.

R. Cartwright, Litho.

Oblong folio. Title + 8 plans (scale 4 inches to 1 mile) + 5 enlarged plans (scale 16 inches to 1 mile), forming 14 sheets each 39 × 71 cm.

BM

1698   *Report on the Various Lines by which a Railway could be carried from London to Porth-Dynllaen in North Wales.*

By Charles Vignoles, Civil Engineer, F.R.A.S., M.R.I.A., M.Inst.C.E.

Signed: Charles Vignoles, Civil Engineer. Trafalgar-Square, London, 29 Nov 1837.

[In]

Second Report of the Commissioners . . . [on] Railways in Ireland.

Dublin, 1838. Appendix A, pp. 35–44.

1669   *Map of North Wales and of portions of the Adjacent English Counties whereon is indicated the General Direction of a Proposed Main Line of Railway . . . from Porth Dynllaen to Bala and thence by the Valley of the*

River Dee, and Eastward to fall into the Grand Junction Railway.

By Charles Vignoles, Civil Engineer.

Engraved map   45 × 93 cm   (scale 1 inch to 5½ miles) with Diagram Section.

[In]

Irish Railway Commission. Plans of the Several Lines laid out under the direction of the Commissioners. 1837.

1700   *Report on the Several Lines of Railway through the South and South-Western Districts of Ireland, as laid out under the direction of the Commissioners.*

By Charles Vignoles, Civil Engineer, F.R.A.S., M.R.I.A., M.Inst.C.E.

Signed: Charles Vignoles. Trafalgar Square, London, March 1838.

[In]

Second Report of the Commissioners . . . [on] Railways in Ireland. Dublin, 1838. Appendix A, pp. 1–31.

The lines run from Dublin to Limerick and Cork with extensions to Tarbet and Bereharen and a branch to Kilkenny.

1701   [Plans of the Lines] *Through the South and South Western Districts of Ireland.*

By Charles Vignoles, Civil Engineer.

Title page + index map + 2 enlarged plans + 11 plans (scale 1 inch to 1 mile), forming 15 sheets each 45 × 90 cm.

[In]

Irish Railway Commission. Plans of the Several Lines laid out under the direction of the Commissioners. 1837.

1702   [Sections of the Lines] *Through the South and South Western Districts of Ireland.*

By Charles Vignoles, Civil Engineer.

Title page + Index + Diagram Section + 46 sections (scale 4 inches to 1 mile) forming 49 sheets each 45 × 90 cm.

[In]

Irish Railway Commission. Sections of the Several Lines laid out under the direction of the Commissioners. 1837.

1703   *Map of Sheffield & Manchester Railway, shewing its connection with other lines particularly with the Line to the Humber. 1838.*

Charles Vignoles, M.R.I.A., F.R.A.S. &c. Engineer. R. Cartwright, Lith.

Litho map   40 × 25 cm   (scale 1 inch to 9
    miles).
                        [In]
*Sheffield & Manchester Railway . . . Report of the
Directors.* June, 1838.
fol 41 cm   (4) pp.
CE

VINCENT, William [d. 1754]

1704 *A Plan of Scarborough. With the Gentlemens
Names of a Committee Appointed to put this Town in a
Posture of Defence against the Rebels 1745 . . . The
whole was conducted under the Direction of the above
committee by Willm. Vincent, Engineer for Building the
new Pier at Scarborough.*
Nath. Hill sc.
Published . . . by W. Vincent 1747.
Engraved plan of town and harbour
    53 × 74 cm   (scale 1 inch to 2 chains = 132
    feet).
BM, RS, (Smeaton)

WALKER, James [1781–1862]
F.R.S.L. & E., M.Soc.C.E., M.Inst.C.E.

See also Reports by WALKER and BURGES,
    1830–1836.

1705 *Plan of the late Improvements in the Port of
London; with the Roads made, or projected, for connecting
the same with the City.*
    Dedicated to the Trustees of the Commercial
    Road, by their obliged Servant, James Walker.
Published 7 Aug 1804 by Robt Wilkinson.
Engraved plan, hand coloured   44 × 142 cm
    (scale 1 inch to 400 feet).
BDL (Gough)

1706 Another issue
As No. 1705 but dated 1 Dec 1805.
PLA (Vaughan)

1707 *Plan of the Commercial Docks at Rotherhithe
with the Intended Improvements.*
    I. Walker Engineer.
Published 1 May 1810 by Robert Wilkinson.
Engraved plan, hand coloured   17 × 33 cm
    (scale 1 inch to 400 feet).
PLA (Vaughan)

1708 *This Plate exhibiting a Picturesque Elevation of
the Iron Bridge erected over the Thames at Vauxhall,
under the Direction of James Walker, Engineer, & completed A.D. 1816.*
    Is respectfully Dedicated to the Vauxhall
    Bridge Company by . . . Sam.! Cossart. Sam.!
    Cossart del.   Js. Basire sc.   Published . . .
    1 Nov 1816.
Engraved elevation with added landscape
    23 × 61 cm   (scale 1 inch to 400 feet).
SML, CE (Rennie)

1709 *Plan of the Commercial Docks at Rotherhithe.*
    J. Walker. January 1820.
Published 12 Jan 1820 by Robert Wilkinson.
Engraved plan, hand coloured   21 × 37 cm
    (scale 1 inch to 400 feet).
PLA

1710 Another issue
As No. 1709 but dated June 1824.
PLA

See WALKER and MYLNE. Report on the Eau
    Brink Cut, 1825.

1711 *Report . . . to the Commissioners of the Haven
and Piers at Great Yarmouth, on the State of the Bar and
Haven; and the Measures advisable to be adopted for their
Improvement.*
Yarmouth: Printed by C. Sloman . . . 1826.
Signed: J. Walker, Civil Engineer. Limehouse, 27
    Aug 1825.
8vo 22 cm 15 pp   Folding engraved plan.
Reduced Plan of the Entrance to the Harbour of Great

Yarmouth and Part of the River Yare, referred to in Mr Walker's Report.
Norfolk R.O., BDL

1712 *Second Report . . . to the Commissioners of the Haven & Piers, of Great Yarmouth, on the State of the Bar and Haven.*
By James Walker, Esq. Civil Engineer.
Yarmouth: Printed by W. Meggy . . . 1826.
Signed: J. Walker. Limehouse, 1 July 1826.
8vo   22 cm   20 pp   Folding engraved plan (as in the previous report).
BM, Norfolk R.O., BDL

1713 *Report . . . to the Corporation of Great Yarmouth, on the Plan for a Ship Navigation from the Sea at Lowestoft to Norwich.*
Yarmouth: Printed by W. Meggy . . . 1826.
Signed: J. Walker. Limehouse 19 Jan 1826.
8vo   21 cm   16 pp.
CE

1714 *Report . . . to the Commissioners of the Haven & Piers, of Great Yarmouth, on the Practicability of making Braydon Navigable for Vessels drawing 10 feet water, &c.*
By James Walker, Esq. Civil Engineer.
Yarmouth: Printed by W. Meggy . . . 1826.
Signed: J. Walker. Limehouse, 4 Sept 1826.
8vo   21 cm   15 pp   Folding litho plan
Plan of the Brayden from Yarmouth Bridge to the Junction of the Yare & Waveney referred to in Mr Walker's Report. Clark & Co. litho.
CE

See WALKER and RASTRICK. Report on comparative merits of locomotive and fixed engines, 1829.

1715 *Liverpool and Manchester Railway. Report to the Directors on the Comparative Merits of Loco-Motive & Fixed Engines, as a Moving Power.*
By James Walker, F.R.S.L. & E. Civil Engineer.
Second Edition, Corrected.
London: Published by John and Arthur Arch . . . 1829.
8vo   23 cm   vii + 50 pp.   Folding plate.
SML, CE, BDL

1716 *Report to the Committee of the Proposed Railway from Leeds to Selby.*
By James Walker, F.R.S.L. & E. Civil Engineer.
London: Printed for John and Arthur Arch . . . 1829.
Signed: J. Walker. Limehouse, 18 July 1829.
8vo   21 cm   24 pp   Folding hand-coloured engraved plan.
Plan & Section of the Proposed Railway from Leeds to Selby in the West Riding of the County of York.
Surveyed under the direction of James Walker . . . Civil Engineer, by A. Comrie, 1829. G. Cruchley sc.
BM (plan only), CE, Elton

See TELFORD and WALKER. Report on the present state of the New London Bridge, 1831.

1717 *Report to the Commissioners of the River Wear on the Formation of Wet Docks at Sunderland.*
By James Walker, F.R.S.L. & E. Civil Engineer.
London: Printed for John and Arthur Arch . . . 1832.
Signed: James Walker. Great George Street, 24 Nov 1832.
8vo   21 cm   17 pp.   2 folding hand-coloured litho plans (R. Martin lith.).
A. *Plan of Docks proposed to be formed in the Port of Sunderland, being part of the extended plan shewn on sheet B.*
B. *Extended Plan of Docks for the Port of Sunderland.*
[Printed with]
Report by George RENNIE, 24 Nov 1832 [q.v.].
[In]
Sunderland. Report on the Formation of Docks; by George Rennie, Esq. Also, Report on the same by James Walker . . . with Plans.
London: Printed for John and Arthur Arch . . . 1832.
8vo   21 cm   (i) + 9 + 17 pp.   2 folding plans (as above).
BM, NLS, CE

See SPECIFICATION and Tender for Repairs to Blackfriars Bridge, 1833.

1718 *Port of Edinburgh. Report on its Improvement . . .*

By James Walker. F.R.S.L. & E.

London: Printed by Ibotson and Palmer . . . 1835.

Signed: Jas. Walker. Great George Street, 25 May 1835.

8vo 21 cm (i) + 16 pp. Folding hand-coloured litho plan.

Plan of Leith Harbour, with the Various Plans Proposed for New Works, referred to in Mr Walker's Report, 1835.

NLS, CE

1719 *Northern and Eastern Railway. Report to the Committee for Promoting a Railway from London to York, with a Branch to Norwich, &c.*

By James Walker, F.R.S.L. & E.

London: W. M. Knight and Co. . . . 1835.

Signed: Jas. Walker. Great George Street, 30 June 1835.

fol 31 cm 20 pp. Folding litho map, the lines in colours.

Northern and Eastern Railway, from London to York &c. 1835.

BDL

For a map and prospectus of an earlier version of this line, with Walker as consulting engineer, *see* CUNDY, 1833.

1720 *Report(s) addressed to the Secretary to the Commissioners for the Loan of Exchequer Bills, on the Improvement of the Navigation of the River Welland.*

By James Walker, F.R.S.L. & E. Civil Engineer.

London: Printed by John Cunningham . . . 1839.

Signed: James Walker. Spalding, 7 Nov 1835 [and] Gt. George St., 21 Nov 1838.

8vo 24 cm 16 pp. Folding litho plan.

River Welland. Plan shewing the Proposed Improvements between Fosdike Bridge and Clay Hole.

Referred to in Mr Walker's Report, 1835. John Grieve, Zincographer.

NLS, CRO, CE

See WALKER and HARTLEY. Report on Whitehaven Harbour, 1836.

1721 *Report on Westminster Bridge, made by Order of the Commissioners.*

By James Walker, Civil Engineer.

London: Printed by Order of the Committee of Commissioners, 1837.

Signed: James Walker. Great George Street, 28 Feb 1837.

4to 27 cm 11 pp.

CE

Issued in brown printed wrappers.

1722 *[Report on the Thames Tunnel works].*

Signed: James Walker. Great George Street, 22 April 1837.

pp. 3–4.

[In]

Copies of Treasury Minutes and Correspondence relating to the . . . Thames Tunnel Company.

House of Commons, 1837.

1723 *Report of James Walker, Esq., Civil Engineer, on the Works of the London and Greenwich Railway.*

London: Printed by A. Macintosh . . . 1837.

Signed: Jas. Walker. Great George Street, 25 Aug 1837.

8vo 23 cm 8 pp. Litho frontispiece.

View of the portion of the London & Greenwich Viaduct which crosses Corbett's Lane.

From Nature & on Zinc by A. R. Grieve.

Printed by Chapman & Co.

BM

1724 *Report of James Walker, Civil Engineer, on Dymchurch Wall in the County of Kent . . . Addressed to the Lord Bailiff and Jurats of Romney March.*

London: Printed by Ibotson and Palmer.

Signed: James Walker. Dymchurch, 22 July 1837.

4to 28 cm 9 pp. Folding litho plan.

Dymchurch Wall, Cross Section for Wall.

R. Martin & Co. lith.

CE

1725 *Copy of Mr Walker's Report to the Treasury, on the Works at the Thames Tunnel.*

House of Commons, Ordered to be printed 30 June 1838.

Signed: James Walker, December 1837.

fol 33 cm 6 pp.

SML

1726 *St. Helier, Jersey. Report of James Walker, Civil Engineer, on the Improvement and Enlargement of the Harbour.*
Jersey: Printed by C. Le Lievre.
Signed: J. Walker. Great George Street, 10 March 1838.
8vo   21 cm   12 pp.   Folding hand-coloured litho plan.
Plan of the Harbour and Small Roads of St. Helier Jersey with Design for Improvements proposed by Mr Walker.
Lith. by S. Le Capelain and C.I. Prignand.
CE

1727 *Aberdeen Harbour. Report of James Walker, Civil Engineer, on the Harbour and Proposed Docks.*
Aberdeen: Printed at the Herald Office . . . 1838.
Signed: J. Walker. Great George Street, April 1838.
8vo   22 cm   12 pp.   Folding hand-coloured litho plan.
Aberdeen Harbour, with Design for Proposed Improvements referred to in Mr Walker's Report.
R. Martin & Co. lith.
BM, NLS, CE
   Issued in yellow printed wrappers.

1728 *Aberdeen Harbour. Report of James Walker, Civil Engineer, on the Plan of Messrs. Hogarth, Pirie, and Read, for Proposed Improvements in the Harbour of Aberdeen.*
Aberdeen: Printed at the Herald Office . . . 1838.
8vo   22 cm   11 pp.
CE

1729   Reprint of the two foregoing reports
[In] *Evidence and Proceedings . . . in regard to the Aberdeen Harbour Bill*, 1839. pp. xxvi–xxxviii. *See* ABERDEEN HARBOUR.

1730 *Copy of Mr Walker's Report to the Board of Treasury on the Caledonian [and Crinan] Canal.*
   House of Commons, 1838.
Signed: J. Walker. Great George Street, 7 June 1838.
fol   33 cm   15 pp.   Docket title on p. (16).

1731 *Leith Harbour. Mr Walker's Report.*
Signed: J. Walker. London, 19 June 1839.
pp. 15–26   with 2 folding litho plans (Hume lithog.).
No. 1. *Leith Harbour, Leith and Newhaven, or Western Design for Improvements.*
No. 2. *Leith Harbour. Leith or Eastern Design for Improvement.*
             [Printed with]
Mr CUBITT's Report, 13 May 1839 [q.v.].
             [In]
Plans and Reports, by J. Walker, Esq., & W. Cubitt, Esq., for the Improvement of Leith Harbour; and Treasury Minutes, and other documents connected therewith.
Edinburgh: Printed by A. Murray . . . 1839.
8vo   22 cm   34 pp.   3 folding litho plans.
NLS, CE

WALKER, James and BURGES, Alfred

1732 *Report of Messrs Walker and Burges, Civil Engineers, on the Improvement of the Harbour of Belfast.*
London: John Weale . . . 1837.
Signed: Walker and Burges. London, 10 July 1830.
4to   29 cm   18 pp.   Large folding litho plan, partly hand coloured.
Map of the Proposed Improvement of the Channel and Harbour of Belfast. (Originally designed by Walker and Burges). With Transverse Sections of the Proposed Channel and Quays.
   By William Bald, F.R.S.E., M.R.I.A. Civil Engineer 1836.   C. F. Cheffins, lithog.
CE

1733 *Report of Messrs Walker and Burges.*
Signed: Jas. Walker, A. Burges. Great George Street, 2 March 1833.
pp. 5–12   3 folding litho plans, one of which is hand coloured. (R. Martin, Lithog.).
[1] *Blackfriars Bridge. Longitudinal Section shewing the depths of the Caissons* [etc.].

[2] *Elevation of Centre Arch with the Piers . . . in their present state.*

[3] *Elevation of Centre Arch with the Piers . . . as proposed to be altered.*

[In]

Report to the Common Council, from the Committee appointed in relation to Blackfriars Bridge.
Corporation of London. April 1833.
fol   33 cm   12 pp.   3 folding plans (as above).
CLRO, CE

1734   *Report to the Subscribers for a Survey of the Part of the Leeds and Hull Junction Railway, between Hull & Selby.*
By Messrs Walker and Burges, Civil Engineers.
Hull: Printed by George Lee.
Signed: J. Walker, A. Burges. Great George Street, 28 July 1834.
8vo   21 cm   8 pp.   Folding engraved plan.
Leeds & Hull Junction Railway. Map and Section of Proposed Line from Selby to Hull, 1834.
   Surveyed under the direction of Messrs Walker & Burges . . . By A. Comrie. Goodwill & Lawson sc.
NLS, CE
   The plan was reissued in *Report of the Directors of the Hull and Selby Railway Company*. Hull: Printed by William Stephenson . . . 1836.

1735   *North Liverpool Railway. Report on the Proposed Line of Railway from the Manchester & Bolton Railway to the North End of Liverpool.*
By Messrs Walker and Burges, Civil Engineers.
Westminster: Vacher & Son . . . 1835.
Signed: Jas. Walker, Alfred Burges. Great George Street, 24 Feb 1835.
8vo   21 cm   8 pp.   2 folding engraved plans.
[1] *Plan of the present lines of Navigation and Railway between Liverpool and Hull, and of the Manchester, Bolton & Bury Railway, with the proposed lines to Liverpool, Middleton, &c.*
[2] *Plan and Section of the Manchester Bolton and Bury Railways, with the proposed Line to the North End of Liverpool.* E. Turrell sc.
B, CE (lacking plans)

1736   *Report by Messrs. Walker and Burges, Civil Engineers, London, regarding the Proposed Wet Docks, &c., at Greenock.*

Greenock: Printed at the Greenock Advertiser Office. 1836.
Signed: Walker & Burges. Great George Street, 9 Feb 1836.
8vo   22 cm   8 pp.
CE

1737   *The Report of Messrs Walker and Burges.*
Signed: Walker and Burgess. Great George Street, 26 Dec 1836.
pp. 3–5   with 2 folding hand-coloured litho plans.
Plan of River Wensum and Proposed Locks. Referred to in the Plan of Messrs Walker and Burges (Parts 1 and 2). Sloman, lithog.
[Printed with]
Report by George EDWARDS, 6 Jan 1838 [q.v.].
[In]
Reports to the Commissioners of the Haven of Great Yarmouth, respecting the Proposed Lock upon the River Wensum, at Norwich.
   By Messrs Walker and Burges, and Mr George Edwards, Civil Engineers.
Yarmouth: Printed by Charles Sloman . . . 1838.
8vo   22 cm   16 pp.   3 folding plans.
CE

1738   *River Thames from Vauxhall Bridge to London Bridge, shewing the line of Proposed Embankments, 1840.*
   Walker & Burges, Engineers. J. Basire lith.
Litho plan, hand coloured   30 × 61 cm   (scale 1 inch to 500 feet).
[and]
[Two Sections] signed: J^s Walker; each 30 × 59 cm   (scale 1½ inches to 100 feet).
[In]
Report from the Select Committee on Thames Embankment.
   House of Commons, 1840.

WALKER, James and HARTLEY, Jesse

1739 *Report of James Walker and Jesse Hartley, Civil Engineers, to the Trustees of Whitehaven Harbour.*
Signed: James Walker, Great George Street, Westminster, Jesse Hartley, Dock Yard, Liverpool. 17 May 1836.
p. 34   with 2 litho plans   (R. Gibson lith.).
[1] *Whitehaven Harbour. Proposed Termination of Piers*, by Messrs Walker & Hartley, 1836.
[2] *Plan for Timber Facing, West Pier.*
[In]
Plans suggested at different periods for the Improvement of Whitehaven Harbour.
Whitehaven, 1836. *See* WHITEHAVEN HARBOUR.
CE, ULG

WALKER, James and MYLNE, William Chadwell

1740 *The Joint Report by Messrs Walker and Mylne, the Engineers appointed in consequence of the late intended Eau Brink Act, 1825.*
Cambridge: Printed at the Independent Press Office.
Signed: J. Walker, William C. Mylne. London, 21 May 1825.
8vo   22 cm   62 pp.
CRO, CE
    Includes also a preliminary report, 25 March and a further note, 10 June 1825.

WALKER, James and RASTRICK, John Urpeth

1741 *Liverpool and Manchester Railway. Report to the Directors on the Comparative Merits of Loco-Motive & Fixed Engines, as a Moving Power.*

By Jas. Walker and J. U. Rastrick, Esqrs. Civil Engineers.
Liverpool: Printed by Wales and Baines . . . 1829.
Signed: J. Walker, Limehouse, 7 March 1829: J. U. Rastrick.
8vo   22 cm   (i) + 80 pp.   3 plates.
BM, NLS, SML, CE, ULG
    Five hundred copies printed. 'Second editions' of separate reports by WALKER and by RASTRICK were published later in 1829 [q.v.].

WALKER, John

1742 *Plan [and Section] of the Proposed Canal from Wakefield to Ferrybridge, in the West Riding of the County of York.*
    John Walker, Surveyor, Wakefield. 1826.
        [Printed on same sheet with]
Map of the Navigations connecting the River Humber & Mersey, exhibiting the Collateral Cuts, Rail Roads, and the Proposed Canal from Wakefield to Ferrybridge.
Franks & Johnson, Engravers, Wakefield.
Engraved maps, 33 × 42 cm.
BM

1743 *Map of the Inland Navigation, Canals and Rail Roads with the Situations of the various Mineral Productions throughout Great Britain, from Actual Surveys. Projected on the Basis of the Trigonometrical Survey made by order of the Honourable Board of Ordnance, by J. Walker, Land and Mineral Surveyor, Wakefield.*
    Accompanied by a Book of Reference, compiled by Joseph Priestley, Esq. of the Aire and Calder Navigation.
Published by Richard Nichols, Bookseller, Wakefield. Longman, Rees . . . and G. & J. Cary . . . London. Jan 1st 1830.
Franks & Johnson, Engravers, Wakefield.
Engraved map   189 × 155 cm   (scale 1 inch to 6 miles).
BM, NLS, SML, RGS, CE, ULG

WALKER, Ralph [c. 1748–1824]

1744 *A Proposed Plan of Wet Docks in Wapping.*
By Ralph Walker.
Engraved plan   46 × 78 cm   (scale 1 inch to 200 feet).
[with]
Reference and Explanation of the Plan produced by Mr Ralph Walker [7 April 1796].
[In]
Report from the Committee on the Port of London.
House of Commons, 1796. pp. 57–70 and pl. 3.

1745 Second printing
[In] *Reports from Committees of the House of Commons* Vol. 14 (1803) pp. 372–376 and pl. 1(c)
See DANCE, JESSOP and WALKER. Plans of the proposed West India docks in the Isle of Dogs, 1797 (two issues).

1746 *A Comparative View of the Wet Docks in Wapping, and in the Isle of Dogs, for the West India Trade: also Hints respecting the bonding of all West-India Produce, with a Plan of the Docks.*
[London] 1797.
8vo   21 cm   27 pp.   Folding hand-coloured engraved plan.
A Plan of the Proposed Docks in the Isle of Dogs for the West India Trade.
Surveyors Office Guildhall. Metcalf sc.
ULG
  With Ralph Walker's autograph signature at end of text on p. 27.

See DANCE, JESSOP and WALKER. Plan of the proposed West India docks in the Isle of Dogs (revised plan). With an explanation of the plan by Ralph Walker, 26 March 1798 and estimate by Jessop & Walker, Nov 1798.

1747 *An Accurate Plan of the Docks for the West India Trade, and Canal, in the Isle of Dogs; Begun 1800.*
Published 7th May 1800 by John Fairburn.
Engraved plan, hand coloured   41 × 54 cm (scale 1 inch to 260 feet).

PLA
  From a drawing by Walker, March 1800, prepared in consultation with Jessop. The plan as finally adopted.

1748 Later edition
As No. 1747 but Published Aug.ᵗ 24th 1801.
PLA

1749 *Plan of the West India Wet Docks.*
Dedicated by Permission to Geo. Hibbert Esq.ʳ Chairman ... & the Directors of the West India Dock Company, by their much obliged and very humble Servant, Ralph Walker, Resident Engineer.
Engraved plan   56 × 83 cm   (scale 1 inch to 200 feet).
PLA (Vaughan)
  Issued about August 1800.

1750 *Plan of a Double Turning Arch Bridge.*
Invented by Ralph Walker, Civil Engineer A.D. 1800. G. Gladwin sc.
Engraved plan and elevation   35 × 63 cm (scale 1 inch to 4 feet).
CE (Rennie)

1751 *Plan of the West India Docks &c.*
Designed and Dedicated to the Directors by their very humble servant Ralph Walker. Published 1 Jan 1802. B. Baker sc.
Engraved plan, hand coloured   27 × 38 cm (scale 1 inch to 400 feet).
PLA (Vaughan)

See RENNIE and WALKER. Plans of the East India docks, 1803 and 1804.

1752 *Plan of the Intended London and Cambridge Junction Canal.*
  Ralph Walker, Engineer. September 15.ᵗʰ 1810.
Engraved plan   21 × 94 cm   (scale 1½ inches to 1 mile).
BM

1753 *Plan and Section of the South Part of the Intended London & Cambridge Canal from the River Lea Navigation to Mile End.*

Ralph Walker, Engineer. W.F. 1810.
Engraved plan   18 × 47 cm   (scale 2½ inches to
   1 mile).
BM

WALTON, William

1754   *Free, and Candid Remarks, on the Reports of
Messrs Watté, Golborne, and Mylne, for, and on the
proposed improvement of the River Ouze ... to which is
subjoined, Observations on the Mode of Jettying proposed
by Mr Hodskinson.*
   By William Walton.
Lynn: Printed by R. Marshall, 1792.
Signed: William Walton. Lynn, 29 March 1792.
4to   28 cm   34 pp.
CUL, Sutro

WARE, Samuel [1781–1860]

1755   *A Treatise on the Properties of Arches, and their
Abutment Piers ... also concerning Bridges, and the
Flying Buttresses of Cathedrals.*
   By Samuel Ware, Architect.
London: Printed by T. Bensley ... 1809.
Dated: Adelphi, London, 1 Jan 1809.
8vo   23 cm   ix + (1) + 62 pp.   19 folding
   engraved plates (Neele sc.).
BM, NLS, CE (Telford), BDL

1756   *Tracts on Vaults and Bridges. Containing observations on the various Forms of Vaults; on the taking
down and rebuilding London Bridge; and on the Principles of Arches ... Also containing the Principles of Pen-
dent Bridges, with reference to the Properties of the
Catenary, applied to the Menai Bridge.*
London: Printed for Thomas and William Boone
   ... 1822.
Signed: Samuel Ware. Adelphi, London.

8vo   25 cm   xx + (5) + 73,   (i) + 71,
   (i) + 177 pp.   20 folding or double-page
   engraved plates (Wm. Alexander sc.).
BM, NLS, SML, CE, RIBA, ULG

WATSON, *Colonel* Justly [*c.* 1710–1757] R.E.

See WATSON and COLLINGWOOD. Report
on Rye New Harbour, 1756.

WATSON, Justly and COLLINGWOOD, Edward

1757   *A Report on the Survey of Rye New Harbour.*
Signed: Justly Watson, Edward Collingwood.
   London, 31 Dec 1756.
pp. 7–24.
                    [In]
*Report from the Committee to whom the Petitions concern-
ing the Harbours of Rye and Dover were referred.*
   [House of Commons]: Printed in the Year
   1757.
fol   33 cm   39 pp.
BM, Sutro
   Based partly on an (unpublished) report by
   Stephen West to the Harbour Commissioners,
   19 Oct 1756.

WATT, James [1736–1819] F.R.S.L. & E., M.Soc.C.E.

See MACKELL and WATT. Report on proposed
   canal from the Forth to the Carron, 1767.

1758 *A Scheme for making a Navigable Canal from the City of Glasgow to the Monkland Coalierys.*
By James Watt.
n.p. [*c.* 1770].
4to   22 cm   12 pp.
BM, NLS

1759 *Report concerning the Harbour of Port-Glasgow, made to the Magistrates of Glasgow.*
By James Watt, Engineer.
n.p.
Signed: James Watt, 9 Aug 1771.
4to   22 cm   7 pp.
BM

1760 *A Report concerning the Navigation of the Upper Part of the River Forth, and communicating it with the Tideway.*
By James Watt [with a Postscript and Answers to Queries].
Signed: James Watt. 4 Dec 1773 [and] Glasgow, 9 Dec and 20 Dec 1773.
pp. 13–53 [and 54–58 and 62–66].
[Printed with]
Mr John GOLBORNE's report, 14 Nov 1773 [q.v.].
[In]
Reports to the Lords Commissioners of Police, relative to the Navigation of the Rivers Forth, Gudie, and Devon.
Glasgow: Printed by Robert and Andrew Foulis . . . 1773.
4to   21 cm   66 + (1) pp.
NLS, ULG

WATT, John [1694–1737]

1761 *The River Clyde. Surveyed by John Watt* [and] *Entry to the river and Frith of Clyde.*
Engraved by Thos. Phinn.
Engraved plan   52 × 72 cm   (scale 1 inch to 1.2 miles).
RS (Smeaton)
Surveyed in 1734, published 1759.

WATTÉ, John [died *c.* 1799] M.Soc.C.E.

1762 *A Map of the Sands as are vested in the Bedford Level Corporation by Act of Parliament of the thirteenth of Geo. the III.*
Reduced from the large Map of an Actual Survey taken by John Watté. 1777.
Engraved plan   59 × 47 cm   (scale $2\frac{3}{4}$ inches to 1 mile).
BM, CRO, Sutro
Shows the recently completed Kinderley's Cut, and marshes of the Nene estuary.

1763 *The Report of John Watté, Surveyor and Engineer, for the better Drainage of the South and Middle Levels of the Fens, and other Lands bordering upon each Side of the River Ouse, and amending the Outfall of the said River, by a New Cut or Channel from Eau-Brink to Lynn.*
n.p.
Signed: John Watté. Wisbich, 21 April 1791.
4to   28 cm   18 pp.   Two folding engraved plans.
[1] *A Sketch or Map of Lynn Haven, and of the River Ouse to St. Mary Magdalen, shewing the Proposed New Cut.*
[2] *A Line and Scale of Levels along the present Channel of the River Ouse from St. German's Bridge to the Crutch, about two Miles below the Port of Lynn,* taken in the year 1791 by John Watté.
BM, CE, CRO, BDL (Gough), CUL

WELCH, Henry   M.Inst.C.E.

See TELFORD. Map of the mail roads from Retford to Morpeth, surveyed by Henry Welch 1826.

See TELFORD. Map of the roads between Carlisle and Edinburgh, surveyed by Henry Welch, 1828.

WELLS, Samuel

1764 *Letter to his Grace the Duke of Bedford, Governor of the Bedford Level Corporation, on the Works in the New Bedford or One Hundred Feet River, the River Ouze, and other matters of the Corporation.*

By the Register.
Cambridge: Printed by Weston Hatfield . . . 1828.
Signed: Samuel Wells. Fen Office . . . 12 Feb 1828.
8vo   22 cm   40 pp.
CE, CRO, CUL
Issued in brown printed wrappers.

1765 *The History of the Drainage of the Great Level of the Fens, called Bedford Level; with the Constitution and Laws of the Bedford Level Corporation.*

By Samuel Wells, Esq., Register of the Corporation. Vol. I.
London: Published for the Author, by R. Pheney . . . 1830.
Preface dated: Fen Office . . . 12 April 1830.
8vo   24 cm   xvii + (i) + 832 pp.
BM, NLS, LCL, CE (Telford), ULG, BDL, CUL

1766 *A Collection of the Laws which form the Constitution of the Bedford Level Corporation with Sundry Documents illustrative of the natural and artificial state of the several rivers, works of drainage . . . and other practical subjects . . . illustrative also of the Ancient and Modern Drainage of that extensive country called The Bedford Level.*

By Samuel Wells, Esq. Register to the Corporation.
London: Published for the Author, by R. Pheney . . . 1828.
Preface dated: Fen Office . . . 1 Aug 1828.
8vo   24 cm   xxii + 802 pp.   Engraved frontispiece. *Salters Lode Sluice.* Mutlow sc.
BM, NLS, LCL, ULG, CUL (lacks frontispiece)

1767   Second issue
The History of the Drainage of the Great Level of the Fens, called Bedford Level.

By Samuel Wells, Esq., Register of the Corporation. Vol. II.

London: Published for the Author, by R. Pheney . . . 1830.
8vo   24 cm   xxii + 802 pp.
CE (Telford), LCL, BDL, CUL
A reissue of No. 1766 with identical text but new title-page and without frontispiece.

1768   *To the most Noble the Governor, the Bailiffs, and Conservators of the Great Level of the Fens called Bedford Level, this map of the said Great Level and parts adjacent is most gratefully dedicated by Samuel Wells. Register.*

Fen Office, 27 March 1829.
London. Published for the Proprietor, by G. & I. Cary . . . 1829.
Engraved map, hand coloured 90×86 cm (scale 1 inch to $1\frac{1}{2}$ miles).
BM, CE, RGS, ULG (Rastrick), CUL
Dissected and mounted on linen, folding into red half-morocco cover with marbled boards and red lettering label, gilt. Issued separately to accompany the *History*.

1769   Another issue
To the most Noble the Governor . . . this map . . . is . . . dedicated by Samuel Wells.

To accompany the Report of Sir John Rennie Dec? 6th 1836.
CE
Map uncoloured, additional drainage channel entered in red. Three instead of seven lines of 'Explanation'.

1770   Another issue
The Great Level of the Fens called Bedford Level.

By Samuel Wells, Register.
London. Published for the Proprietor, by G. & J. Cary . . . 27 March 1829.
BM, BDL
Map coloured and apparently identical to No. 1768 but with changed title and three lines of 'Explanation'.

WEST, Stephen

See WATSON and COLLINGWOOD. Report on Rye New Harbour, 1756.

WESTMINSTER BRIDGE

1771 *A Short Narrative of the Proceedings of the Gentlemen, concerned in obtaining the Act, for building a Bridge at Westminster; and of the Steps, which the Honourable the Commissioners . . . have taken to carry it into Execution. In a Letter to a Member of Parliament . . . together with his Answer.*
London: Printed for T. Cooper . . . 1738.
Dated: 10 Nov and 24 Nov 1737.
8vo   21 cm   (i) + 70 pp.
CE, ULG

1772 *Reports by Messrs Telford, Cubitt, and Swinburne, Civil Engineers, as to the State of the Foundations, &c. of Westminster Bridge.*
London: Printed by Order of the Committee of Commissioners, 1836.
8vo   22 cm   48 pp.
CE
Ten reports dating from 1823 to 1836.

WESTON, Robert Harcourt

1773 *Letters and Important Documents relative to the Edystone Lighthouse, selected chiefly from the Correspondence of the late Robert Weston, Esq . . . to which is added a Report made to the Lords of the Treasury in 1809, by the Trinity Corporation; with some Observations upon that Report.*
By Robert Harcourt Weston.
London: Printed by C. Baldwin . . . 1811.
4to   28 cm   (xxiii) + 308 pp.   2 engraved plates (H. Mutlow sc.).

BM, CE, IC
Includes many letters written by Josias Jessop, John Smeaton, Robert Weston and others during the years 1755–1759.

WESTON, William [c. 1752–1833]

1774 *A Plan of the Line of the proposed London and Western Canal for Hampton Gay to Isleworth.*
By William Weston, Engineer.
Engraved by B. Baker.
Engraved plan   51 × 148 cm   (scale 1 inch to 1 mile).
BM
Scheme prepared in 1792.

WEY, River Navigation

1775 *A Plan of the River Wey from Guildford to Godalming in the County of Surrey; with a Scheme for making the same navigable.*
R. W. Seale sc. [1758].
Engraved plan   19 × 80 cm   (scale 8 inches to 1 mile).
MCL, RS (Smeaton)
Note by John Farey on the RS copy: 'This appears by Mr Smeaton's letter book to have been made in 1758'. Smeaton was consulted on this navigation.

1776 *Plan of the River Wey, Surveyed by Order of the Proprietors 1782.*
Engraved plan   16 × 79 cm   (scale 2 inches to 1 mile).
BM
The navigation from Weybridge to Godalming.

WHIDBEY, Joseph [c. 1755–1833] F.R.S., M.Soc.C.E.

See RENNIE and WHIDBEY. Report on Plymouth Breakwater, 1806.

See WHIDBEY and RENNIE. Report on Whitehaven Harbour, 1823.

WHIDBEY, Joseph and RENNIE, Sir John

1777 *Report of Messrs. Whidbey and Rennie on the Means of Improving the Harbour of Whitehaven.*
Signed: J. Whidbey, John Rennie. London, 30 May 1823.
p. 18 with litho plan.
Design for the Extension of Whitehaven Harbour.
By Messrs Whidbey & Rennie. Civil Engineers, 1823.
[In]
Plans suggested at different periods for the Improvement of Whitehaven Harbour.
Whitehaven, 1836. *See* WHITEHAVEN HARBOUR.
CE, ULG

WHISHAW, Francis [1804–1856] M.Inst.C.E.

See WHISHAW and THOMAS. Report on proposed railway between Perranporth and Truro, 1831.

1778 *Holborn Hill Improvements. Report on the Projected Viaduct from Fetter Lane to the Old Bailey, and Improvement of Farrington Market.*
By Francis Whishaw, C.E. Mem.Inst.Civ.Eng.

London: Published by Effingham Wilson . . . 1835.
Signed: Francis Whishaw. Gray's Inn, Aug 1835.
8vo   22 cm   8 pp.   2 folding litho plans.
[1] *Holborn Hill Improvements* [plan]. Francis Whishaw. R. Martin, Lithog.
[2] *Elevation of Part of the Proposed Viaduct* (scale 1 inch to 50 feet).
BM, CE

1779 *Hertfordshire Grand Union Railway. Report to the Promoters of a Proposed Line of Railway, between the Northern Railway and the London and Birmingham Railway . . . between the Towns of Watford, St. Alban's, Hatfield, Hertford and Ware . . .* [etc].
By Francis Whishaw, Esq., Civil Engineer, M.Inst.C.E.
Hertford: Published by St. Austin and Sons . . . 1836.
Signed: Francis Whishaw. Gray's Inn, 2 Dec 1836.
8vo   22 cm   15 pp.   Folding litho plan.
Hertfordshire Grand Union Railway. Plan of the District through which the Proposed Line is intended to pass . . .
By Francis Whishaw . . . 1836.
CE

1780 *Analysis of Railways: consisting of a series of reports on the twelve hundred miles of Projected Railways in England and Wales, now before Parliament . . . with a Glossary; and other useful information.*
By Francis Whishaw, Esq., Civil Engineer, M.Inst.C.E.
London: Published by John Weale . . . 1837.
8vo   23 cm   (i) + xv + 296 pp. and errata leaf.
BM, ULG, BDL

1781   Second edition
Analysis of Railways: consisting of a series of reports on the Railways Projected in England and Wales in the Year 1837 . . . with a Copious Glossary and several useful tables.
By Francis Whishaw . . . Second Edition, with Additions and Corrections.
London: John Weale . . . 1838.
Signed: Francis Whishaw. Gray's Inn, 31 Oct 1837.
8vo   23 cm   (i) + xv + 298 pp.   Engraved frontispiece.
SML, ULG, Elton

1782 *Lancaster and Penrith Proposed Railway.*
Report of Francis Whishaw, Civil Engineer, M.Inst.C.E.

n.p.

Signed: Francis Whishaw. Gray's Inn, 12 June 1840.

4to   28 cm   8 pp.   Folding litho plan, hand-coloured (R. Martin lith.).

Lancaster and Penrith Proposed Railway. Section of Gate Scarth with the Summit Tunnel in the Diverted Line as proposed by and taken under the directions of Mr Whishaw by Mr Bintley, Surveyor to the Kendal Committee.

CE

1783 *The Railways of Great Britain and Ireland practically described and illustrated.*
By   Francis   Whishaw,   Civil   Engineer. Mem.Inst.Civ.Eng.

London: Simpkin, Marshall and Co . . . 1840.

4to   28 cm   xxvi + 500 + lxiv pp.   Engraved dedication leaf + 16 engraved plates, mostly double-page or folding (F. Mansell and W. A. Beever sc.) + large folding engraved map.

Plan of the Railways of Great Britain and Ireland. 1840. Francis Whishaw.

Engraved by J. Dower.

BM, NLS, CE, ULG, BDL

1784   Additional plates

Three plates, published by Simpkin, Marshall and Co . . . 1841, with title page (as No. 1783 with *Additional Plates*) and explanatory leaf. The plates include a folding plan, hand-coloured:

Lengths, Summits and General Levels of the Main English Lines of Railway Communication. Datum, Trinity High Water, London. Francis Whishaw. H. P. Hughes del.   Engraved by J. Dower.

CE

A second edition of *The Railways of Great Britain and Ireland* was published in 1842.

WHISHAW, Francis and THOMAS, Richard

1785   *Report on Two Proposed Lines of Railway, between Perran Porth and Truro, in the County of Cornwall.*

By Francis Whishaw and Richard Thomas, Civil Engineers.

London: Published by Josiah Taylor . . . 1831.

Dated: 7 January 1831.

8vo   22 cm   14 pp.   Folding litho plan, hand-coloured.

Plan of the Intended Railway from Perran-Porth to Truro, exhibiting also another proposed Line by Zeala . . . January 1831. Francis Whishaw, Richard Thomas, Civil Engineers. R. Martin, Lithographer.

BM, CE

1786   Another edition

Report on the proposed lines of Railway from Perran-Porth to Truro, by Perran Alms'-House.

By Messrs Whishaw and Thomas.

Truro: Printed by John Brokenshir . . . 1831.

Dated: 7 January 1831.

8vo   22 cm   12 pp.   Folding engraved plan; title as in No. 1785 but Engraved by Mitchell, Truro.

BM

Only minor textual variations in the two editions.

WHITE, John [d. 1850]

1787   *On Cementitious Architecture, as applicable to the Construction of Bridges.*

By John White, Architect.

With a Prefatory Notice of the First Introduction of Iron as the Constituent Material for Arches of a Large Span, by Thomas Farnolls Pritchard, in 1773.

London: Printed by Richard Taylor . . . 1832.

8vo   24 cm   (iii) + 27 pp.   Woodcuts in

text.   2 folding engraved plates (G. Gladwin sc.)
CE, RIBA, ISE, BDL

1788   *An Essay on the formation of Harbours of Refuge, and the improvement of the Navigation of Rivers and Sea Ports, by the adoption of Moored Floating Constructions as Breakwaters of the Force of the Sea* [etc].
By John White.
London: J. Weale . . . 1840.
Dated: 1 Jan 1840.
8vo   23 cm   xiv + (1) + 59 pp.   7 litho plates (J. R. Jobbins).
NLS, CE, RIBA, BDL

WHITE, William [*c.* 1749–1817]

See JESSOP. Plan of part of the rivers Avon and Frome, surveyed by White, 1792.

1789   *A Plan of Kings-sedgemoor Drains and Parochial Allotments.*
Wm. White Surveyor.
Engraved plan   32 × 48 cm.
Somerset R.O., Sutro
   The Somerset R.O. copy endorsed Wm. White 1795.

See JESSOP. Designs for improving the harbour of Bristol, 1802 and 1803.

WHITEHAVEN HARBOUR

1790   *Plans suggested at different periods for the Improvement of Whitehaven Harbour, with Reports and Memorials connected therewith.*
Printed by Order of the Trustees.
Whitehaven: Printed by R. Gibson . . . 1836.
fol   46 cm   (38) pp.   19 litho plans.
CE, ULG

WHITTENBURY, William [d. 1757]

1791   *A Map of the River Lee, from Hertford Toll-Bridge to Ware Toll-Bridge.*
By Wm. Whittenbury 1733.
Engraved map   30 × 20 cm   (scale 4 inches to 1 mile).
BDL

1792   *A Plan of the River Lee from Hartford to Bow Bridge with a Profile of the Fall.* [*c.* 1740].
Engraved map   23 × 80 cm   (scale 1½ inches to 1 mile).
BM, RS (Smeaton)
   Minute Book, River Lee Trustees, August 1741: Whittenbury paid £25 for preparing the map as part of an account for work by him in the period June 1740 to June 1741.

WHITWORTH, Richard [1734–1811]

1793   *The Advantages of Inland Navigation; or, some Observations offered to the Public, to shew that an Inland Navigation may be easily effected between the three great Ports of Bristol, Liverpool, and Hull.*
By R. Whitworth, Esq.
London: Printed for R. Baldwin . . . 1766.
Signed: Rich. Whitworth. Batchacre Grange, 20 Feb 1766.
8vo   22 cm   vi + 73 + (1) pp.   4 tables
   Engraved   frontispiece   (p. i)   Folding engraved plan, the lines entered in colour.
A Plan for a Navigation from the River Severn, near Tern Bridge, in the County of Salop; By . . . Stafford to Burton, on the River Trent . . . Thence to Wilden Ferry in the County of Derby, and from or near Bridgeford . . . in the County of Stafford, to Winsford in the County of Chester to unite with the Wever, near the River Mersey.
   C. Grignion sc.
BM, CE, ULG
   This is the second issue with corrected text.

The first issue (only one copy seen, lacking frontispiece and plan) has identical title-page and pagination but includes a list of errata.

WHITWORTH, Robert [1734–1799] M.Soc.C.E.

See BRINDLEY. Plans of Coventry Canal 1767, Droitwich Canal 1767 and Oxford Canal 1768. Drawn and probably surveyed by Whitworth.

1794 *A Plan and Estimates of the intended Navigation from Lough-Neagh to Belfast, as surveyed by Mr Robert Whitworth: Together with his Report, concerning the best Method of executing the Work.*
> Presented to the local Committee ... and approved by Mr James Brindley, Engineer.
Belfast: Printed by Henry and Robert Joy. 1770.
Signed: Rob.ͭ Whitworth. Lisburn, 24 Aug 1768.
8vo 21 cm 35 pp. Folding engraved plan.
A Plan of part of the River Lagan, and of the intended Navigable Canal from Lough Neagh to Belfast.
> Surveyed ... in 1768. By Rob.ͭ Whitworth. Bowen sc.
CE, BDL (Gough, plan only)

See BRINDLEY and WHITWORTH. Report on a proposed canal from Stockton by Darlington to Winston, 1768–69. Printed 1770.

See BRINDLEY. Plan of the navigable canals now making ... for opening a communication to the ports of London, Bristol, Liverpool and Hull, 1769. Drawn by Whitworth.

1795 *A Plan for a Navigable Canal from Taunton in the County of Somerset to Tiverton Exeter and Topsham in the County of Devon.*
> Surveyed in 1769 by Rob.ͭ Whitworth. J. Cary sc.
Engraved plan 31 × 78 cm (scale 1 inch to 1 mile).
BM

See BRINDLEY. Plan of the intended canal in Berkshire from Sunning to Monkey Island, 1770. Drawn and surveyed by Whitworth.

1796 *A Plan of the intended Navigable Canal from Andover to Redbridge in the County of Southampton.*
> Survey'd in March 1770. Rob.ͭ Whitworth del.
Engraved plan 26 × 62 cm (scale 1 inch to 1 mile).
BM

See BRINDLEY. Plan of the Thames from Kennets Mouth to London shewing the intended canal from Sunning Lock ... to Isleworth, 1770. Drawn by Whitworth. Also the revised scheme from Reading. 1770.

See BRINDLEY. Profile and Plan of the River Thames from Boulters Lock to Mortlake, 1770. Drawn and surveyed by Whitworth.

1797 *A Plan of the intended Navigable Canal from the Leeds and Liverpool Canal near Eccleston in the County Palatine of Lancaster to Kendall in Westmoreland.*
> Surveyed by Robert Whitworth. Engraved by Thomas Jeffreys ... 1772.
Engraved plan 31 × 90 cm (scale 1 inch to 1⅓ miles).
BM

1798 *A Report and Survey of the Canal, proposed to be made on one Level, from Waltham-Abbey to Moorfields. Also a Report and Survey, of a Line, which may be continued from Marybone to the said proposed Canal ... By Robert Whitworth. To which is subjoin'd, an Address to the Right Honourable the Lord-Mayor ... By James Sharp.*
> Corporation of London, 1773.
Signed: Rob.ͭ Whitworth. London, 5 Oct 1773.
James Sharp. Leadenhall-Street, 22 Dec 1773.
fol 32 cm (i) + 8 + (i) + 16 pp. 2 folding engraved plans Survey'd by Order of the City of London. By Rob.ͭ Whitworth. Engrav'd by Faden and Jefferys 1773.
[1] *Plan and Profile of the Intended Navigable Canal, from Moor-Fields, into the River Lee at Waltham Abby.*

[2] *Plan and Profile of the Intended Navigable Canal from Mary-le-Bone to Moor-Fields: Being a Continuation of the Propos'd Canal from Uxbridge.*

BM, CLRO, ULG, Elton

Issued in pink boards. City Navigation Committee, 5 Oct 1773: Thos. Jefferys to make an engraving of Mr Whitworth's plan and 1,000 copies to be printed. Mr Costin to print the reports.

1799 Second edition

Title as No. 1798.

Corporation of London, 1774.

Text as in first edition with additional pages by James Sharp dated 29 Jan 1774.

fol 33 cm (i) + 8 + (i) + 18 pp. 2 folding engraved plans (as in first edition).

BM, CE, ULG (lacking plans), BDL (Gough)

1800 *A Plan of the River Thames from Boulter's Lock to Mortlake. Surveyed by order of the City of London in 1770 by James Brindley Engineer. Revised and continued to London Bridge in 1774 by Rob.t Whitworth.*

Engraved plan 66 × 143 cm (scale 2 inches to 1 mile).

BM

1801 Later edition

As No. 1800 but with notes on *Some Account of Improvements made in the Navigation at the sole Expence of the City of London from August 1774 to April 1777.*

BM, SML, BDL (Gough)

1802 *A Table of Distances of the several Towns and Bridges upon the River Thames between London Bridge and Boulters Lock near Maidenhead; together with the usual time of Navigating loaded Barges, both upwards and downwards; supposing that there is no uncommon hindrance by Winds, Floods, or want of Water.*

By Robert Whitworth. Engraved by William Faden 1779.

Engraved table 24 × 49 cm

CE (Page)

1803 [Report] *To the Worshipful Committee for Thames and Canal Navigation.*

Corporation of London, 1780.

Signed: Robert Whitworth, 10 Oct 1780.

fol 33 cm 9 pp. 2 folding engraved plans, both 'Survey'd by Order of the City of London in 1779 and 1780 by Robert Whitworth, Engineer. Engraved by Wm. Faden'.

[1] *A Plan of the propos'd Navigable Canal, from Bishops Stortford to Cambridge.*

[2] *A Profile of the Line of the proposed Navigable Canal from Bishops Stortford to Cambridge.*

BM, CLRO, CE (lacking plans), BDL (Gough, plans only)

1804 *A Plan of the intended Navigable Canal from the Coventry Canal near Griff in the County of Warwick, to the Coal Mines at Measham, Oakthorp, Donisthorp and Ashby-Woulds, in the Counties of Leicester and Derby.*

Survey'd in 1781 by Rob.t Whitworth, Eng.r Engraved by W. Faden.

Engraved plan 27 × 56 cm (scale 1 inch to 1 mile).

BM, SML, CE

1805 *A Plan of the Thames and Severn Canal Navigation.*

Laid down from Actual Surveys. By Rob.t Whitworth Engin.r 1783. Engraved by Wm. Faden.

Engraved plan, the line in colour 27 × 80 cm (scale 1 inch to 1 mile).

BM, SML, RS (Smeaton), CE (Page), BDL

1806 *A Plan and Profile of the proposed Navigable Canal from the Line of the Thames and Severn Canal in the Parish of Kempsford and County of Gloucester, to the River Thames at Abingdon in the County of Berks.*

Survey'd in 1784 by Rob.t Whitworth, Eng.r Engraved by Wm. Faden.

Engraved plan 34 × 63 cm (scale 1 inch to 1 mile).

BM, CE (Page), BDL

1807 *Report of Robert Whitworth, Esq; Engineer; to the Company of Proprietors of the Forth and Clyde Navigation. Relative to the Tract of the Intended Canal, from Stockingfield westward, and different Places of Entry into the River Clyde.*

n.p.

Signed: Rob.! Whitworth. Edinburgh, 2 Aug
  1785.
4to   25 cm   (i) + 20 + 5 pp.
NLS

1808   *A Plan [and Profile] of the Great Canal from
Forth to Clyde.*
  By Rob.! Whitworth Esq.! and Mr John
Laurie.
[with inset] *Plan for a Reservoir proposed to be made in
Dolater Bogg.*
  Drawn and Engraved by John Ainslie Land
  Surveyor 1785.
Engraved plan   58 × 186 cm   (scale 2 inches to
  1 mile).
NLS

1809   *Report of Robert Whitworth, respecting the
South and North Lines for the Forth and Clyde Navigation, to the Westward of the River Kelvin.*
n.p.
Signed: Rob.! Whitworth. Glasgow, 28 Sept 1785.
4to   26 cm   4 pp.
Scottish Record Office

See WHITWORTH and JESSOP. Plan of
  intended canal from Thrinkston Bridge to
  Leicester, 1785.

1810   *A Plan and Profile of the proposed Navigable
Canal from Melton Mowbray in the County of Leicester,
to Oakham in the County of Rutland; and Stamford in
the County of Lincoln.*
  Surveyed in 1786. By Rob.! Whitworth,
  Engineer.
Engraved plan   40 × 81 cm   (scale 1 inch to 1
  mile).
BM, Sutro

1811   *Report of Robert Whitworth, Engineer, to the
Company of Proprietors of the Forth and Clyde Navigation; relative to the Supplies of Water necessary to be
taken into the Canal . . . Also, of the Junction of the
Great Canal with the Monkland Canal.*
Glasgow: Printed by Robert Chapman and Alexander Duncan. 1786.
Signed: Rob.! Whitworth. Edinburgh, 1 Aug
  1786.

4to   28 cm   (i) + 20 pp.
Scottish Record Office

1812   *A Plan [and Profile] of the Great Canal from
Forth to Clyde.*
  By Rob.! Whitworth Engin.! and Mr John
Laurie.
[with inset] *Plan of the Monkland Canal and the
several Reservoirs Lochs & Feeders to the Great Canal.*
[1786].
  Engraved plan, the lines entered in colours
  58 × 186 cm   (scale 2 inches to 1 mile).
  BM
  Revised (second state) of No. 1808.

1813   *Report and Estimates, relative to the Enlarging
of the Harbour of Leith, with Plans of a Bason above
Leith-mills, and another between the North Pier and the
Citadel, with Sections of parts of the Work.*
  By Robert Whitworth, Esq; Engineer.
n.p.
Signed: Rob.!. Whitworth, 7 Nov 1786.
4to   26 cm   (i) + 18 pp.   Folding engraved
  plan.
*A Plan for the improvement of the Harbour of Leith either
by Making a Bason or Wet Dock . . . South of Leith
Mills or . . . upon North Leith Sands.*
  By Rob.!. Whitworth. Reduced from the Origi-
  nal Drawing & Engraved by J. Ainslie.
BM, Leith Docks Office

1814   *Observations by Robert Whitworth, Esq.
Engineer, on Several Schemes and Plans proposed for
Enlarging and Improving the Harbour of Leith.*
n.p.
Signed: Rob.!. Whitworth. Edinburgh, 8 Jan 1787.
4to   26 cm   (i) + 25 pp.
BM

1815   *Mr Whitworth's Report to the Company of
Proprietors of the Leeds and Liverpool Canal.*
Signed: Robert Whitworth. Liverpool, 9 Oct
  1789.
p. 2.

[With]
Report of John HUSTLER, William BIRK-
  BECK and Jo. PRIESTLEY, 9 Oct 1789
[q.v.].

Leeds & Liverpool Canal. At a General Assembly of the Proprietors . . . held at Liverpool 9th and 10th October 1789 [etc].

n.p.

fol   32 cm   3 pp.

BM, Lancashire R.O.

1816   [Report] *To the Company of Proprietors of the Borrowstounness Canal.*

n.p.

Signed: Rob.! Whitworth. Glasgow, 28 Dec 1789.

4to   28 cm   15 pp.   Folding engraved plan.

A Plan of the proposed Navigable Canal from the Forth & Clyde Canal to Borrowstounness.

Surveyed in Dec.! 1789 by Rob.! Whitworth.

NLS

Issued in grey wrappers.

1817   *A Plan of the Great Canal from Forth to Clyde.*

As finished by Rob.! Whitworth Eng.! Anno 1790.

Engraved plan   58 × 187 cm   (scale 2 inches to 1 mile).

NLS

Further revised (third state) of Nos. 1808 and 1812.

1818   *Report to the Heritors of the County of Roxburgh, respecting the Practicability and Expence of making a Navigable Canal from Berwick to Kelso and Ancrum-Bridge.*

Kelso: Printed by James Palmer. Anno 1792.

Signed: Robert Whitworth. Kelso, 4 Feb 1790 and 22 April 1791.

4to   27 cm   27 pp.

NLS

See WHITWORTH, Robert and William. Plan of the proposed deviation line of the Leeds and Liverpool Canal, 1792.

1819   *A Plan of the intended Navigable Canal from Ashby de la Zouch to join and communicate with the Coventry Canal at or near Griff, with the Proposed Navigable Cuts or Branches from Ashby de la Zouch to or near the Lime works at Ticknal; Lime works, Lead mines and Coal Mines at Staunton Harold; Lime works at Cloudhill and Coal mines at Coleorton.*

Surveyed in 1792, by Robert Whitworth, Engineer; John Smith, Surveyor. Downes sc.

Engraved plan   46 × 66 cm   (scale 1 inch to 1 mile).

BM

1820   Later state

As No. 1819 but with table of lengths and falls.

BM, CE (Page)

See WHITWORTH, Robert and William. Plans of the Wilts and Berks Canal, 1793 and 1794.

See WHITWORTH and JESSOP. Plan of the Ashby de la Zouch canal, 1794.

See WHITWORTH and MYLNE. On the proposed London Canal from Boulter's Lock to Isleworth, 1794.

1821   *Report on the Proposed Line of Navigation from Stella to Hexham, and from Hexham to Haydon-Bridge, on the South Side of the River Tyne.*

By Rob.! Whitworth, Engineer.

Newcastle: Printed for John Bell . . . 1797.

Signed: Rob.! Whitworth. Burnley, 23 Feb 1797.

8vo   20 cm   9 pp.

CE (Vaughan), ULG (Rastrick), BDL

1822   [Report] *To The Committee of Management of the proposed Canal from Newcastle to Haydon Bridge, upon the North Side of the River Tyne.*

Signed: Robt. Whitworth, Sen. Burnley, 23 Feb 1797.

pp. 3–16.

[In]

Mr Whitworth's Report on the Proposed Canal on the North Side of the River Tyne, with Observations on that Report; together with various Publications and Petitions.

Newcastle: Printed by Joseph Whitfield . . . 1797.

8vo   20 cm   (i) + 101 pp.

CE (Vaughan), BDL

WHITWORTH, Robert and JESSOP, William

1823 *A Plan of the intended canal and River Navigation from Thrinkston Bridge to Leicester: shewing its communication with the River Wreak to Melton . . .*
Surveyed in 1785.
Engraved plan   37 × 54 cm   (scale 1 inch to 1 mile).
BM, CE (Page), Leicestershire R.O., BDL
Whitworth and Jessop reported on this scheme in January 1786.

1824 *Messrs Whitworth & Jessop's Report and Estimate for improving the River Trent Navigation.*
n.p.
Signed: Robert Whitworth, William Jessop. Nottingham, 8 July 1793.
fol   33 cm   4 pp.
PRO

1825 *A Plan of the Ashby de la Zouch Canal with the Cuts or Branches therefrom.*
R. Whitworth, W. Jessop Engineers. J. Smith Surveyor. 1794. Downes sc.
Engraved plan   42 × 62 cm   (scale 1 inch to 1 mile).
BM

WHITWORTH, Robert and MYLNE, Robert

1826 [Report] *To the Gentlemen Subscribers for the London Canal, from Boulter's Lock to Isleworth on the River Thames.*
Signed: Robert Whitworth, Robert Mylne. 17 July 1794.
pp. (5)–15.
[In]
Extract from the Minutes of the Proceedings of the Subscribers to the intended London Canal, presented to the Commissioners of the Thames Navigation . . . at Henley

on Thames . . . the 1st day of August 1794. And also a Report of the Engineers appointed . . . to survey and determine in what Manner, the Navigation may be best improved from Reading to Isleworth.
[London]: Printed by Order of the Commissioners 1794.
8vo   20 cm   15 pp.
CE (Page), ULG

1827 *Plan of the proposed London Canal from a Place call'd Hog Hole in the River Thames in the Parish of Datchet . . . to a place in the said River call'd the Rails Head in the Parish of Isleworth . . .*
Surveyed 1794. Engraved by W. Woodthorpe.
Engraved plan, hand coloured   42 × 77 cm (scale 2 inches to 1 mile).
CE
An abbreviated form of the Boulters–Isleworth canal.

WHITWORTH, Robert and William

1828 *Plan of the proposed Deviation Line of the Leeds and Liverpool Canal from near Colne, to Wigan.*
By Rob! & Wm Whitworth 1792. Engraved by W. Faden.
Engraved plan, the line entered in colour   29 × 78 cm   (scale 1 inch to 1 mile).
BM

1829 *A Plan of the Wilts and Berks Canal from Trowbridge in the County of Wilts to Abingdon in the County of Berks, with Branches to the Towns of Chippenham, Calne, and Wantage.*
By Rob! & Willm Whitworth, Engineers 1793. Engraved by W. Faden 1795.
Engraved plan, the line entered in colour   42 × 121 cm   (scale 1 inch to 1 mile).
BM, BDL

1830 *A Plan of the Wiltshire and Berkshire Canal, with a Branch to the Thames & Severn Canal, also Branches to the Towns of Calne and Chippenham.*
By R. & W. W.   W.F. 1794.

Engraved plan   24 × 40 cm   (scale 1 inch to 4
  miles).
BM, CE (Page)

WHITWORTH, Robert *junior*

See AINSLIE and WHITWORTH. Report on
  proposed canal betwixt Edinburgh and Glasgow, [1794].

WHITWORTH, William

See WHITWORTH, Robert and William. Plan of
  the proposed deviation line of the Leeds and
  Liverpool Canal, 1792.

See WHITWORTH, Robert and William. Plans
  of the Wilts and Berks Canal, 1793 and 1794.

1831   *A Plan of the intended Severn Junction Canal.*
  W.ᵐ Whitworth Engineer 1810. W.F. 1810.
Engraved plan, the line entered in colour
  36 × 83 cm   (scale 4 inches to 1 mile).
BM

1832   [Report] *To the Chairman of the Committee of
Management of the Wilts and Berks Canal.*
Signed: W. Whitworth. Watchfield, 9 March
  1811.
p. 2.
                [In]
*Proposed Severn Junction Canal, from the Wilts and
Berks Canal, near Swindon, to the Thames and Severn
Canal, at Latton.*
  Single sheet   38 × 22 cm   2 pp.
BM

1833   *A Plan of the intended Bristol Junction Canal,
from the Wilts and Berks Canal near Wotten Basset in
the County of Wilts to the Bristol Dock Companys Canal
near the City of Bristol.*

By Wm. Whitworth Civil Engineer 1811.
W.F. 1811.
Engraved plan, the line entered in colour
  53 × 164 cm   (scale 2 inches to 1 mile).
BM

WICKINGS, William

1834   *Plan of a road intended to be made from the West
India Docks in the Isle of Dogs to communicate with
Aldgate High Street in the City of London: to be called
the Commercial Road.*
  Surveyed by W. Wickings. Oct.ʳ 1801.
Engraved plan, the line entered in colour
  30 × 62 cm   (scale 8 inches to 1 mile).
PLA

WICKSTEED, Thomas [1806–1871]
M.Inst.C.E.

1835   *Observations on the Past and Present Supply of
Water to the Metropolis.*
  By Thomas Wicksteed, Civil Engineer.
London: Printed by Richard Taylor . . . 1835.
Dated: 12 May 1835.
4to   30 cm   (i) + 20 pp.
CE

WILLIAMS, John [d. 1811]

See HASSALL and WILLIAMS. Road from Milford to Gloucester, 1792.

WILLIAMS, Philip

1836 *Plan of an Intended Dram Road from or near Carno Mill in the Parish of Browellty & County of Monmouth, to or near the Sea Lock below the Town of Cardiff, with a Branch from the same to the Limestone Rocks in the Parish of Merthyr Tydvill & County of Glamorgan.*

By P. Williams 1799.

Engraved plan 30×78 cm   (scale $1\frac{1}{4}$ inches to 1 mile).

BM, CE, RGS

WILSON, Thomas

1837 *Mr Wilson's Estimate or Specification of a Cast Iron Bridge over the Thames.*

Signed: T. Wilson.

p. 76   with folding engraved plate.

Mr Wilson's Design for a Cast Iron Bridge over the Thames instead of the Present.

(Laurie & Whittle sc.).

[In]

Third Report from the Select Committee upon the Improvement of the Port of London.

House of Commons, 1800.

The plate (No. 8) issued in a separate volume.

1838 Second printing

In *Reports from Committees of the House of Commons* Vol. 14 (1803) p. 574 and pl. 37.

1839 *Observations by Mr T. Wilson [on the design of a Cast Iron Bridge of three arches over the Thames].*

Signed: Tho.s Wilson. Bentinck-street [London], 13 April 1801.

[In]

Report from the Select Committee upon Improvement of the Port of London.

1801. pp. 83–85.

1840 Second printing

In *Reports from Committees of the House of Commons* Vol. 14 (1803) p. 635.

WING, John [1723–1780]

1841 *An Actual Survey of the North Level, Part of the Great Level of the Fens call'd Bedford level. Also of Crowland, Great Porsand, and Part of South Holland in the County of Lincoln. And of Wisbeach North-side in the Isle of Ely . . . Wherein is Described the Several Drains, Sewers, Sluices &c by which the Lands contain'd in this Survey Drain to their Outfalls at Sea.*

Taken Aug.t 1749 by Jn.o Wing.

Engraved   map,   partly   hand   coloured 52×72 cm   (scale $1\frac{1}{3}$ inches to 1 mile).

BM, LCL, RS (Smeaton)

WING, John *junior* [1752 –1812]

1842 *An Inquiry into the State of the Revenues of the North Level, part of the Great Level of the Fens called Bedford Level, and the Causes of the Distresses of that Level during the last Twenty-Five Years. With an Inquiry into the Probable Means of Preventing the Like in Future.*

By J. Wing.

Peterborough: Printed by J. Jacob. 1788.

Signed: J. Wing, 17 Sept 1788.

4to   22 cm   19 pp.

BM, CRO, CUL

WING, Tycho [1794–1851]

1843 *Considerations on the Principles of Mr Rennie's Plan for the Drainage of the North Level, South Holland, &c. With a View to their Practical Adoption.*

By Tycho Wing.

Peterborough: Printed by C. Jacob . . . 1820.

Dated: August 1820.

8vo   23 cm   (i) + 30 pp.

BM, CE, CRO, LCL

1844 *Memoir of the Nene Outfall and the North Level Drainage.*

By Tycho Wing, Esq.

Dated: 1837.

[In] TELFORD. *Life of Thomas Telford*, London, 1838 [q.v.]. pp. 320–323.

WINGROVE, Benjamin

1845 *Report of Benjamin Wingrove, to the Trustees of the Taunton Turnpike Roads.*

Taunton: Printed by C. H. Drake . . . [1819].

Signed: Benjamin Wingrove. Roads Office, Bath, 9 July 1819.

4to   27 cm   10 pp.

Somerset R.O.

1846 *Remarks on a Bill now before Parliament, to amend the General Laws for regulating Turnpike-Roads; in which are introduced, Strictures on the Opinions of Mr M'Adam on the subject of roads* [etc].

By Benjamin Wingrove, Surveyor of the Bath Turnpike-Roads.

Bath: Printed by Richard Cruttwell . . . 1821. Price 1/6^d.

8vo   21 cm   35 pp.

ULG

WOOD, Nicholas [1795–1865] F.R.S., M.Inst.C.E.

1847 *A Practical Treatise on Rail-Roads, and Interior Communication in General; with Original Experiments, and Tables of the comparative value of Canals and Rail-Roads.*

By Nicholas Wood, Colliery Viewer.

London: Printed for Knight and Lacey . . . 1825.

Dated: Killingworth, April 1825.

8vo   22 cm   (iii) + 314 pp.   6 folding engraved plates.

BM, SML, CE, ULG

Issued in boards, with printed paper label.

1848   Second edition

A Practical Treatise on Rail-Roads . . . containing an account of the performances of the different Locomotive Engines at and subsequent to the Liverpool Contest; upwards of two hundred and sixty experiments; with Tables . . . [etc].

By Nicholas Wood, Colliery Viewer, Mem. Inst.Civ.Eng.

London: Hurst, Chance & Co . . . 1831.

Dated: Killingworth, 22 Feb 1831.

8vo   22 cm   xxiii + 530 pp. + errata leaf and directions to binder   8 folding plates   + 3 wood engraved plates.

BM, SML, CE, ULG

Issued in boards, price 18s.

1849   Second edition, second issue

As No. 1848 but *Second Edition* and London: Printed for Longman, Rees [etc]. 1832.

BM, ULG

Text and illustrations unchanged from the 1831 issue.

1850   Third edition

A Practical Treatise on Rail-Roads . . . containing numerous experiments on the powers of the improved Locomotive Engines: and Tables of the comparative cost of conveyance on Canals, Railways, and Turnpike Roads.

Third Edition, with Additions. By Nicholas Wood, Colliery Viewer, Mem.Inst.Civ.Eng.

London: Printed for Longman, Orme [etc] . . . 1838.

Dated: Killingworth, 8 June 1838.

8vo   23 cm   xxvii + 760 pp.   13 folding engraved plates (H. Adlard sc.)

BM, SML, CE, ULG (Rastrick)

Issued in publisher's cloth, embossed, with gilt lettering.

See THOMPSON, JOHNSON and WOOD. Reports on Newcastle & Carlisle Railway, 1834–1841.

1851 *Report of Nicholas Wood, Esq. To the Directors of the Great Western Railway.*

13. Part of a plan of the Caledonian Canal by Telford and Downie, 1805.

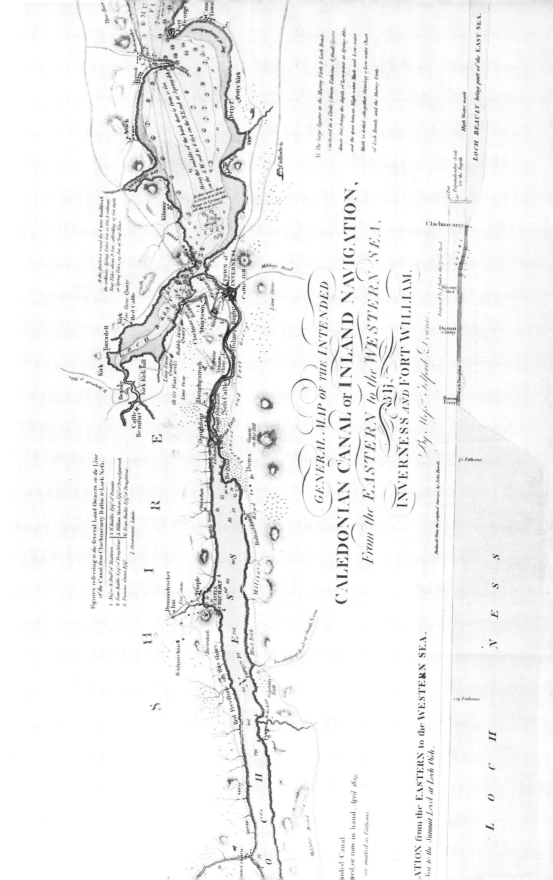

REPORT AND ESTIMATE

ON THE IMPROVEMENT

OF THE

Drainage and Navigation

OF THE

SOUTH & MIDDLE LEVELS

OF THE

Great Level of the Fens.

IN CONSEQUENCE OF A REFERENCE

FROM THE HONORABLE CORPORATION

OF THE

BEDFORD LEVEL,

Made the 24th May 1809,

By JOHN RENNIE, CIVIL ENGINEER.

***

LONDON:

PRINTED BY E. BLACKADER, TOOK'S COURT, CHANCERY LANE.

1810.

14.   Report by Rennie on the South and Middle Levels, 1810.

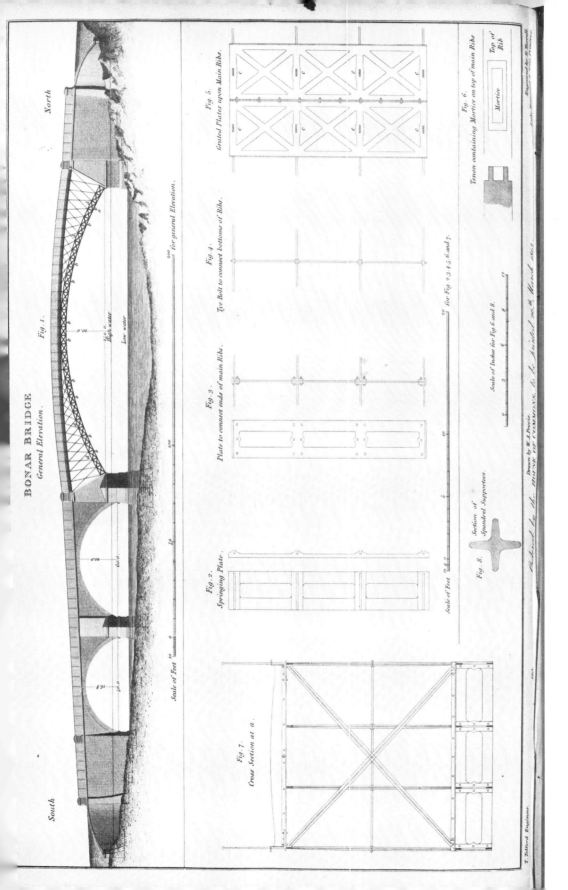

15. Elevation and details of Bonar Bridge by Telford, 1813.

REPORT

ON THE

IMPROVEMENT

OF THE

HARBOUR OF DUNDEE,

BY

ROBERT STEVENSON, Esq.

PRINTED BY ORDER OF THE MAGISTRATES AND
TOWN-COUNCIL OF DUNDEE.

1814.

16.   Report by Stevenson on Dundee Harbour, 1814.

Holyhead Road.

REPORTS OF MR. TELFORD

TO THE

Commissioners for the Improvement of the Holyhead Road,

UPON THE STATE OF

THE ROAD BETWEEN LONDON AND SHREWSBURY;

WITH MAPS, AND ESTIMATES.

1. REPORT - - - dated 5 June 1820.
2. REPORT - - - dated 30 June 1819.

Ordered, by The House of Commons, *to be Printed,*
5 *June* 1820.

17.   Reports by Telford on the Holyhead Road, 1820

REPORT

BY

Messrs GRAINGER & MILLER, Civil Engineers

To the Committee of the Subscribers to the Proposed Railway between the Cities of Edinburgh and Glasgow, and the Port of Leith.

Gentlemen,

By our "Observations" on the formation of a Railway between the Cities of Edinburgh and Glasgow, and Port of Leith, published on the 13th of October last, we had the honour of directing your attention to this important communication; and, in accordance with your instructions, we have since completed a *Particular Survey* of the line we then generally pointed out, and have now to lay the Plans and Sections of the same before you.

Improvement in the means of conveyance between the Cities of Edinburgh and Glasgow, and their respective Ports, must always be regarded as an object of the greatest importance, not only to these Cities, but also to the Agricultural, Commercial, and Manufacturing interests on both sides of the Island, and more especially, when the intercourse with these Ports, has been so perfectly established by means of Steam Navigation. This means of intercourse is now carried on to a greater extent on the River Clyde, than on any other river in Europe, and insures a regular and expeditious means of conveyance between Glasgow and Ireland; and the west coast of England. By it, also, a like regular and expeditious means of conveyance has been established between Leith, and the east coast of Scotland; also to London, and the whole east coast of England.

When conveyance is thus found to be in such a perfect state on the Forth and Clyde, the establishment of such a connexion between their two chief Ports as would ensure a regular, cheap, expeditious and uninterrupted means of carrying on the trade already established with them, is certainly an object of the greatest importance to the country.

To a certain extent, this has already been accomplished by means of the Forth and

d

Clyde

ABERDEEN HARBOUR.

REPORT

OF

JAMES WALKER,

CIVIL ENGINEER,

ON THE

HARBOUR AND PROPOSED DOCKS.

PRINTED AND CIRCULATED

BY ORDER OF

THE HARBOUR TRUSTEES.

April, 1838.

ABERDEEN :
PRINTED AT THE HERALD OFFICE,
BY J. FINLAYSON.

MDCCCXXXVIII.

19.   Printed wrapper of Walker's report on Aberdeen Harbour, 1838.

0.   Warrington Viaduct: engraved plate from Roscoe *Grand Junction Railway,* 1839.

21.   New Cross Cutting: lithograph plate from Trotter *Croydon Railway,* 1839.

EXPERIMENTAL RESEARCHES

ON

THE STRENGTH

OF

PILLARS OF CAST IRON,

AND OTHER MATERIALS.

BY

EATON HODGKINSON, Esq.

From the PHILOSOPHICAL TRANSACTIONS.—PART II. FOR 1840.

LONDON:

PRINTED BY R. AND J. E. TAYLOR, RED LION COURT, FLEET STREET.

1840.

22.   Offprint of Eaton Hodgkinson's paper on the strength
of iron columns, 1840.

Newcastle: Printed by John Hernaman.
Dated: Killingworth, 10 Dec 1838.
8vo   22 cm   82 pp.
BM, SML, CE, ULG
  Issued with reports by Hawkshaw and Brunel,
  and an Introductory Letter by Wood, dated 5
  Oct 1838.

WOODHOUSE, Thomas Jackson [1793–1855] M.Inst.C.E.

1852   *First (and Second) Report on the Improvement of Belfast Harbour.*
  By Thomas J. Woodhouse, Esq. C.E.
Belfast: George Harrison . . . 1834.
Signed: Thos. J. Woodhouse. Belfast 10 Sept 1834
  [and] 22 Oct 1834.
8vo   22 cm   9 [and] 4 pp.   Folding litho plan,
  hand-coloured.
Plan for the Improvement of Belfast Harbour.
  By T. J. Woodhouse, Civil Engineer. Belfast,
  1834. McBrair Lithog.
CE

WOOLER, John

See SMEATON and WOOLER. Report on Tyne
  Bridge, 1772.

1853   *To Sir Walter Blackett, Bart. Mayor, the
Aldermen, and Common Council of Newcastle upon
Tyne.*
n.p.
Signed: John Wooler. Newcastle 17 March 1772.
8vo   21 cm   8 pp.
P
  Observations on Robert MYLNE's report of 12
  March 1772 on Tyne Bridge [q.v.].

WORTHINGTON, *Lieut.* Benjamin, R.N.

1854   *Proposed Plan for Improving Dover Harbour, by
an Extension of the South Pier Head.*
  By Lieut. B. Worthington, R.N.
Dover: Printed by W. Batcheller . . . 1838.
Signed: B. Worthington. Dover, 14 April 1838.
8vo   23 cm   (i) + 174 pp.   7 plans including:
Plan of Dover Harbour 1838. J. Netherclift lith.
East View of Breakwater, Piers, and Sea Wall. Selater
  sc.
BM, SML, CE, ULG
  Includes extracts from reports by Perry (1718),
  Smeaton (1769), Nickalls (1782), Ralph
  Walker (1812).

WYATT, Robert Harvey [1769–1836]

1855   *Plan of the River Trent from Burton to Caven-
dish Bridge.*
  Rob.! H. Wyatt, Surveyor.   Jn.º Cary,
  Engraver. [*c.* 1792].
Engraved map   23 × 60 cm   (scale 1½ inches to
  1 mile).
BM

1856   *Plan of intended Canal from Burton upon Trent
to Fradley Heath to join the Coventry Canal.*
  Rob.! H. Wyatt, Surveyor.   Jn.º Cary,
  Engraver. [*c.* 1793].
Engraved map   22 × 37 cm   (scale 1½ inches to
  1 mile).
BM, CE (Page)

WYATT, Samuel [1737–1807] M.Soc.C.E.

1857   *Explanations, shewing the peculiar Advantages
to be attained by Adoption of the Docks at the Isle of*

Dogs, and extending the Limits of the Port of London.
Signed: Sam. Wyatt.
With 2 folding engraved plans.
[1] *Mr S. Wyatt's Plan. The Proposed London Docks compared with those proposed at the Isle of Dogs*. J. Cary sc.
[2] *A Design shewing the manner of bringing the Kings Beam to the ships, by means of a floating Platform*.
[In]
Report from the Committee on the Port of London.
House of Commons, 1796. Appendix Oo. pp. 103–109 and pls. 6 and 7.

1858 Second printing
In *Reports from Committees of the House of Commons* Vol. 14 (1803) pp. 400–402 and pls. 6 and 7.

YARRANTON, Andrew [1616–c. 1684]

1859 *England's Improvement by Sea and Land. To Out-do the Dutch without Fighting . . . With the Advantages of making the Great Rivers of England Navigable* [etc].
By Andrew Yarranton, Gent.
London: Printed by R. Everingham for the Author . . . 1677.
Dated: 4 Oct 1676.
4to   19 cm   (xx) + 195 pp.   8 folding engraved plates.
BM, NLS, SML, ULG, BDL (Gough)
The numbers 73–96 are missing from the pagination. In some copies (presumably a first issue) pp. 193–195 are incorrectly numbered 177–179.

1860 *England's Improvement by Sea & Land. The Second Part. Containing I. An Account of Its Scituation, and the Growths and Manufactures thereof . . . VI. The Way to Make New-haven in Sussex, fit to Receive Ships of Burthen* [etc].
By Andrew Yarranton.
London: Printed for the Author . . . 1681.

Signed: Andrew Yarranton, 2 Feb 1680/1.
4to   19 cm   (viii) + 212 pp.   7 folding engraved plates.
BM, SML, ULG, BDL
The numbers 121–128 are missing from the pagination.

YEOMAN, Thomas [?1708–1781] F.R.S., M.Soc.C.E.

1861 *A Plan of the River Nen, from Thrapston to Northampton with the Mills and Locks necessary for the Navigation, together with the Towns near the River. The Dotted Lines shew on which side of the River the Locks are to be fixed*.
Engraved plan   30 × 19 cm   (scale 1 inch to 3 miles).
Northampton Central Library
This plan is referred to in the *Northampton Mercury* 30 Dec 1754. Yeoman gave evidence on the Bill in Parliament in January 1756.

See DALLAWAY. A Scheme to make the River Stroudwater Navigable, 1755. With an Estimate of the General Expenses by John Willets, Tho. Yeoman and others.

1862 *A Plan of the Road from Towcester in the County of Northampton, to Weston Gate in the County of Oxford*.
Surveyed by T. Yeoman.
Engraved plan   21 × 36 cm   (scale 1 inch to 1 mile).
Northants R.O.
J.H. of C., 23 Dec 1756; Thomas Yeoman has surveyed the road, and gives evidence. The Act (30 Geo II, c.48) was obtained in 1757.

1863 [Report on] *A Survey of the River Chelmer, from Chelmsford to Maldon*.
By Thomas Yeoman.
pp. 94–102.
[In]
A New and Complete History of Essex.

By a Gentleman [Peter Muilman]. Vol. 1.
Chelmsford: Printed and Sold by Lionel Hassall. 1771.

8vo   21 cm   xviii + 465 pp.   Plates + 2 folding engraved maps.

BM, Essex R.O., BDL (Gough)

The report, with estimates, describes the 1762 scheme for making the river navigable as shown in the following plan. The book, however, includes a copy of the 1765 plan [q.v.].

1864   *A Plan of the River Chelmer from Chelmsford to Maldon in the County of Essex.*

Surveyed by Thomas Yeoman, 1762.

Engraved plan   22 × 38 cm   (scale $1\frac{3}{4}$ inches to 1 mile).

BM

Shows the proposed locks and cuts, totalling 5 miles.

1865   Revised scheme

A Plan of the River Chelmer from Chelmsford to Maldon in the County of Essex.

Surveyed by Thomas Yeoman, 1765.

Engraved plan   22 × 38 cm   (scale $1\frac{3}{4}$ inches to 1 mile).

BM

Total length of cuts now reduced to just over $3\frac{1}{2}$ miles.

1866   *A Report of the State of the River Stour in the County of Kent, in October 1765, and the Means of improving the Drainage of the Marshland or Levels near the said River, and also of preserving the Port of Sandwich.*

By Tho. Yeoman.

Dated: Westminster, 8 Oct 1765.

p. (2).

[Printed with]

Report by Richard DUNTHORNE, 8 Sept 1774 [q.v.]

[and]

Tables of levels by Thomas and Henry HOGBEN, 1773 [q.v.].

n.p.   [Printed 1775].

fol   34 cm   (4) pp.

BM

1867   *A Report of the State of the Level of Ancholme, and the Methods proposed for the more effectual Draining the same, together with making the River Ancholme Navigable, from Ferriby Sluice to Bishop Brigg.*

By Thomas Yeoman.

n.p.

Signed: Thomas Yeoman. Brigg, 17 Sept 1766.

fol   32 cm   2 + (1) pp.   with engraved plan.

A Section of the River Ancholme in Profile.

BM, Sutro

1868   *An Appendix: containing an Estimate for making a navigable Cutt from Limehouse-Hole to near the Four-Mills at Bromley, Middlesex, for Barges navigating upon the River Lee; proposed as an Addition to Mr Smeaton's Plan.*

By Tho. Yeoman, Engineer.

p. 12.

[In]

The Report of John Smeaton, Engineer, upon the New-making and completing the Navigation of the River Lee, from the River Thames . . . to the Town of Hertford.

See SMEATON, 1766.

fol   32 cm   12 pp.   Folding engraved plan with inset showing the proposed Limehouse Cut.

P, B

1869   [Report relating to the Petition of the Proprietors of the London-Bridge Waterworks] *To the Right Honourable the Lord-Mayor, Aldermen, and Commons . . . in Common-Council Assembled.*

Signed: Tho. Yeoman. Westminster, 5 Feb 1767.

pp. 4–5.

[In]

Kite, Mayor. A Common-Council holden . . . on Wednesday the 25th Day of February 1767.

Corporation of London, 1767.

fol   33 cm   7 pp.

CLRO

1870   *Answers to Questions relative to a Navigable Communication between the Friths of Forth and Clyde, and Observations thereupon.*

By Thomas Yeoman, F.R.S.

Dated: Edinburgh, 23 Sept 1768.

pp. 16–18.

[In]

Reports by James Brindley Engineer, Thomas Yeoman Engineer, and F.R.S. and John Golborne Engineer, relative to a Navigable Communication betwixt the Friths of Forth and Clyde . . . With Observations.

Edinburgh: Printed by Balfour, Auld, and Smellie, 1768.

4to   25 cm   (iii) + 44 pp.   Folding engraved plan.

NLS, CE, ULG

1871   *The Report of Thomas Yeoman, Engineer, concerning the Drainage of the North Level of the Fens, and the Outfall of the Wisbeach River.*

n.p.

Signed: Thomas   Yeoman.   Castle-street, Leicester-fields, 15 Nov 1769.

4to   24 cm   12 pp.   Folding engraved plan.

A Chain and Scale of Levels along Wisbeach River and Channel from Peterborough Bridge down to the Eye at Sea.

Taken in 1767 by William Elstobb. Vere sc.

CE, CRO, BDL (Gough), CUL

Issued in grey wrappers.

1872   Second edition

Title as No. 1871.

Printed in the Year 1769.

4to   24 cm   8 pp.   [no plan].

P

1873   *A Survey and Estimate for Extending the Navigation [of the River Medway] from Tunbridge to Edenbridge and Six Miles above it.*

By Tho. Yeoman, Engineer.

n.p.

Signed: Tho. Yeoman. Castle-street, Leicester-fields, 20 June 1771.

fol   32 cm   3 pp.   Docket title on p. (4).

ULG

1874   *A Report of the State of the River Stour, in the County of Kent, in July 1775. And the Means of improving the Drainage of the Marsh Land, or Levels, near the said River, and also of preserving the Port of Sandwich.*

By Thomas Yeoman, F.R.S.

n.p.

Signed: Tho. Yeoman, 15 July 1775.

fol   32 cm   3 pp.   Folding engraved plan.

A Plan of the River Stour, in the County of Kent, from Fordwich Bridge to the Sea: 1775.

BM

1875   *Mr Yeoman's Strictures and Observations on Mr M'Kenzie's Report, Remarks, &c.*

n.p.

Signed: Tho. Yeoman. Castle-street, Leicester-fields, 24 Aug 1775.

fol   32 cm   4 pp.

BM

Refers to MACKENZIE's report on the River Stour, 27 July 1775 [q.v.].

YOUNG, George [1750–1820] M.Soc.C.E.

1876   *Plan of the River Severn from the Meadow Wharf near Coalbrookdale, to the City of Gloucester.*

By G. Young, 1786.

Engraved plan   18 × 72 cm   (scale 1 inch to 2 miles).

BM, Staffordshire R.O.

1877   *Plan of the Intended Shrewsbury Canal.*

By George Young, 1793.

Engraved plan   21 × 19 cm   (scale 1 inch to 2 miles).

BM, CE (Page)

YOUNG, Thomas [1773–1829] F.R.S.

1878   *A Summary of the most useful parts of Hydraulics, chiefly extracted and abridged from Eytelwein's Handbuch der Mechanik und der Hydraulik.* [Berlin, 1801].

By Thomas Young, M.D. F.R.S.

pp. 185–219.

[In]

Tracts on Hydraulics; edited by Thomas TRED-
GOLD. London, 1826. [q.v.]

Reprinted from *Journals of the Royal Institution*
Vol. 1 (1802) with editorial notes by Tredgold.

1879 *A Course of Lectures on Natural Philosophy and
the Mechanical Arts.*

By Thomas Young, M.D. For.Sec.R.S.

London: Printed for Joseph Johnson . . . 1807.

Dated: Welbeck Street, 30 March 1807.

2 vols   4to   28 cm   **1**, xxiv + (1) + 796 pp.
43 engraved plates. **2**, xii + (1) + 738 pp.   15
engraved plates (Joseph Skelton sc.).

BM, NLS, RS (from the author), CE, BDL

1880 *Hydraulic Investigations, subservient to an
intended Croonian Lecture* . . .

By Thomas Young, M.D. For.Sec.R.S.

London: Printed by W. Bulmer . . . 1808.

Dated: 5 May 1808.

4to   28 cm   25 pp.

CE

Offprint, with new title-page, from *Phil. Trans.*
Vol. 98 (1808).

ADDENDA: *Bridge Prints, Railway Guides and Additional Entries*

Bridge Prints

Between 1819 and 1826 Josiah Taylor of High Holborn, London, published nine large engravings by Matthew Dubourg of bridges shown in perspective, or more often in strict elevation, with a landscape setting. Taylor published and sold other bridge prints, and his 1826 catalogue lists a total of fifteen. In February 1986 the Institution of Civil Engineers acquired by gift a folio in grey paper wrappers, which, according to a contemporary manuscript list pasted on the front cover, originally contained thirteen of these prints. One is missing (a perspective of the Menai suspension bridge by W. A. Provis, 1825) and four are of wooden bridges in America and Switzerland. Two have already been recorded from other sources (Nos 173 and 187). The rest are described below over the symbol CE.

Several bridge prints are among recent acquisitions in the Science Museum Library. They include, by coincidence, three of the Dubourg engravings and a set of London Bridge lithographed drawings which also exist in the CE (Rennie) collection, recently returned after conservation; as well as four other prints.

BROWN, *Sir* Samuel

1881   *The Chain Pier at Brighton. Designed and Erected by Cap.! S. Brown.*
   Drawn from actual admeasurement by T. W. Clisby, Architect. Engraved by M. Dubourg.
London: Published by J. Taylor . . . Jan.! 1825.
Engraved plan, elevation with aquatint landscape, and details   32 × 59 cm   (scale 1 inch to 60 feet and $1\frac{1}{2}$ inches to 1 foot).
CE

1882   *View of a Bridge of Suspension to be erected at Welney in the County of Norfolk. Designed by Cap.! S. Brown, R.N.*
[London]: D. Clark and Co . . . [1825].
Litho elevation   30 × 55 cm   (scale 1 inch to 18 feet).

SML
The design dates from 1825; the bridge was opened in August 1826.

HAYWARD, William [c. 1740–1782]

1883 *To the Mayor . . . of the Town of Henley upon Thames . . . [and] Commissioners appointed by Act of Parliament, this Plate exhibiting a Geometrical View of the North Elevation & Plan of the Superstructure of the Bridge at Henley, with a Section of the River, is . . . dedicated by the Delineator.*
Designed by William Hayward 1781. Measured & Delineated by Robert Shennan 1793.
Engraved plan and elevation   30 × 56 cm (scale 1 inch to 18 feet).
SML

IRONBRIDGE

1884 *This Elevation represents one set of the Ribs of the Iron Bridge cast at Coalbrookdale, and erected over the River Severn, near that place, in the Year 1799 . . . [and] a Section of the Bridge.*
[London]: Published by James Phillips . . . 1782.
Engraved elevation and section   45 × 60 cm (scale 1 inch to 6 feet).
SML
Issued with *View of the Cast-Iron Bridge near Coalbrook-Dale.* Drawn by M. A. Rooker. Engraved by Wm. Ellis. Published by James Phillips, June 1782.

1885 *The Cast Iron Bridge over the River Severn near Coalbrookdale.*
Eng.ᵈ by M. Dubourg.
London: Published by J. Taylor . . . 1823.
Half elevation, details and aquatint view showing structural form   39 × 56 cm   (scale 1 inch to 12 feet).
SML, CE, Elton

RENNIE, John

1886 *The Southwark Cast Iron Bridge erecting over the River Thames at London under the direction of John Rennie Esq.ʳ Civil Engineer, F.R.S.*
Drawn by H. Gardner. Engraved by G. Hawkins [senior].
London: Published by H. Gardner . . . Oct.ʳ 1818 [and sold by J. Taylor].
Engraved plan and elevation   41 × 81 cm (scale 1 inch to 18 feet).
CE

1887 *Plan and Elevation of the Waterloo Bridge over the River Thames.*
Engraved by M. Dubourg.
London: Published by J. Taylor . . . 1822.
Engraved plan, elevation with aquatint landscape, and detail of one span   25 × 61 cm (scales 1 inch to 60 feet and 1 inch to 30 feet).
SML, CE

1888 *The New London Bridge . . . designed by the late John Rennie Esq and . . . now executing by Mess.ʳˢ I. and G. Rennie, Engineers.*
Eng.ᵈ by M. Dubourg.
London: Published by J. Taylor . . . 1825.
Engraved plan, and elevation with aquatint landscape   26 × 60 cm   (scale 1 inch to 40 feet).
SML, CE

RENNIE, *Sir* John

1889 *New London Bridge.*
London: R. Martin & Co . . . [1831].
Signed: John Rennie [autograph].
Set of three lithographs each 41 × 65 cm with subtitles:
I.   *Elevation and Plan* (scale 1 inch to 50 feet).
II. *Longitudinal Section, Transverse Section, Plan of Cofferdams and Foundations* (scale 1 inch to 50 feet).

III. *General Plan and Elevation* of the bridge and approaches (scale 1 inch to 100 feet).
SML, CE (Rennie)
Lithographed from engineering drawings. Issued in printed wrappers titled *A View of Old London Bridge, in 1823 . . . from a drawing by Major G. Yates. And Three Views of New London Bridge, from drawings by Sir John Rennie.* Probably published shortly after the opening of the bridge in August 1831. The Yates print is missing from SML and CE, and the SML set lacks wrappers.

TELFORD, Thomas

1890   *A Plan & View of a Chain Bridge, erecting over the Menai at Bangor Ferry 1820.*
[London]: Published by J. Taylor . . . 1820.
Engraved plan, elevation, and perspective end view with aquatint landscape   20 × 34 cm (scale of elevation 1 inch to 80 feet).
SML
The preliminary design of 1818–19.

WALKER, James

1891   *View of the Cast Iron Bridge over the River Thames at Vauxhall . . . executed from the design & under the superintendence of James Walker, Esq.*
Drawn by E. Gyfford, Architect. Engraved by M. Dubourg.
London: Published by J. Taylor . . . Oct.ʳ 1819.
Engraved perspective view showing exact structural form, with aquatint landscape   26 × 87 cm.
CE

Railway Guides

Many popular guide books were published for travellers on the early railways. Chiefly concerned with the countryside and towns along the line, and often accompanied by pages of advertisements, the guides usually devote some attention to engineering matters and may include views of the railway itself. The examples selected here supplement the works of greater technical and artistic merit (and far higher cost) listed in the bibliography, such as those of Bourne and Britton, Bury, Roscoe, Brees, and Whishaw. Short titles only are given and the pagination ignores advertisements.

LIVERPOOL & MANCHESTER RAILWAY

1892   *An Accurate Description of the Liverpool and Manchester Railway.*
By James Scott Walker.
Liverpool: Printed & Published by J. F. Cannell . . . [1830].
8vo   21 cm   46 pp   Folding litho map with gradient profile and elevation of viaduct.
SML, ULG (Rastrick)
Price 1s. 6d. in printed wrappers. A second edition was published later in 1830 with 52 pp. and the addition of a litho plate.

1893   Third edition
As No. 1892 but Third Edition, 1831 (and 1832).
8vo   21 cm   52 pp.   2 folding engraved plates (drawn and engraved by I. Shaw).
BM, SML, CE, ULG
There were two issues in 1831 and one in 1832, the latter with 53 pp. The plates are second states of two of those published by SHAW [q.v.].

1894   *The Railway Companion, describing an Excursion along the Liverpool Line, accompanied with a succinct and popular history of the Rise and Progress of Rail-Roads.*
By a Tourist.
London: Published for the Author, by Effingham Wilson . . . 1833.

8vo 22 cm 46 pp. 5 folding litho plates each with two views (E. Colyer lith.).
BM, SML

NEWCASTLE & CARLISLE RAILWAY

1895 *Scott's Railway Companion, describing all the Scenery on and contiguous to the Newcastle and Carlisle Railway.*
Carlisle: Printed and Published by H. Scott . . . 1837.
8vo 15 cm vii + 105 pp. Folding map with gradient profile.
ULG (Rastrick)

GRAND JUNCTION RAILWAY

1896 *The Grand Junction Railway Companion to Liverpool, Manchester, and Birmingham . . . Containing an account of every thing worthy the attention of the Traveller upon the Line.*
By Arthur Freeling.
Liverpool: Published by Henry Lacey . . . 1837.
12mo 15 cm (i) + iv + 192 + viii pp.
Folding map with gradient and ground profiles.
BM, SML, CE, ULG
Price 2s. 6d. in gilt-stamped cloth; 3,000 copies printed. Another edition was published in 1838. Freeling, in this and his other *Railway Companions*, makes a point of giving dimensions of embankments, cuttings, bridges and viaducts.

1897 *Osborne's Guide to the Grand Junction, or Birmingham, Liverpool, and Manchester Railway, with the Topography of the Country through which the Line passes.*
Birmingham: E. C. & W. Osborne . . . 1838.
12mo 17 cm iv + 378 pp. Folding map

with gradient profile and engraving of train, 2 plans, geological map, 15 text figures and 26 wood-engraved vignettes. 3 plates.
BM, ULG
Price 3s. 6d. in cloth. Pages 1–57 give a 'History of Railways'; the guide to the GJR occupies pp. 59–306. A 'corrected' 2nd edn appeared later in 1838.

LONDON & BIRMINGHAM RAILWAY

1898 *The London and Birmingham Railway Companion, containing a complete description of every thing worthy of attention on the Line.*
By Arthur Freeling.
London: Published by Whittaker and Company . . . [1838].
12mo 15 cm xi + (1) + 204 pp Folding map with gradient and ground profiles.
BM, ULG
A new edition was published later in 1838 combined with 2nd edn of the Grand Junction guide, under the title *The Railway Companion from London to Birmingham, Liverpool, and Manchester*. Price 5s. in gilt-stamped cloth.

1899 *Osborne's London & Birmingham Railway Guide, illustrated with numerous Engravings & Maps.*
Birmingham: E. C. & W. Osborne . . . [1840].
12mo 17 cm Engraved vignette title page + viii + 270 pp. Folding map with gradient profile and engraving of train, 2 plans, 15 text figures and 16 vignettes. 18 wood-engraved plates (Samuel Williams sc.).
BM, ULG
Includes a 'History of Railways' as in No. 1897.

LONDON & BIRMINGHAM and GRAND JUNCTION RAILWAYS

1900 *Drake's Road Book of the London and Birmingham and Grand Junction Railways, being a Complete Guide to the entire Line of Railway from London to Liverpool and Manchester.*
London: Hayward and Moore . . . [1839].
8vo   16 cm   [and 19 cm 'large paper']   vi + (Part I London to Birmingham) 112 pp. Folding map with gradient profile and engraving (J. B. Wallace del.   J. Archer sc.) 6 wood-engraved vignettes, 6 steel-engraved plates + (Part II Birmingham to Liverpool and Manchester) 147 pp. Folding map with gradient profile and engraving of train, 9 vignettes, 9 plates (H. Harris, W. Green, J. Wallace del.   E. Roberts, J. Grieg sc.).
BM, ULG (no plates), B (with plates)
Issued in gilt-stamped cloth. Price 4s. 6d. in small 8vo or 9s. in large paper with steel engravings.

LONDON & SOUTHAMPTON RAILWAY

1901 *The London and Southampton Railway Companion, containing a complete description of every thing worthy of attention on the Line.*
By Arthur Freeling.
London: Printed for J. T. Norris . . . 1839.
12mo   15 cm   191 pp.   Folding map with gradient and ground profiles. 20 wood-engraved vignettes. Folding hand-coloured litho plan. *Chart of Southampton Water and Site of the intended Dock.*
Francis Giles, Engineer.
BM, CE, ULG
A second edition appeared in 1840 with the title *London and South Western Railway Companion.*

MIDLAND COUNTIES RAILWAY

1902 *The Midland Counties Railway Companion, with Topographical Descriptions.*
Published by . . . R. Allen, Nottingham and E. Allen, Leicester . . . 1840.
12mo   18 cm   Engraved vignette title page + xii) + 135 pp.   Folding map with gradient profile, 23 wood-engraved vignettes (Branston sc.), 4 steel-engraved plates (W. Radclyffe after A. B. Johnson, J. C. Sammons).
BM, SML, CE, ULG

1903 *A Guide or Companion to the Midland Counties Railway, containing its Parliamentary history, engineering facts, with a description of . . . the Line.*
Leicester: Printed and Sold by R. Tebbutt . . . 1840.
12mo   18 cm   110 pp.   Folding map with gradient profile. 16 vignettes, 4 wood-engraved plates (J. Burton sc.).
BM, ULG

Additional Entries:

Items and cross-references omitted from the main bibliography.

ADAM, John [1721–1792]

See   SMEATON, ADAM and BAXTER. Report on the North Bridge, Edinburgh, 1769.

BARLOW, Peter William [1809–1885] F.R.S., M.Inst.C.E., M.Soc.C.E.

See   PALMER. South Eastern Railway, survey by Barlow, 1835.

**BATEMAN, John Frederic
[1810–1889] F.R.S., M.Inst.C.E.,
M.Soc.C.E.**

See FAIRBAIRN. Reservoirs on the River
Bann, survey by Bateman, 1835.

BAXTER, John [d. 1798]

See SMEATON, ADAM and BAXTER. Report
on the North Bridge, Edinburgh, 1769.

BELCHER, Henry

See DODGSON. Illustrations of the Whitby
and Pickering Railway, with a description by
Henry Belcher, 1836.

BRIDGES, James

1904 *Four designs for rebuilding Bristol Bridge.
Humbly inscrib'd to the Gentlemen appointed Com-
missioners for building the same.*
   By James Bridges, Architect.
Bristol: Printed by E. Ward . . . 1760.
8vo   19 cm   59 pp.
BM, Bristol R.O.

CARTER, John Richard

1905 *Observations on the Proposed Improvement of the
Nene Outfall in reference to the Drainage of South
Holland.*

Spalding: Printed by T. Albin, 1822.
Signed: Jnᵒ. Rᵈ. Carter. Spalding, 23 Jan 1822.
8vo   21 cm   16 pp.
CE (Telford)

CARTWRIGHT, William [d. 1804]

1906 *Report to the Committee of the Lancaster Canal
Navigation.*
London: Printed by S. Gosnell . . . [1799].
Signed: William Cartwright.
fol   37 cm   3 pp.   Folding engraved plan.
A Plan of the Lancaster Canal.
ULG
   General Meeting of Lancaster Canal Prop-
   rietors 2 July 1799: resolved that the report,
   with plan annexed, should be printed.

**COMRIE, Alexander [1786–1855]
Assoc.Inst.C.E.**

See WALKER, Plan and section of railway from
Leeds to Selby, survey by Comrie, 1829.

See WALKER and BURGES. Plan and section
of railway from Selby to Hull, survey by Comrie, 1834.

**DYSON, John, *of Bawtry* [died *c.*
1805]**

See HUDSON and DYSON. Report on draining Witham fens, 1784.

EASTBURN, Henry [1753–1821] M.Soc.C.E.

See SMEATON. Plan for improving the drainage of Hatfield Chase, survey by Eastburn, 1776.

FOULDS, John [*c.* 1743–1815] M.Soc.C.E.

1907 *A Statement of the Soundings of the River Thames, from London Bridge to Blackwall.*
By J. Foulds, March 1796.
Appendix Cc (2 tables).
[In]
Report from the Committee on the Port of London.
House of Commons, 1796.
Reprinted in *Reports from Committees of the House of Commons* Vol. 14 (1803) pp. 378–383. *See also* in this volume pl. 18: Soundings of the Great Arch of London Bridge, 1799.

1908 *Borings of the River betwixt London and Blackfriars Bridge; performed betwixt the 19th May and 16th of June 1800.*
By John Foulds and Assistants.
p. 39.
[In]
Third Report from the Select Committee upon the Improvement of the Port of London.
House of Commons, 1800
Reprinted in *Reports from Committees of the House of Commons* Vol. 14 (1803) p. 555. Eighteen borings, each to a depth of 12 feet below low water level.

FROUDE, William [1810–1879] F.R.S., M. Inst.C.E.

See PALMER. South Eastern Railway, survey by Froude, 1835.

GLADSTONE, Murray

See STEPHENSON, G. Chester & Crewe Railway. Plan and section by Stephenson and Gladstone, 1836.

GOTT, John [1720–1793] M.Soc.C.E.

See JESSOP, GOTT and WRIGHT. Plan of intended canal from River Calder to Barnby Bridge, 1792.

GRAVATT, William [1806–1866] F.R.S., M.Inst.C.E., M.Soc.C.E.

See BRUNEL, I. K. Bristol & Exeter Railway, survey by Gravatt, 1835.

HENNET, George [1799–1857] Assoc.Inst.C.E.

See STEPHENSON, G. and R. London & Birmingham Railway, survey by Hennet, 1833.

See STEPHENSON, G. and SWANWICK, F. North Midland Railway, survey by Hennet, 1835.

HOGARD, Thomas [d. 1783] M.Soc.C.E.

See HODSKINSON. Report by Grundy, Hogard and Hodskinson on Wells Harbour. 1782.

LECOUNT, *Lieut.* Peter

1909 *General Results of the traffic returns, &c. between London and Birmingham for one Year; also the expenses of travelling and carriage by the present means and by the Railway.*
[London] J. B. Nicholls and Son . . . [1832].
Signed: Peter Lecount.
fol   33 cm   6 pp.
ULG
> Pages 2–5 reprinted in *London & Birmingham Railway. Preamble to Bill*, May 1832 (House of Commons).

NICHOLLS, *Captain* George

See   TELFORD. Reports on ship canal for the junction of the English and Bristol Channels, by Telford and Nicholls, 1824.

PARKINSON, John [d. 1818]

See   HUDSON and PARKINSON. Report on draining Witham fens, 1788.

PRITCHARD, Thomas Farnolls [1723–1777]

See WHITE, J. Early designs for Ironbridge by Pritchard, 1773 (published 1832).

SALUSBURY, Thomas [died *c.* 1665]

1910 *Mathematical Collections and Transactions: The Second Tome. The Second Part, containing, I. Benedictus Castellus, his Discourse on the Mensuration of Running Waters. II. His Geometrical Demonstrations of the Measure of Running Waters. III. His Letters and Considerations touching the Draining of Fenns, Diversions of Rivers, &c. IV. D. Corsinus, His Relation of the state of the Inundations, &c in the Territories of Bologna, and Ferrara.*
By Thomas Salusbury, Esq.
London: Printed by William Leybourne, 1661.
fol   32 cm   (xiii) + 118 + (5) pp.   Diagrams in text.
BM, BDL, CUL

SHEASBY, Thomas

1911 *Plan (and Section) of an Intended Navigable Canal from Swansea to Pentrecrybarth in the Counties of Glamorgan and Brecon.*
By T. Sheasby, Engineer. 1793.
Engraved by J. Cary.
Engraved plan   34 × 100 cm   (scale 2 inches to 1 mile).
BDL (Gough)

SMITH, John Thomas [1805–1882] R.E., F.R.S., Assoc.Inst.C.E.

1912 *A Practical and Scientific Treatise on calcareous Mortars and Cements, artificial and natural . . . by L. J. Vicat . . . Translated, with the addition of explanatory notes, embracing remarks upon the results of various new experiments.*
By Captain J. T. Smith, Madras Engineers, F.R.S. Associate Member of the Civil Engineers Institution.
London: John Weale . . . 1837.

8vo   23 cm   (ii) + xii + xiv + 302 pp.
3 engraved plates.
BM, SML, IC

STEPHENSON, George and SWANWICK, Frederick

1913   Deposited plan
Plan and Section of an Intended Railway, to be called the North Midland Railway, commencing at or near the Town of Leeds . . . and terminating at or near the Town of Derby.

George Stephenson, Engineer. Frederick Swanwick, Assistant Engineer [1835].
Engraved plans and sections, hand coloured, title page + 14 sheets each 45 × 67 cm   (scales 4 inches to 1 mile and 1 inch to 100 feet).
HLRO
With autograph signature of George Hennet (surveyor). Swanwick gave the engineering evidence on the Bill: House of Lords, June 1836.

STRATFORD, Ferdinando

1914   *Observations on a Letter . . . To which is added, an Explanation of a certain Method of Building a Single-Arch Bridge. Humbly offered to the . . . Commissioners for re-building a Stone Bridge in the City of Bristol.*
By Ferdinando Stratford, Engineer.
Glocester: Printed in the Year 1762.
8vo   19 cm   22 pp.   Folding engraved plate.
The Elevation and Section of a single Arch-bridge [*and*] *Plan of the Foundations of the Arch and the Batterdeaux.*
Ferd? Stratford del.   R. Hancock sc.
BM, Bristol, R.O.

WRIGHT, Elias

See   JESSOP, GOTT and WRIGHT. Plan of an intended canal from River Calder to Barnby Bridge, 1792.

Name Index to Addenda

Subject Index

Market Harborough to Stamford
1810 Bevan 88
Market Weighton
1772 Grundy 628
1772 Smith 1355
Middlewich to Altringham
1837 Provis 1098
Monkland
1770 Watt 1758
1786 Millar 926
1786 Whitworth 1812
Monmouthshire
1792 Dadford 339
1793 Dadford 340
Newcastle under Lyme Junction
1797 Roberts 1258
Newcastle upon Tyne to Maryport
1795 Chapman 222, 223, 224
1795 Dodd 225, 226, 227, 229
1795 Jessop 373
1795 Thompson 758
1796 Chapman 1634
1796 Sutcliffe 1476
1797 Chapman 232
1797 Sutcliffe 1477
1797 Whitworth 1821, 1822
1810 Dodd 369
North London
1802 Dodd 389
Nottingham
1791 Green 584
Oakham
1786 Whitworth 1810
1792 Jessop and Staveley 790
Oxford
1768 Brindley 149
1828 Vignoles 1694
Paisley
1803 Rennie 1163, 1164
1804 Telford 1494
1805 Telford 1495, 1496
Pocklington
1814 Leather 832
Polbrock
1796 Rennie 1141
Portsmouth & Arundel
1816 Rennie 1205
1817 Rennie 1206
Regent's
1802 Rennie 1156
1812 Morgan 950
1818 Morgan 951
Rochdale
1791 Rennie 1129
1793 Papers 1268
1794 Rennie 1130
1799 Townshend 1644

Royal (Ireland)
1803 Specifications 1370
Selby
1772 Jessop 730
Severn Junction
1810 Whitworth 1831
1811 Whitworth 1832
Severn & Wye
1810 Bowdler 103
Sheffield
1813 Chapman 255
1814 Chapman 256
Shrewsbury
1793 Young 1877
1800 Telford 1488
Shropshire
1800 Telford 1488
Sir John Ramsden's
1774 Holt 679
Somersetshire Coal
1800 Sutcliffe 1478
Stainforth & Keadby
1772 Kelk 798
1772 Tofield 1639
1793 Fairbank 501
Staffordshire & Worcestershire
1766 Brindley 161
Stamford Junction Navigation
1810 Telford 1505, 1506
Stirling
1835 Macneill 892
Stockton & Darlington
1770 Brindley and
        Whitworth 159
1796 Dodd 377
Stratford-upon-Avon
1792 Snape 1366
Strathmore
1817 Stevenson 1434
1820 Stevenson 1443
1821 Stevenson 1453
Swansea
1793 Sheasby 1911
Thames & Medway
1799 Dodd 381
1799 Dodd 382
Thames & Severn
1783 Whitworth 1805
Trent & Mersey
1760 Brindley 140
1765 Bentley 85
1765 Brindley 143
1766 Brindley 160
Uxbridge to Marylebone
1776 Erskine 454
Wakefield to River Went
1772 Smith 1356

Waltham Abbey to Moorfields
1773 Whitworth 1798
Warwick & Birmingham
1792 Sherriff 1286
Warwick & Braunston
1793 Sherriff 1289
Weald of Kent
1802 Sutherland 1480
1809 Rennie 1177
1811 Rennie 1193
Wey & Arun
1813 Jessop 725
Whitby to Pickering
1794 Crosley 309
Wilts & Berks
1795 Whitworth 1829
Witham & Ancholme
1828 Padley 1010
Worcester & Birmingham
1789 Snape 1365
1810 Smith 1353
1815 Giles 537
Wyrley & Essington
1793 Pitt 1077

CIVIL ENGINEERING (general)
1763 Vallancey 1669
1797 Smeaton 1345
1812 Smeaton 1346
1826 Strickland 1475
1837 Brees 131
1837 Mahan 904
1838 Simms 1292
1838 Stevenson 1428
1838 Telford 1556
1839 Brees 133
1840 Brees 134, 135

COASTAL SURVEYS
1737 Labelye 810
1792 Pickernell 1073
1809 Telford and McKerlie
        1605
1817 Stevenson 1431
1819 Telford 1584
1823 Clegram 290
1832 Nimmo 1003
1832 Stevenson 1457

DOCKS
General
1716 Perry 1054
Bristol
1763 Smeaton 1305
1767 Mylne 961
1793 Jessop 166
1793 Jessop and White 752

294

Blaenavon and Beaufort
(Monmouthshire)
  1792 Dadford 339
  1793 Dadford 340
Clydach (Brecknock &
Abergavenny)
  1793 Dadford 340
Little Eaton (Derby)
  1792 Outram 1004
Nanpantan (Leicester Navigation)
  1790 Jessop and Staveley 789
Newcastle under Lyme Junction
  1797 Roberts 1258
Penderyn (Aberdare)
  1792 Dadford 336

Horse Railways
Berwick & Kelso
  1810 Rennie 1178
Brechin & Montrose
  1819 Stevenson 1437
East Lothian
  1826 Stevenson 1452
Edinburgh & Glasgow
  1826 Jardine 722
Glasgow & Berwick
  1810 Telford 1502, 1503
Knaresborough
  1818 Telford 1515
London & Portsmouth
  1803 Marshall 907
Mansfield & Pinxton
  1817 Jessop 726
Merthyr (and Carno Mill) to
Cardiff
  1799 Williams 1836
Mid Lothian
  1819 Stevenson 1436
Monkland & Kirkintilloch
  1823 Grainger 559
Plymouth & Dartmoor
  1819 Tyrnhitt 1666
Prior Park, Bath
  1734 Labelye 807
Roxburgh & Selkirk
  1821 Stevenson 1445
Severn & Wye (Lydney &
Lydbrook)
  1801 Outram 1008
  1810 Bowdler 103
Sirhowy
  1801 Hodgkinson 662
Stirling to Forth & Clyde Canal
  1827 Steedman 1391
Stratford & Moreton
  1820 Baylis 74
Strathmore
  1821 Stevenson 1443

  1827 Stevenson 1453
Surrey Iron
  1800 Jessop 767
  1801 Jessop 768
  1802 Jessop 773, 774
West Lothian
  1824 Baird 48
Whitby & Pickering
  1836 Dodgson 393

Railways with Stationary Engines
Cromford & High Peak
  1824 Jessop 727
  1832 Page 1016
Dundee & Newtyle
  1825 Dunn 405
  1825 Landale 822
Edinburgh & Dalkeith
  1824 Grieve and McLaren
        599
  1825 Jardine 720
  1828 Jardine 721
Stockton & Darlington
(preliminary)
  1818 Meynell 923
  1821 Meynell 924

Steam Railways
Basingstoke & Bath
  1836 Giles and Brunton 536
Birmingham & Gloucester
  1835 Moorsom 683
Bodmin & Wadebridge
  1831 Hopkins 683
Bristol & Exeter
  1835 Brunel 185
Carlisle, Glasgow & Edinburgh
  1837 Locke 847
Central Kent
  1836 Braithwaite 130
Central Kentish
  1836 Cundy 334
  1836 Rennie 1126
  1837 Rennie 1122
  1837 Rennie 1127
Chester & Crewe
  1836 Stephenson and
        Gladstone 1404
Chester & Holyhead
  1839 Stephenson 1409
Cork to Cove
  1837 Macneill 896
Dublin & Kingstown
  1836 Nicholl 982
  1836 Stevenson 1427
Dundee & Perth
  1835 Findlater 505
Eastern Counties
  1834 Braithwaite 128

  1835 Braithwaite 129
Edinburgh & Glasgow
  1830 Grainger and Miller
        569, 570
  1831 Stephenson 1397
  1835 Stevenson 1461
  1838 Evidence 490
Edinburgh, Haddington & Dunbar
  1835 Grainger and Miller
        578
Edinburgh, Leith & Newhaven
  1835 Grainger and Miller
        577
Edinburgh to Newcastle
  1838 Stephenson 1408
Edinburgh & Northern
  1836 Geddes 519
  1836 Stevenson 1463
  1841 Grainger 563
Glasgow
  1829 Glasgow 538
  1829 Grainger and Miller
        568
  1829 Telford 1552
Glasgow & Garnkirk
  1825 Grainger 560, 561
  1828 Grainger and Miller
        564
  1829 Grainger and Miller
        567
  1832 Buchanan 201
Glasgow, Paisley & Greenock
  1840 Brees 134
Glasgow, Paisley, Kilmarnock &
Ayr
  1837 Evidence 489
  1839 Miller 928
Gloucester & Hereford
  1836 Price and Laxton 1087
Grand Junction
  1831 Stephenson 1399
  1832 Rastrick 1104
  1833 Rastrick 1105
  1836 Locke 844
  1837 Freeling 1896
  1838 Osborne 1897
  1839 Drake 1900
  1839 Roscoe 1270
Grand Northern
  1833 Cundy 333
  1834 Cundy 332
Grand Southern
  1834 Cundy 332
Grand Surrey
  1835 Price 1085
Great Inland Junction
  1838 Blackmore 99

Great North of England
1836 Storey 1471, 1472
1837 Storey 1473
Great Northern
1835 Gibbs 522
Great Western
1833 Britton 170
1833 Brunel 181
1834 Brunel 182
1834 Evidence 484
1835 Brunel 184
1835 Evidence 485
1837 Specifications 1389
1838 Brunel 186
1838 Hawkshaw 657
1838 Wood 1851
1839 Brees 133
Hertfordshire Grand Union
1836 Whishaw 1779
Hull, Lincoln & Nottingham
1836 Laxton 828
Hull & Selby
1834 Walker and Burges
1734
Ireland (general)
1837 Drummond *et al.* 401
1837 Macneill 898, 899
1837 Vignoles 1701, 1702
1838 Drummond *et al.* 402
1838 Macneill 897
1838 Vignoles 1700
Lancaster & Carlisle
1837 Stephenson 1407
1840 Whishaw 1782
Launceston & Victoria
1836 Hopkins 684
Leeds & Selby
1829 Walker 1216
1837 Brees 131
Limerick & Waterford
1825 Nimmo 999
Liverpool & Manchester
1824 Sandars 1274
1824 Stephenson 1395
1825 Evidence 473
1825 Stephenson 1396
1825 Sylvester 1484
1826 Evidence 474
1826 Rennie 1123, 1124
1829 Telford 1548
1829 Walker and Rastrick
1741
1830 Booth 100
1830 Stephenson and
Locke 1421
1830 Walker 1892
1831 Bury 211

1831 Shaw 1284
1833 Tourist 1894
1836 Stevenson 1427
London & Birmingham
1832 Evidence 480, 481
1832 Lecount 1909
1833 Stephenson 1411
1835 Stephenson 1413, 1414
1837 Brees 131
1838 Freeling 1898
1839 Bourne 102
1839 Brees 133
1839 Drake 1900
1839 Lecount 838
1839 Roscoe 1271
1840 Osborne 1899
London & Blackwall
1836 Cubitt 322
1836 Stephenson 1403
1838 Stephenson and
Bidder 1410
London & Brighton
1835 Gibbs 524
1836 Gibbs 525
1836 Stephenson 1412
1837 Evidence 488
1837 Locke 845
1837 Rastrick 1108
1837 Stephenson 1416, 1417
London & Croydon
1835 Gibbs 523
1839 Trotter 1661
1840 Brees 134
London to Glasgow
1836 Locke 843
London & Greenwich
1832 Landmann 823
1833 Landmann 824
1837 Walker 1723
1838 Simms 1292
London to Porth Dynllaen
1837 Vignoles 1698, 1699
London & Southampton
1832 Giles 531
1834 Evidence 482, 483
1834 Giles 532
1836 Giles 534
1839 Freeling 1901
Manchester & Cheshire Junction
1836 Rastrick 1106, 1107
Manchester & Leeds
1837 Specifications 1390
Maryport & Carlisle
1836 Stephenson 1405, 1406
Midland Counties
1832 Jessop 795

1833 Glynn 541
1833 Rennie 1119
1836 Rennie 1120
1837 Brees 131
1840 Allen 1902
1840 Tebbutt 1903
Newcastle & Carlisle
1824 Chapman 269, 270
1825 Jessop 728
1825 Thomas 1630
1828 Thompson 1631, 1632
1829 Evidence 478
1829 Leather 833
1830 Giles 528, 529, 530
1830 Specifications 1384
1834 Thompson *et al.* 1633
1836 Blackmore 98
1837 Scott 1895
Newcastle, Edinburgh & Glasgow
1836 Richardson 1248
1837 Richardson 1249
Newcastle & North Shields
1831 Richardson 1244
Newtyle, Coupar & Glamis
1833 Blackadder 93
North Liverpool (to Manchester)
1835 Locke 842
1835 Vignoles 1696
1835 Walker and Burges 1735
North Midland
1836 Stephenson and
Swanwick 1913
Northern & Eastern
1835 Walker 1719
1839 Stephenson 1418
Northumberland
1835 Dunn 406
Paisley & Renfrew
1835 Grainger and Miller
576
Perranporth & Truro
1831 Whishaw 1785
Portsmouth Junction
1836 Giles 533
Preston & Wigan
1831 Vignoles 1695
Preston & Wyre
1835 Landmann 825
Sheffield, Ashton-under-Lyme &
Manchester
1836 Vignoles 1697
1836 Vignoles 1703
Sheffield & Manchester
1831 Stephenson 1400
1832 Page 1016
South Durham
1836 Nicholson 985

298

299

1813 Edgeworth 422
1816 McAdam 865, 869
1817 McAdam 866
1819 Paterson 1048
1819 Evidence 470
1819 McAdam 868
1820 Rules 1255
1820 Chambers 215
1821 Wingrove 1846
1822 Paterson 1049
1823 Evidence 472
1824 Paterson 1050
1824 Stevenson 1450
1825 McAdam 878
1825 MacLaren 886
1833 Macneill 888
1833 Parnell 1041
1834 Grahame 557
1835 Gordon 552
1836 Evidence 487
1837 Gordon 553
1837 Simms 1291

Birmingham to Bromsgrove
1786 Sherriff 1285

Carlisle to Edinburgh
1828 Telford 1551
1829 Telford 1549

Carlisle to Glasgow
1815 Telford 1511

Carlisle to Portpatrick
1811 Rennie 1189
1811 Telford 1508

Edinburgh
1824 Jardine 719
1824 Stevenson 1451

Edinburgh to Newcastle (via Carter Fell)
1820 Telford 1516
1829 McAdam 879
1831 Grainger and Miller 573

Exeter to Plymouth
1819 Green 585, 586

Glasgow
1828 Macquisten 901
1829 Cleland 293
1829 Grainger and Miller 566
1829 Macquisten 902

Glasgow, Hamilton & Elvanfoot
1813 Abercrombie 1
1813 Macquiston 903
1814 Inglis 713

Glasgow to Kilmarnock
1829 Grainger and Miller 565

Glasgow to Port-Patrick
1824 Telford 1526

Highgate Archway
1838 Sims 1292

Holyhead
1811 Telford 1582
1813 Fulton 516
1817 Huddart 692
1817 Provis 1095
1819 Telford 1585, 1586
1820 Rules 1255
1820 Telford 1587
1821 Provis 1096
1821 Telford 1588
1822 Easton 413
1822 Telford 1589–1592
1823 Telford 1593
1824 Telford 1594
1825–
1834 Telford 1595
1828 Provis 1093
1828 Telford 1544
1830 Evidence 479
1830 Macneill 887
1830 Telford 1596
1833 Parnell 1041
1835 Macneill 891
1836 Provis 1094
1838 Simms 1292

Inverness to Nairn
1835 Mitchell 944

Inverness to Perth
1810 Telford 1504
1828 Mitchell 939
1832 Flint 506
1836 Mitchell 945

Ketley Ironworks to Chirk
1820 Telford 1553

London, Commercial Road
1801 Wickings 1834
1804 Walker 1705

London to Edinburgh
1821 Telford 1519
1824 Telford 1532
1825 Telford 1533
1826 Telford 1541, 1542
1827 Telford 1540, 1543

London to Liverpool
1827 Telford 1538
1828 Telford 1544
1829 Telford 1547

Milford to Gloucester
1792 Hassall and Williams 655

Plymouth & Devonport to St. Austell
1835 Rendel 1112

Scotland, Highland Roads
1803 Brown 172
1803 Telford 1491, 1557

1804 Telford 1558
1807 Rickman 1251
1807 Telford 1560
1810 Telford 1565
1814 Rickman 1252
1828 Mitchell 940
1829 Telford 1569
1830 Mitchell 941
1833 Mitchell 943

South Wales, Mail Roads
1824 Telford 1535
1825 Telford 1536

Staffordshire
1825 Hamilton 640
1826 Hamilton 641

Surrey & Sussex Turnpike Trust
1838 McAdam 863

Taunton Turnpike
1819 Wingrove 1845

Towcester to Weston Gate
1756 Yeoman 1862

Warwickshire
1760 Tomlinson 1642

West Wycombe to Chilworth
1824 Telford 1534

Whetstone & St. Albans Turnpike
1828 Evidence 477

STRENGTH OF MATERIALS
General
1803 Banks 59
1817 Barlow 60
1820 Tredgold 1649
1822 Robison 1265
1837 Barlow 67
1840 Hodgkinson 661

Cement and Concrete
1796 Telford 1487
1830 Pasley 1046
1832 White 1787
1836 Godwin 543
1837 Smith 1912
1838 Pasley 1047

Cordage
1804 Huddart 689
1807 Chapman 250

Iron
1803 Banks 59
1822 Tredgold 1653
1824 Cumming 328
1831 Turnbull 1662
1832 Drewry 400
1832 Turnbull 1663
1835 Barlow 64
1836 Lecount 837
1837 Fairbairn 498
1838 Fairbairn 499